Student Solutions Manual
to Accompany

# Finite Mathematics
## An Applied Approach

**Eighth Edition**

Abe Mizrahi
*Indiana University Northwest*

Michael Sullivan
*Chicago State University*

*Prepared by*

Sharon O'Donnell
*Chicago State University*

Iztock Hozo
*Indiana University Northwest*

*With assistance from*

**Stephen L. Davis**
*Davidson College*

**William L. Hosch**
*Indiana University Northwest*

**John Wiley & Sons, Inc.**
*New York / Chichester / Weinheim / Brisbane / Singapore / Toronto*

Cover Art: Marjory Dressler

To order books or for customer service call 1-800-CALL-WILEY (225-5945).



ISBN 0-471-35508-9

Printed in the United States of America

10 9 8 7 6 5 4 3

Printed and bound by Bradford & Bigelow, Inc.

# Table of Contents

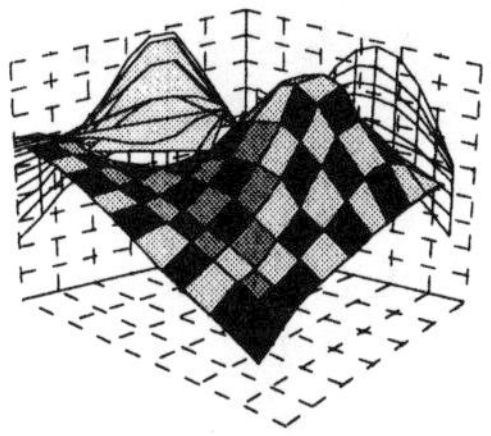

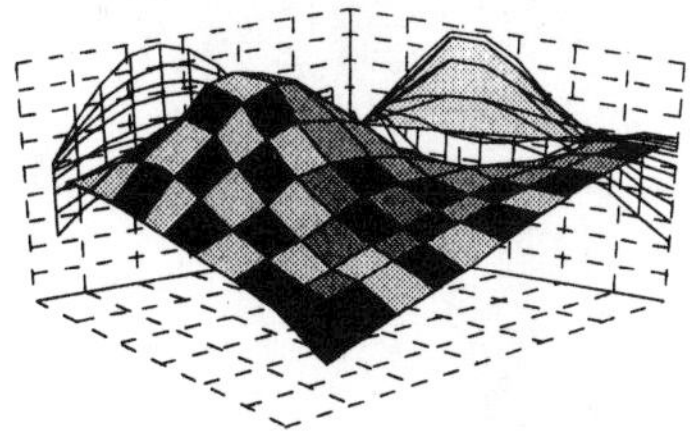

# Chapter 1

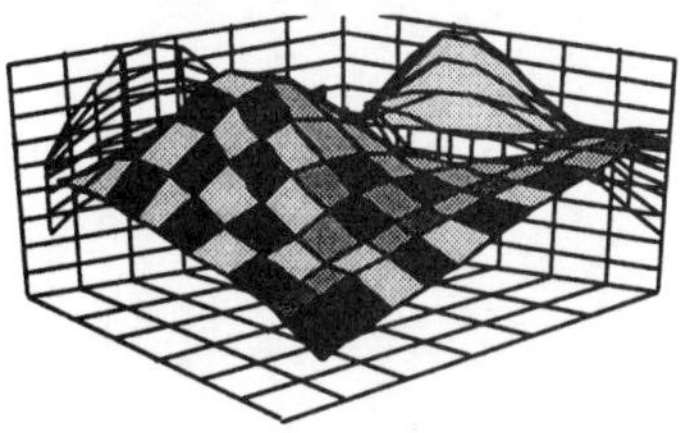

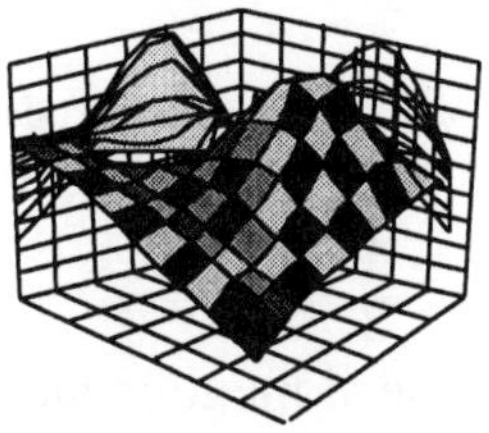

# Linear Equations

## 1.1 Rectangular Coordinates; Lines

**1.** $A=(4,\ 2)$; $B=(6,\ 2)$; $C=(5,\ 3)$; $D=(-2,\ 1)$; $E=(-2,\ -3)$; $F=(3,\ -2)$; $G=(6,\ -2)$; $H=(5,\ 0)$

**3.** The set of points of the form $(2,\ y)$, where $y$ is a real number, is a vertical line passing through 2 on the $x$-axis.

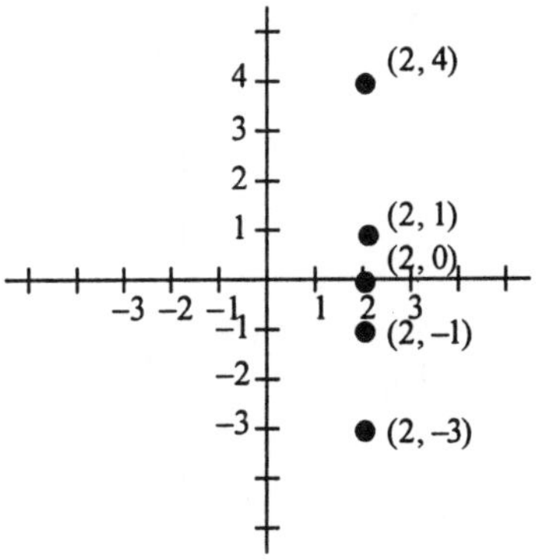


**5.** $y=2x+4$

| $x$ | 0 | –2 | 2 | –2 | 4 | –4 |
|---|---|---|---|---|---|---|
| $y$ | 4 | 0 | 8 | 0 | 12 | –4 |

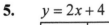

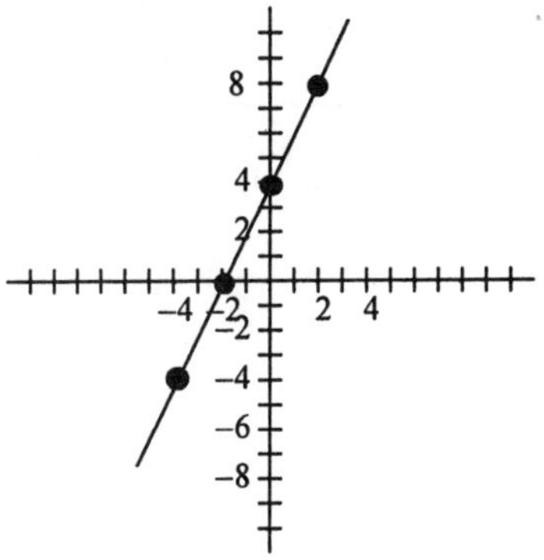

**7.** $2x - y = 6$

| $x$ | 0 | 3 | 2 | –2 | 4 | –4 |
|---|---|---|---|---|---|---|
| $y$ | –6 | 0 | –2 | –10 | 2 | –14 |

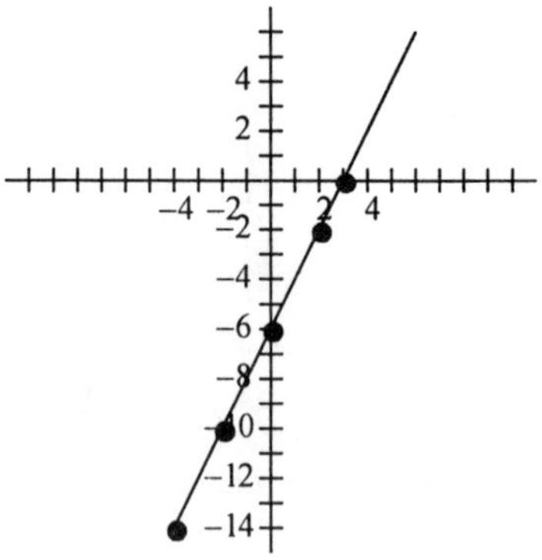

**9.** $\text{slope} = \dfrac{1-0}{2-0} = \dfrac{1}{2}$

We interpret the slope to mean that for every 2 unit change in $x$, $y$ will change by 1 unit.

**11.** $\text{slope} = \dfrac{3-1}{-1-1} = -1$

We interpret the slope to mean that for every 1 unit change in $x$, $y$ will change by –1 unit.

**13.** $\text{slope} = \dfrac{3-0}{2-1} = 3$

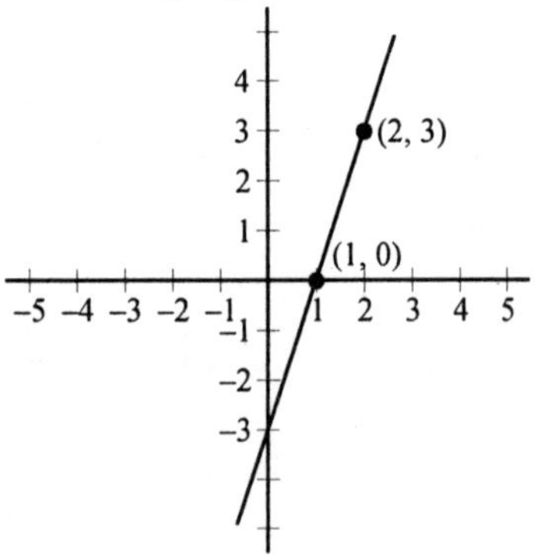


**15.** $\text{slope} = \dfrac{3-1}{-2-2} = -\dfrac{1}{2}$

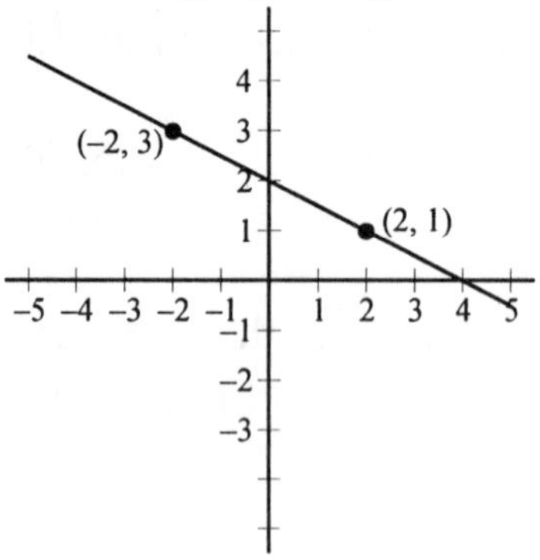


**17.** $\text{slope} = \dfrac{-1-(-1)}{-3-2} = 0$

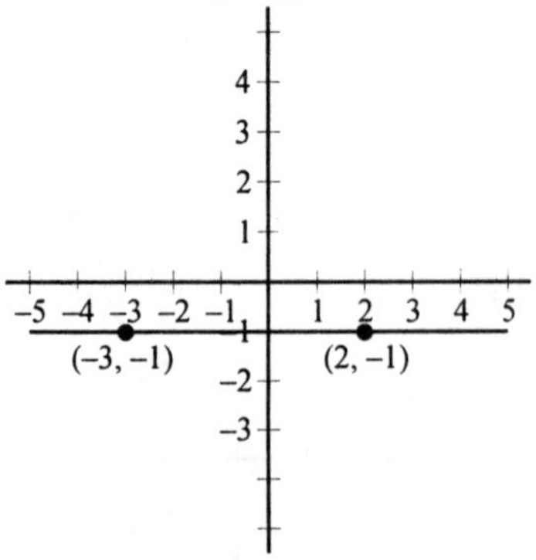


**19.** The slope is undefined.

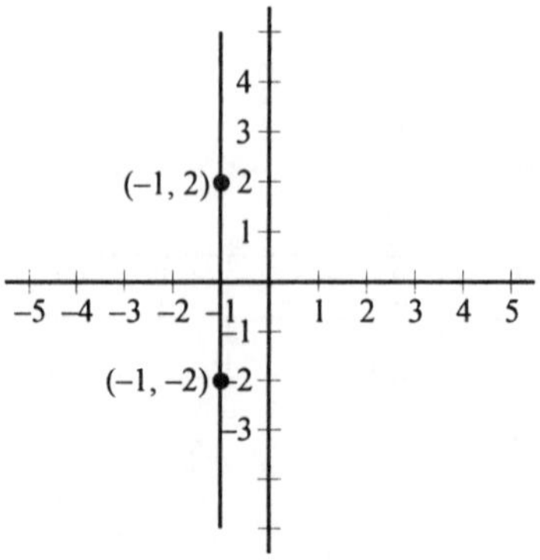

**21.** slope $= \dfrac{3-\sqrt{3}}{\sqrt{2}-1} \approx 3.0611$

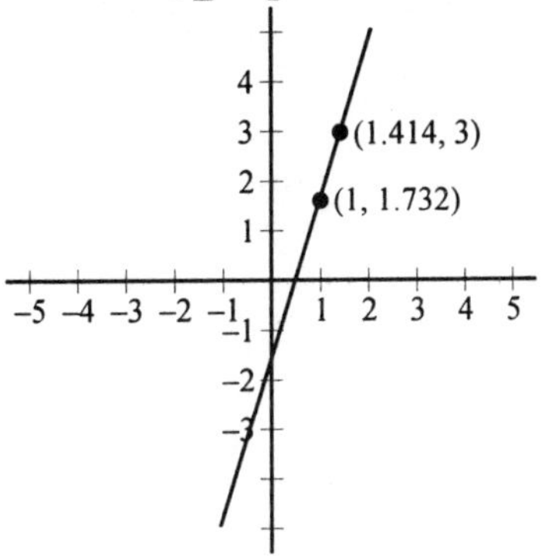


**23.**

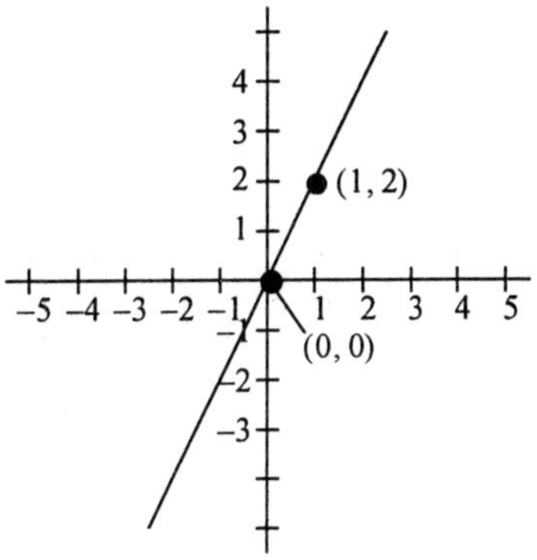


**25.**

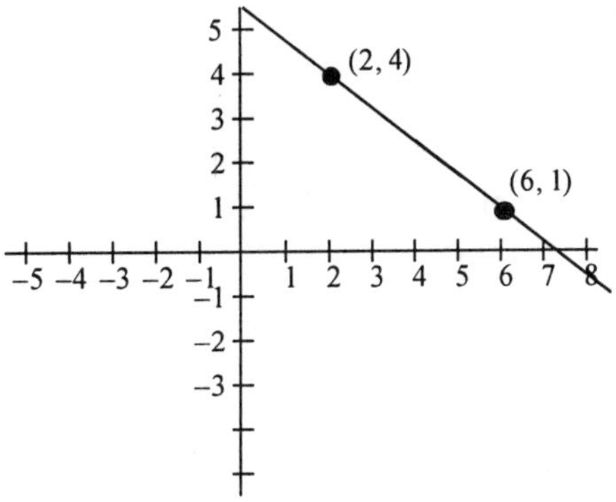


**27.**

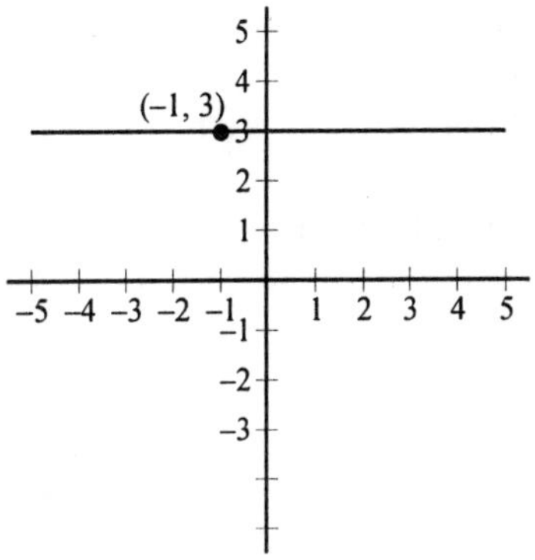


**29.** the $y$-axis

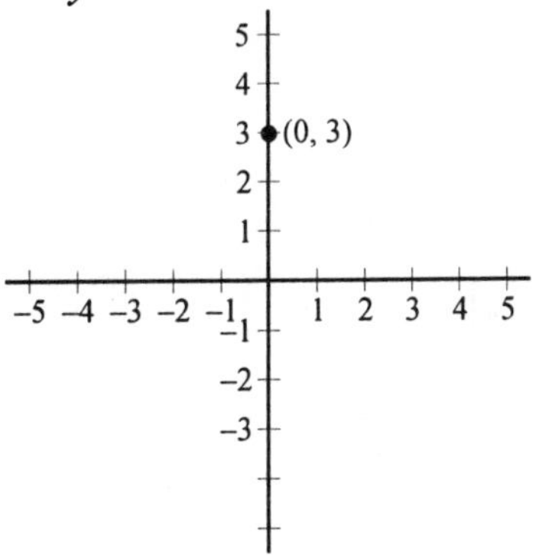


**31.** slope $= \dfrac{1-0}{2-0} = \dfrac{1}{2}$. Equation: (0, 0) is the $y$-intercept. $y = \dfrac{1}{2}x + 0$; $y = \dfrac{1}{2}x$; $2y = x$ or $x - 2y = 0$.

**33.** slope $= \dfrac{3-1}{-1-1} = -1$. Equation: $y - 3 = (-1)(x - (-1))$; $y - 3 = -x - 1$, or $x + y = 2$.

**35.**
$$y-1=2(x-(-4))$$
$$y-1=2x+8$$
$$2x-y=-9$$

**37.**
$$y-(-1)=-\frac{2}{3}(x-1)$$
$$3y+3=-2(x-1)$$
$$3y+3=-2x+2$$
$$2x+3y=-1$$

**39.** $\text{slope}=\dfrac{3-2}{1-(-1)}=\dfrac{1}{2}$
$$y-3=\frac{1}{2}(x-1)$$
$$2y-6=x-1$$
$$x-2y=-5$$

**41.**
$$y=-2x+3$$
$$2x+y=3$$

**43.**
$$y-0=3(x-(-4))$$
$$y=3x+12$$
$$3x-y=-12$$

**45.** (0,0) is the $y$-intercept
$$y=\frac{4}{5}x+0$$
$$5y=4x$$
$$4x-5y=0$$

**47.** $\text{slope}=\dfrac{0-(-1)}{2-0}=\dfrac{1}{2}$
$$y=\frac{1}{2}x-1$$
$$2y=x-2$$
$$x-2y=2$$

**49.** vertical line containing (1, 4); $x=1$

**51.** horizontal line containing (1, 4); $y=4$

**53.** slope $=2$; $y$-intercept $=(0,\ 3)$

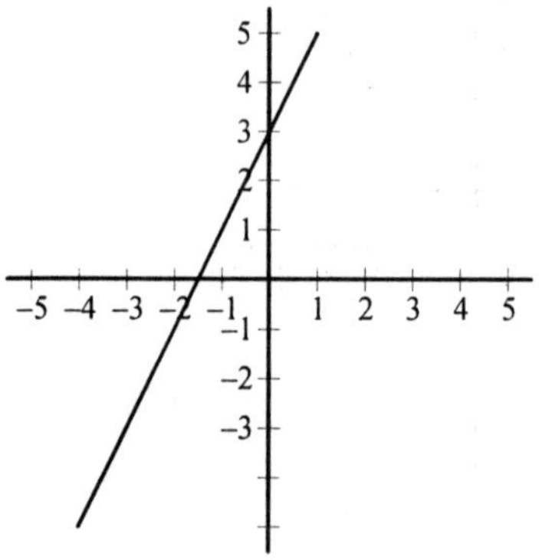


**55.** $y=2x-2$; slope $=2$;
$y$-intercept $=(0,\ -2)$

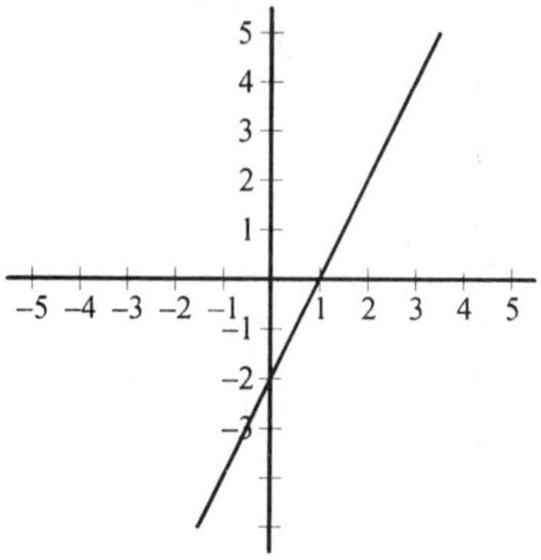

**57.** $-3y = -2x + 6$

$y = \frac{2}{3}x - 2$; slope $= \frac{2}{3}$;

$y$-intercept $= (0,\ -2)$

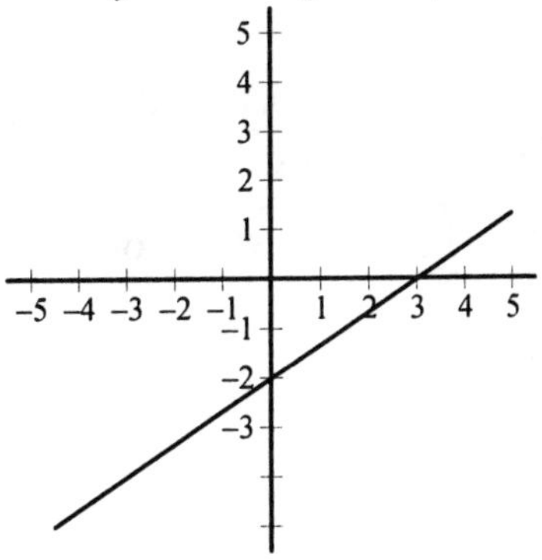

**59.** $y = -x + 1$; slope $= -1$;

$y$-intercept $= (0,\ 1)$

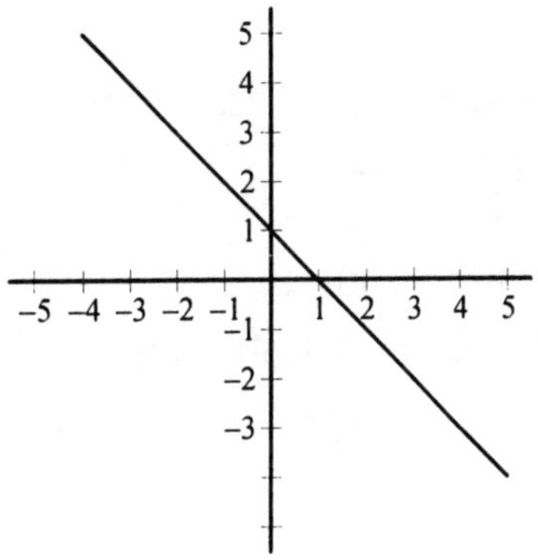

**61.** $x = -4$; slope undefined (vertical line);

no $y$-intercept

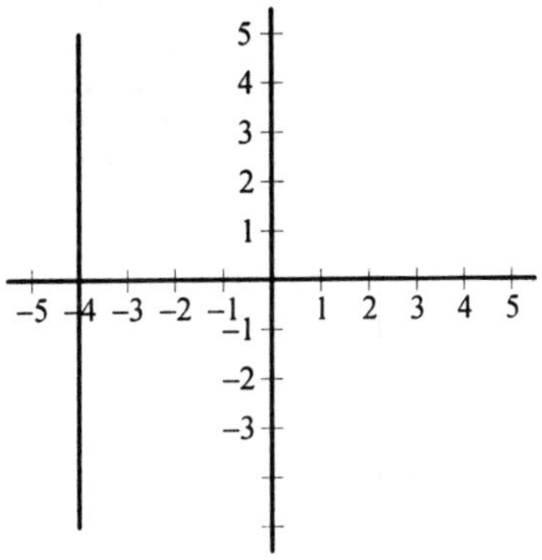

**63.** $y = 5$; slope $= 0$ (horizontal line);

$y$ – intercept $= (0,\ 5)$

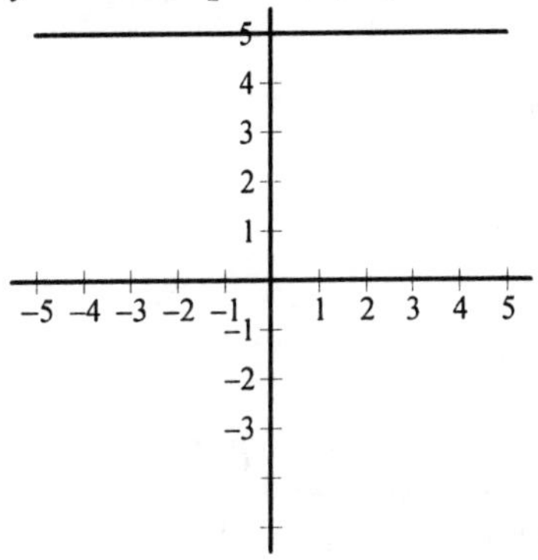

**65.** $y = x$; slope $= 1$; $y$-intercept $= (0,\ 0)$

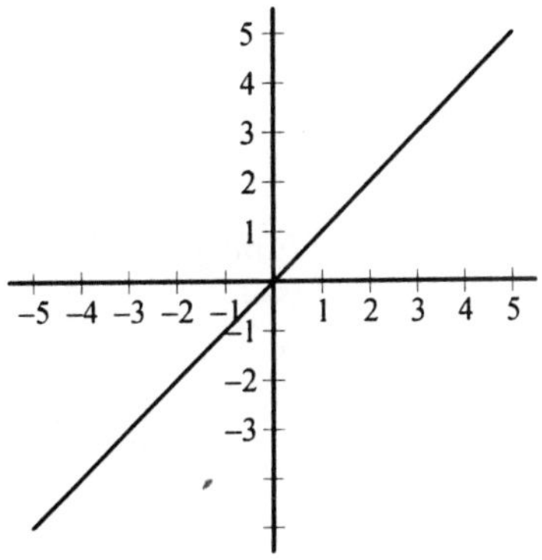

**67.** $2y = 3x$; $y = \frac{3}{2}x$; slope $= \frac{3}{2}$;

$y$-intercept $= (0,\ 0)$

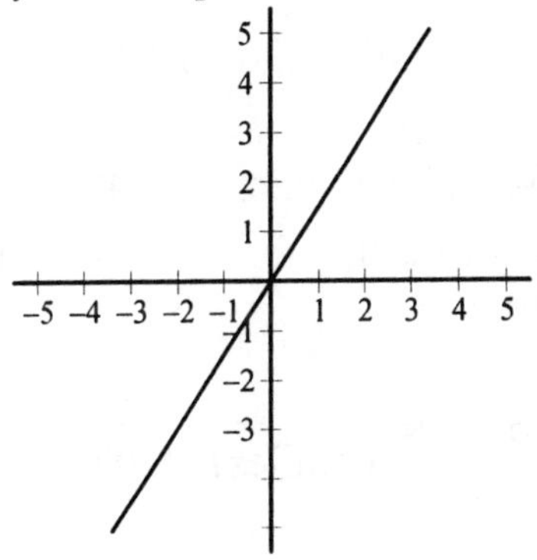

**69.** $x$-axis: a horizontal line containing points (1, 0), (2, 0), (3, 0), . . . . Equation: $y = 0$

**71.** The $y$-coordinate at any point on a horizontal line is always the same. Since (−1, −3) is a point on the horizontal line, its equation is $y = -3$.

**73.** Using $C$ and $F$ for degrees Celsius and Fahrenheit, respectively, we seek an equation for the line through points $(C, F) = (0, 32)$ and $(C, F) = (100, 212)$. The slope is $\frac{\text{change in } F}{\text{change in } C} = \frac{212-32}{100-0} = \frac{9}{5}$. Thus, an equation is $F = \frac{9}{5}C + 32$; $5F = 9C + 160$ or $9C - 5F = -160$. Substituting $F = 70$: $9C - 350 = -160$; $9C = 190$; $C = \frac{190}{9} = 21\frac{1}{9} \approx 21.1111$. 70° F corresponds to $21\frac{1}{9}$ °C.

**75.** Revenue is $x$ dollars and cost is $0.5x + 100$ dollars, so profit $P = x - (0.5x + 100)$, or $P = 0.5x - 100$ dollars.

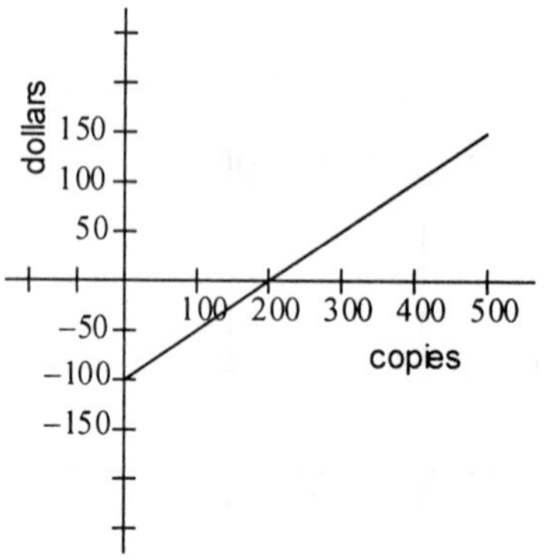


**77.** $10.494\frac{\text{cents}}{\text{kwh}} = 0.10494\frac{\text{dollars}}{\text{kwh}}$

$C = \left(0.10494\frac{\text{dollars}}{\text{kwh}}\right)(x \text{ kwh}) + \$9.36$,

$C = 0.10494x + 9.36$ dollars, $0 \le x \le 400$.

When $x = 100$,

$C = 0.10494(100) + 9.36 = 10.494 + 9.36 = \$19.85$.

When $x = 300$,

$C = 0.10494(300) + 9.36 = 31.482 + 9.36 = \$40.84$.

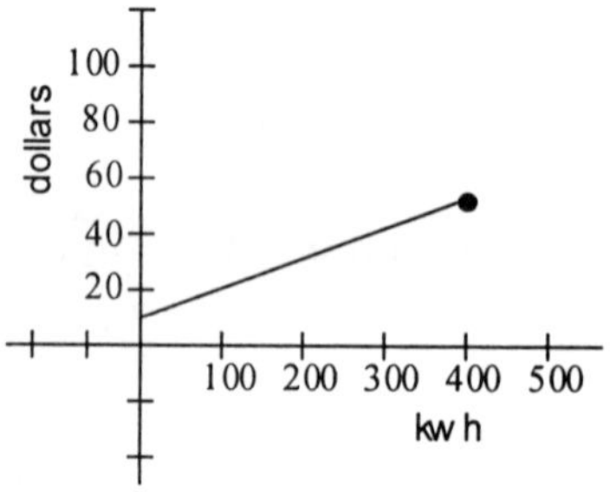


The slope means that for each additional kwh of usage the customer is charged \$0.10494, or 10.494¢.

**79.** $C = \left(0.2\frac{\text{dollars}}{\text{mile}}\right)(x \text{ miles}) + \$280$; $C = 0.2x + 280$ dollars

**81.** Using $(h, w) = (62, 120)$ and $(h, w) = (72, 170)$, slope $= \frac{170-120}{72-62} = 5$;

$w - 120 = 5(h - 62)$; $w = 5h - 190$.

**83.** (a) Using $(t, w) = (10,300)$ and $(t, w) = (18,262)$, where $t$ is the day of the month, and $w$ is millions of gallons of water in the reservoir, the slope is $\frac{262-300}{18-10} = -\frac{19}{4} = -4.75$. $w$ and $t$ are related by $w - 300 = -4.75(t-10)$; $w = -4.75t + 347.5$

(b) $w = -4.75(14) + 347.5 = 281$ million gallons were in the reservoir on the 14th of the month.

(c) The slope means that each day the reservoir loses 4.75 million gallons of water.

## Technology Exercises

**1.** $1.2x + 0.8y = 20$;
$x$-intercept = (16.67, 0)
$y$-intercept = (0, 25)

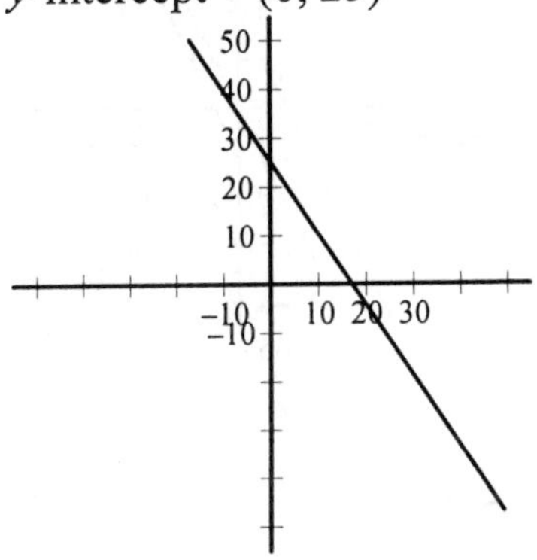


**3.** $215x - 0.1y = 53$;
$x$-intercept = (0.25, 0)
$y$-intercept = (0, –530)

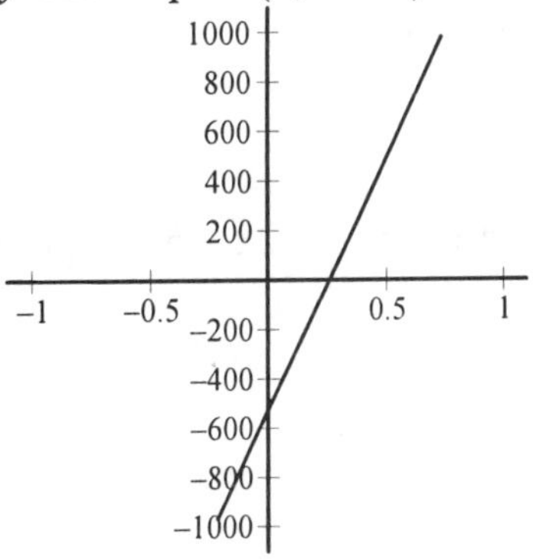


**5.** $\frac{4}{17}x + \frac{6}{23}y = \frac{2}{3}$;
$x$-intercept = (2.83, 0)
$y$-intercept = (0, 2.56)

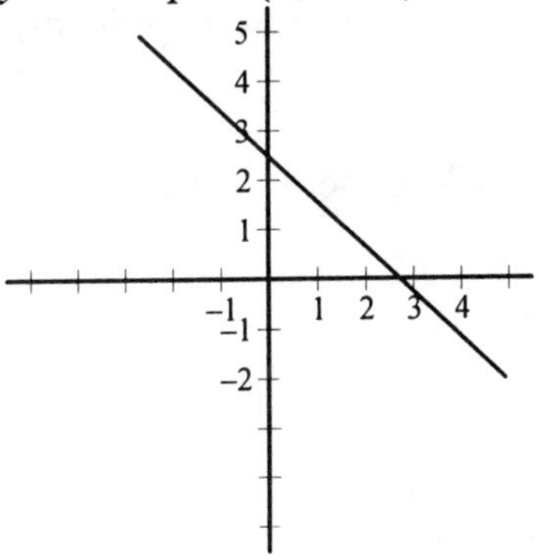


**7.** $\pi x - \sqrt{3}y = \sqrt{6}$;
$x$-intercept = (0.78, 0)
$y$-intercept = (0, –1.41)

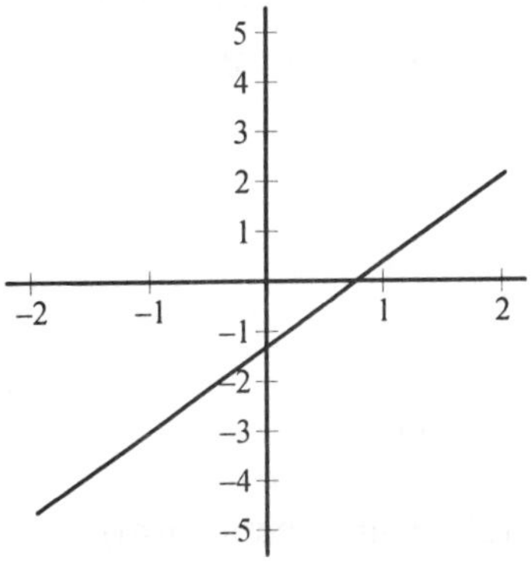

**9.** (a) $y = 0$ (slope 0)

(b) $y = \frac{1}{2}x \left(\text{slope } \frac{1}{2}\right)$

(c) $y = x$ (slope 1)

(d) $y = 2x$ (slope 2)

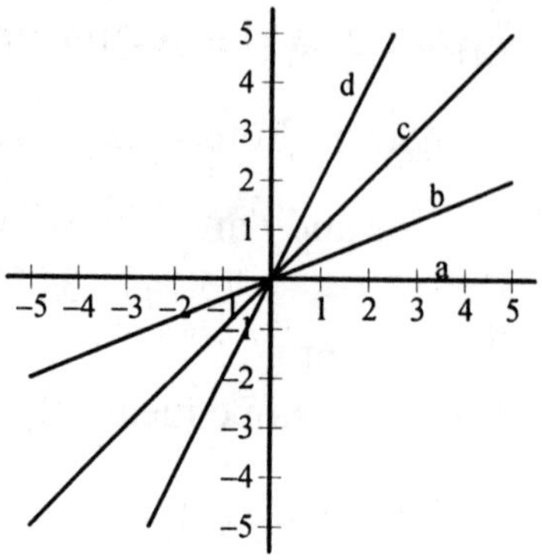


**11.** (a) $y = 2x - 3$

(b) $y = 2x - 1$

(c) $y = 2x$

(d) $y = 2x + 2$

Lines of the form $y = 2x + b$ are parallel.

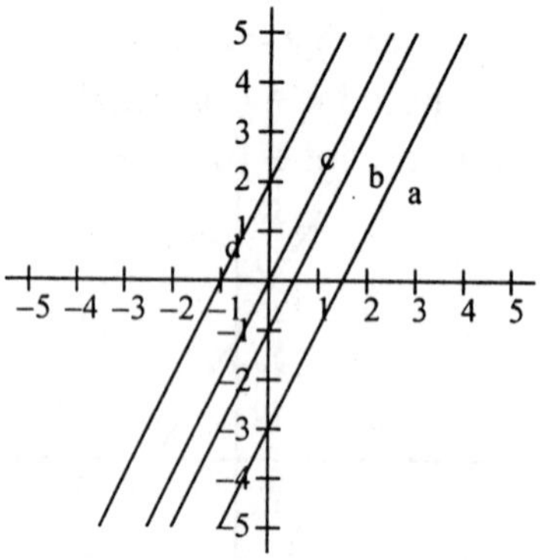


**13.** $\text{slope} = \frac{2}{1} = 2$ (b)

**15.** $\text{slope} = \frac{4}{1} = 4$ (d)

**17.** $\text{slope} = \frac{-2}{-2} = 1$

$y$-intercept = (0,2)

$y = x + 2$ (slope-intercept form)

or $x - y = -2$ (general form)

**19.** $\text{slope} = \frac{-1}{3}$

$y$-intercept = (0,1)

$y = -\frac{1}{3}x + 1$ (slope-intercept form)

$3y = -x + 3$

or $x + 3y = 3$ (general form)

**21.** $\text{slope} = \frac{-2}{3}$

$y$-intercept = (0,0)

$y = -\frac{2}{3}x$ (slope-intercept form)

$3y = -2x$

or $2x + 3y = 0$ (general form)

## 1.2 Parallel and Intersecting Lines

**1.** In slope-intercept form: $L$: $y=-x+10$, $M$: $y=-x+2$. The lines have the same slope $(-1)$, but *different* $y$-intercepts (10 & 2, respectively), so $L$ and $M$ are *parallel.*

**3.** In slope-intercept form: $L: y=-2x+4$, $M: y=2x-8$. The lines have *different* slopes (–2 & 2, respectively), so $L$ and $M$ *intersect* in exactly one point.

**5.** In slope-intercept form: $L$: $y=x+2$, $M$: $y=x+2$. The lines have the same slope (1) and the same $y$-intercept (2), so $L$ and $M$ are *coincident.*

**7.** In slope-intercept form: $L$: $y=\frac{2}{3}x+\frac{8}{3}$, $M$: $y=\frac{2}{3}x+\frac{2}{9}$. The lines have the same slope $\left(\frac{2}{3}\right)$, but *different* $y$-intercepts $\left(\frac{8}{3} \ \& \ \frac{2}{9}, \text{respectively}\right)$, so $L$ and $M$ are *parallel.*

**9.** In slope-intercept form: $L: y=\frac{3}{4}x-\frac{1}{4}$, $M$: $y=\frac{1}{2}x+2$. The lines have *different* slopes $\left(\frac{3}{4} \ \& \ \frac{1}{2}, \text{respectively}\right)$, so $L$ and $M$ *intersect* in exactly one point.

**11.** $L$ is a vertical line (slope undefined) and $M$ is a horizontal line (slope = 0), so $L$ and $M$ intersect.

**13.** In slope-intercept form: $L$: $y=-x+5$, $M$: $y=3x-7$.

$-x_0+5=3x_0-7$

$-4x_0=-12$

$x_0=3;\ y_0=-(3)+5=2$

Point of intersection: $(x_0, y_0)=(3,\ 2)$

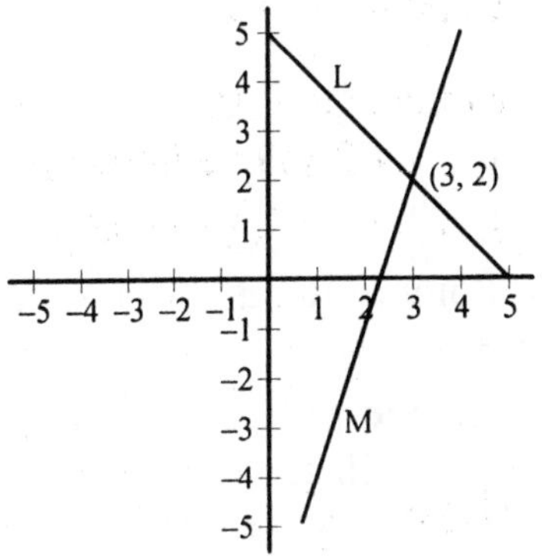


**15.** In slope-intercept form: $L$: $y=x-2$, $M$: $y=-2x+7$.

$x_0-2=-2x_0+7$

$3x_0=9$

$x_0=3;\ y_0=3-2=1$

Point of intersection: $(x_0, y_0)=(3,\ 1)$

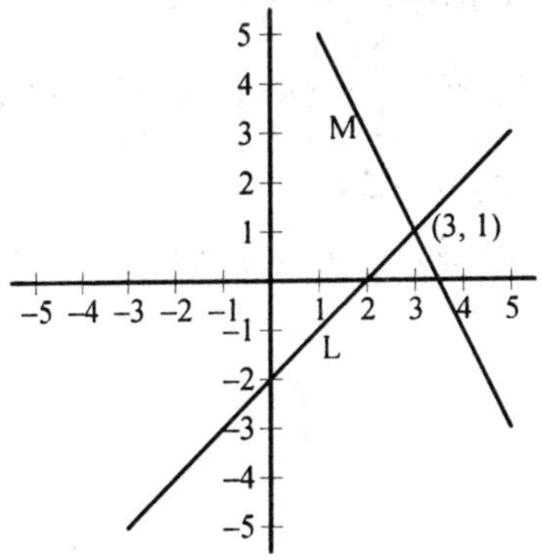

**17.** In slope-intercept form: $L$: $y = -2x + 2$, $M$: $y = 2x - 2$.

$-2x_0 + 2 = 2x_0 - 2$

$-4x_0 = -4$

$x_0 = 1$; $y_0 = -2(1) + 2 = 0$

Point of intersection: $(x_0, y_0) = (1, 0)$

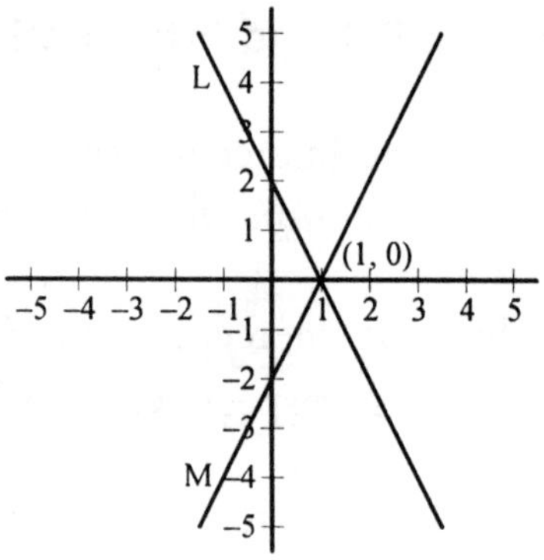


**19.** In slope-intercept form: $L$: $y = \frac{3}{4}x - \frac{1}{2}$, $M$: $y = -\frac{1}{2}x + 2$.

$\frac{3}{4}x_0 - \frac{1}{2} = -\frac{1}{2}x_0 + 2$

$\frac{5}{4}x_0 = \frac{5}{2}$

$x_0 = \frac{4}{5} \cdot \frac{5}{2} = 2$; $y_0 = \frac{3}{4} \cdot 2 - \frac{1}{2} = 1$

Point of intersection: $(x_0, y_0) = (2, 1)$

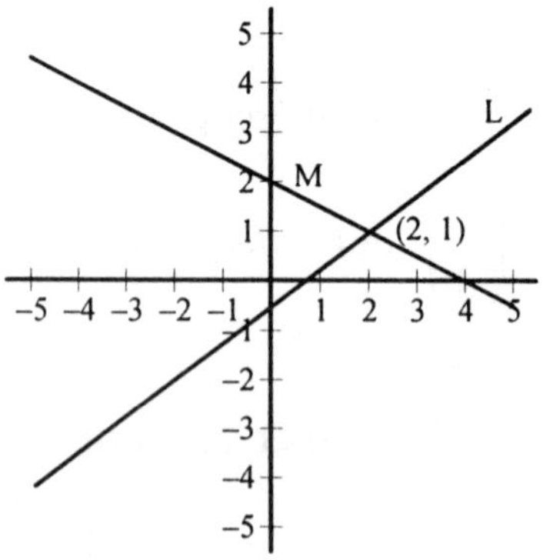


**21.** In slope-intercept form: $L$: $y = \frac{3}{2}x + \frac{5}{2}$, $M$: $y = -3x - 2$.

$\frac{3}{2}x_0 + \frac{5}{2} = -3x_0 - 2$

$\frac{9}{2}x_0 = -\frac{9}{2}$

$x_0 = -1$; $y_0 = \frac{3}{2} \cdot (-1) + \frac{5}{2} = 1$

Point of intersection: $(x_0, y_0) = (-1, 1)$

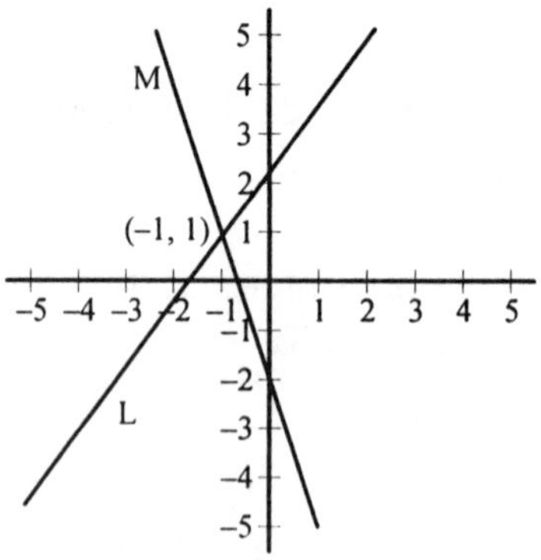


**23.** $L$ is the vertical line that contains all points with $x$-coordinate 4.
$M$ is the horizontal line that contains all points with $y$-coordinate −2.
$L$ and $M$ intersect at (4, −2)

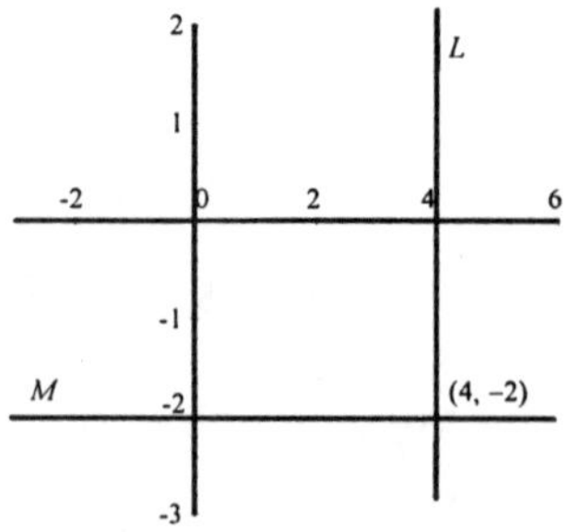

**25.** $-x+3y=2 \Rightarrow y=\frac{1}{3}x+\frac{2}{3}$; $m_1=\frac{1}{3}$; $6x+2y=5 \Rightarrow y=-3x+\frac{5}{2}$; $m_2=-3$. Since $m_1m_2=\frac{1}{3}\cdot(-3)=-1$, the lines are perpendicular.

**27.** $x+2y=7 \Rightarrow y=-\frac{1}{2}x+\frac{7}{2}$; $m_1=-\frac{1}{2}$; $-2x+y=15 \Rightarrow y=2x+15$; $m_2=2$. Since $m_1m_2=-\frac{1}{2}\cdot 2=-1$, the lines are perpendicular.

**29.** $3x+12y=2 \Rightarrow y=-\frac{1}{4}x+\frac{1}{6}$; $m_1=-\frac{1}{4}$; $4x-y=-2 \Rightarrow y=4x+2$; $m_2=4$. Since $m_1m_2=-\frac{1}{4}\cdot 4=-1$, the lines are perpendicular.

**31.** Parallel to $y=2x$, so the slope is 2. Point-slope form: $y-3=2(x-3)=2x-6$. Slope-intercept form $y=2x-3$. General form: $2x-y=3$.

**33.** Perpendicular to $y=2x$, so the slope is $-\frac{1}{2}$. Point-slope form: $y-3=-\frac{1}{2}(x-3) \Rightarrow$ $y-3=-\frac{1}{2}x+\frac{3}{2}$. Slope-intercept form: $y=-\frac{1}{2}x+\frac{9}{2}$. Multiply both sides by 2: $2y=-x+9$ General form: $x+2y=9$.

**35.** Parallel to $y=4x$, so the slope is 4. Point-slope form: $y-2=4(x-(-1)) \Rightarrow y-2=4x+4$. Slope-intercept form: $y=4x+6$. General form: $4x-y=-6$.

**37.** The slope-intercept form of $2x-y+2=0$ is $y=2x+2$, so the slope is 2. Slope-intercept form: $y=2x$. General form: $2x-y=0$.

**39.** Vertical line through $(4, 2)$: $x=4$.

**41.** Perpendicular to $y=2x-5$, so the slope is $-\frac{1}{2}$. Point-slope form: $y-(-2)=-\frac{1}{2}(x-(-1))$ $\Rightarrow y+2=-\frac{1}{2}x-\frac{1}{2}$. Slope-intercept form: $y=-\frac{1}{2}x-\frac{5}{2}$. Multiply both sides by 2: $2y=-x-5$. General form: $x+2y=-5$.

**43.** Perpendicular to $y = 2x - 5$, so the slope is $-\frac{1}{2}$. Point-slope form:

$y - \frac{4}{5} = -\frac{1}{2}\left(x - \left(-\frac{1}{3}\right)\right) \Rightarrow y - \frac{4}{5} = -\frac{1}{2}x - \frac{1}{6}$. Slope-intercept form: $y = -\frac{1}{2}x + \frac{19}{30}$.

Multiply both sides by 30: $30y = -15x + 19$. General form: $15x + 30y = 19$.

**45.** The slope-intercept form of $tx - 4y + 3 = 0$ is $y = \frac{t}{4}x + \frac{3}{4}$; slope is $\frac{t}{4}$. The slope-intercept form of $2x + 2y - 5 = 0$ is $y = -x + \frac{5}{2}$; slope is $-1$. Since the product of the slopes $\frac{t}{4}\cdot(-1)$ must equal $-1$, $t = 4$.

**47.** The slope of the line containing $(-2,\ 9)$ and $(3,\ -10)$ is $m = \frac{9-(-10)}{-2-3} = -\frac{19}{5}$. The requested line is perpendicular to this one, so its slope is $\frac{5}{19}$. Point-slope form:

$y-(-5) = \frac{5}{19}(x-(-2)) \Rightarrow y + 5 = \frac{5}{19}x + \frac{10}{19}$. Slope-intercept form: $y = \frac{5}{19}x - \frac{85}{19}$.

Multiply both sides by 19: $19y = 5x - 85$. General form: $5x - 19y = 85$.

## 1.3 Applications

**1.** (a) $S = 5000\cdot 0 + 80{,}000 = \$80{,}000$

(b) $S = 5000\cdot 3 + 80{,}000 = \$95{,}000$

(c) $S = 5000\cdot(2003-1998) + 80{,}000 = \$105{,}000$

(d) $S = 5000\cdot(2006-1998) + 80{,}000 = \$120{,}000$

**3.** Let $t$ represent the number of years since 1995, and let $C$ represent the average cost of a compact car. Then the data provided yields the points $(t, C) = (0, 8000)$ and $(t, C) = (3, 9500)$. Slope $= \frac{9500-8000}{3-0} = \frac{1500}{3} = 500$. Using the slope-intercept form yields $C = 500t + 8000$. The average cost of a compact car $t$ years after 1995 is $500t + 8000$ dollars. In the year 2000 ($t = 2000 - 1995 = 5$), the average cost should be approximately $C = 500\cdot 5 + 8000 = 2500 + 8000 = \$10{,}500$. The slope means that each year the average cost of a compact car will increase by \$500.

**5.** (a) Let $t$ represent the number of years since 1994, and let $SAT$ represent the average SAT score. Then the data provided yields the points $(t, SAT) = (0,592)$ and $(t, SAT) = (4,564)$. Slope $= \frac{564-592}{4-0} = \frac{-28}{4} = -7$. Using the slope-intercept form yields $SAT = -7t + 592$.

(b) If the trend continues, in the year 2000 ($t = 2000 - 1994 = 6$) the average *SAT* score of incoming freshmen will be $SAT = -7 \cdot 6 + 592 = 550$.

**7.** Revenue is $30x$ dollars and cost is $10x + 600$ dollars, so break-even occurs when $30x = 10x + 600$, $20x = 600$, $x = 30$ items.

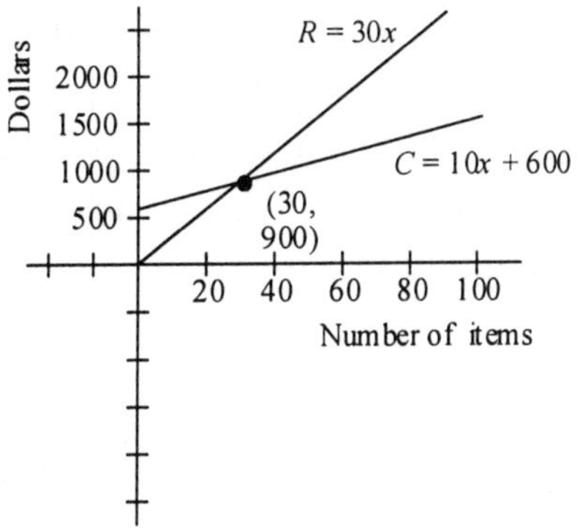


**9.** Revenue is $0.3x$ dollars and cost is $0.2x + 50$ dollars, so break-even occurs when $0.3x = 0.2x + 50$, $0.1x = 50$, $x = 500$ items.

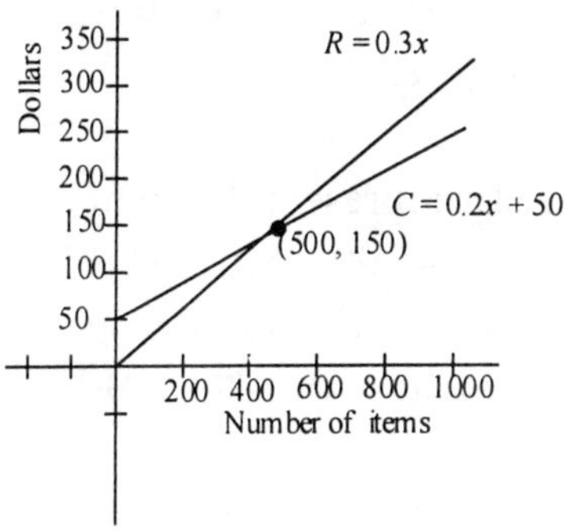


**11.** The cost $C$ of production is the operational overhead ($300) plus the variable cost of producing $x$ items at $0.75 per item.

$C = \$0.75x + \$300 = 0.75x + 300$

$R = \$1x = x$ Break-even occurs when

$x = 0.75x + 300$

$0.25x = 300$

$x = 1200$ items

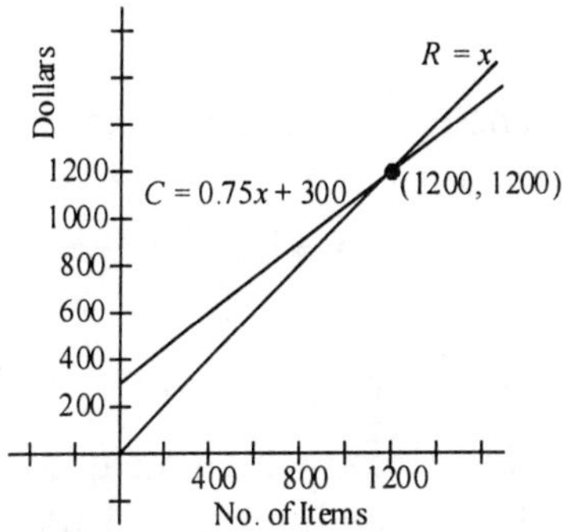


**13.** Let $x$ represent the number of Sunday papers sold. Then revenue is described by $R = 2x$ dollars and cost by $C = 1 \cdot x + 200$ dollars. The break-even point occurs when $R = C$, or $2x = x + 200$; $x = 200$. To break-even, sell 200 Sunday newspapers.

**15.** Let $x$ represent the number of caramels per box and $y$ the number of creams per box. Then $x + y = 50$, or $y = -x + 50$. The cost for the candy in each box is $0.10x + 0.20y$ dollars, and the revenue for each box is $8, so the company breaks-even when $0.10x + 0.20y = 8$

or $y=-0.5x+40$. Solve $y=-x+50$ and $y=-0.5x+40$ ; $-x+50=-0.5x+40$; $-0.5x=-10$; $x=20$; $y=-(20)+50=30$. Each box should contain 20 caramels and 30 creams for no profit or loss. Increasing the number of caramels yields a profit (since the caramels cost less than the creams.)

**17.** Let $x$ represent the dollars invested in AA bonds and $y$ the dollars invested in S&L Certificates. Then $x+y=150{,}000$, or $y=-x+150{,}000$. His annual return on these investments is $0.10x+0.05y$, so he requires $0.10x+0.05y=10{,}000$, or $y=-2x+200{,}000$. Solve $y=-x+150{,}000$ and $y=-2x+200{,}000$; $-x+150{,}000=-2x+200{,}000$; $x=50{,}000$, $y=-50{,}000+150{,}000$; $y=100{,}000$. He should invest \$50,000 in AA bonds and \$100,000 in S&L Certificates.

**19.** Let $x$ represent the number of adult patrons and $y$ the number of child patrons. From the attendance figures, $x+y=2600$, or $y=-x+2600$. From the receipts, $8x+4y=16{,}440$, or $y=-2x+4110$. Solve $y=-x+2600$ and $y=-2x+4110$; $-x+2600=-2x+4110$; $x=1510$, $y=-1510+2600=1090$. There were 1510 adult patrons (and 1090 child patrons).

**21.** Let $x$ represent the cc of 15% acid and $y$ the cc of 5% acid in a 100 cc mixture. Then $x+y=100$, or $y=-x+100$ and $0.15x+0.05y=0.08(100)$, or $y=-3x+160$. Solve $y=-x+100$ and $y=-3x+160$; $-x+100=-3x+160$; $2x=60$, $x=30$; $y=-30+100=70$. Mix 30 cc of the 15% acid solution with 70 cc of the 5% acid solution.

**23.** When $S=D$, $p+1=3-p$; $2p=2$; $p=1$. The market price is \$1.00.

**25.** When $S=D$, $20p+500=1000-30p$; $50p=500$; $p=10$. The market price is \$10.00.

**27.** The market price occurs when $S=D$, or

$$0.7p+0.4=-0.5p+1.6$$
$$1.2p=1.2$$
$$p=1$$

(when $p=1$, $S=0.7(1)+0.4=1.1=D$). The market price is \$1.00, at which the quantity supplied and the quantity demanded both are 1.1 units. Graphically, this corresponds to the point of intersection of the supply ($S$) and demand ($D$) lines.

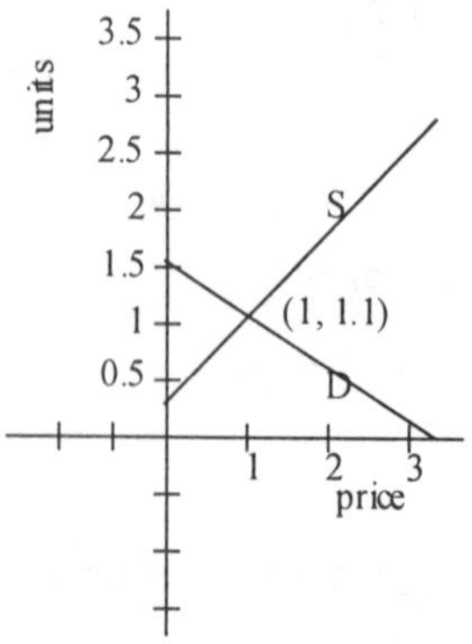


**29.** $S(3)=2(3)+5=11$, so the demand equation satisfies $(p,\ D)=(3,\ 11)$ (at market price, supply and demand are equal) and $(p,\ D)=(1,\ 19)$. The slope of the demand line is $\dfrac{11-19}{3-1}=-4$. The point-slope form of the demand equation is $D-11=-4(p-3)$, or $D=-4p+23$.

## Chapter 1 Review

### True or False

**1.** False **3.** True **5.** False **7.** False

**9.** False

### Fill in the Blank

**1.** *x*-coordinate; *y*-coordinate **3.** negative

**5.** coincident **7.** intersecting

### Review Exercises

**1.** $y = -2x + 3$;

| $x$ | $-2$ | $0$ | $2$ |
|---|---|---|---|
| $y$ | $7$ | $3$ | $-1$ |

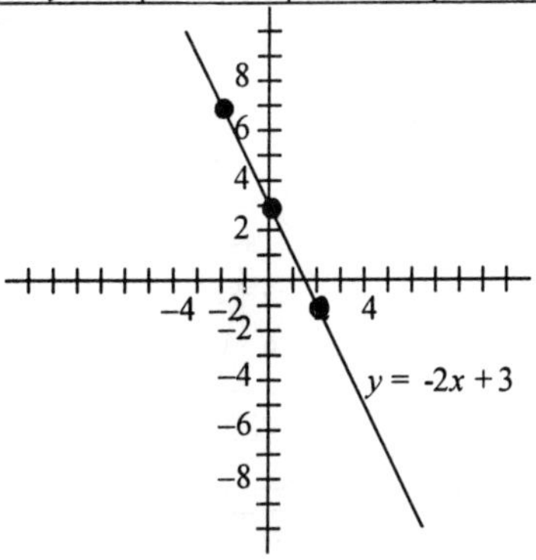


**3.** $2y = 3x + 6$, or $y = \frac{3}{2}x + 3$.

| $x$ | $-2$ | $0$ | $2$ |
|---|---|---|---|
| $y$ | $0$ | $3$ | $6$ |

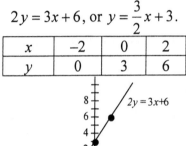

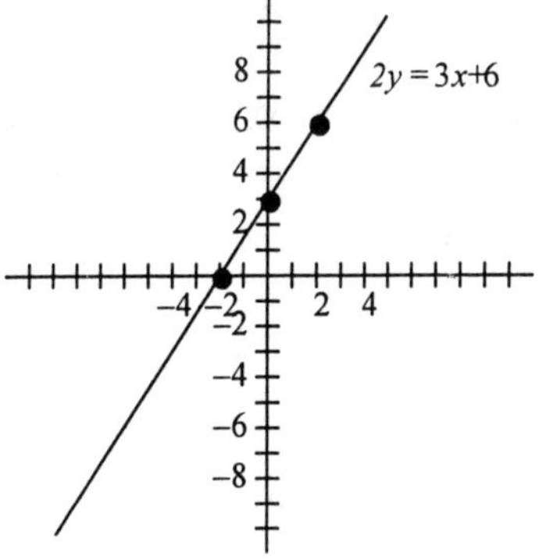


**5.** slope $= \frac{2-4}{1-(-3)} = -\frac{1}{2}$. Point-slope equation: $y - 2 = -\frac{1}{2}(x-1) \Rightarrow 2y - 4 = -(x-1) \Rightarrow$ $2y - 4 = -x + 1 \Rightarrow$ General equation: $x + 2y = 5$.

**7.** slope $= \frac{0-3}{0-(-2)} = -\frac{3}{2}$. Slope-intercept equation: $y = -\frac{3}{2}x \Rightarrow 2y = -3x \Rightarrow$ General equation: $3x + 2y = 0$.

**9.** Point-slope form: $y - (-1) = -3(x-2) \Rightarrow y + 1 = -3x + 6 \Rightarrow$ Slope-intercept form: $y = -3x + 5 \Rightarrow$ General form: $3x + y = 5$.

**11.** Lines with slope = 0 are horizontal. The horizontal line containing (–3, 4) has equation $y = 4$.

**13.** Slope $= \frac{0-(-5)}{2-4} = -\frac{5}{2}$; Point-slope form: $y-0=-\frac{5}{2}(x-2) \Rightarrow$ Slope-intercept form: $y=-\frac{5}{2}x+5$. Multiply both sides by 2: $2y=-5x+10 \Rightarrow$ General form: $5x+2y=10$.

**15.** Slope $= \frac{0-(-4)}{-3-0} = -\frac{4}{3}$. Slope-intercept form: $y=-\frac{4}{3}x-4$. Multiply both sides by 3: $3y=-4x-12 \Rightarrow$ General form: $4x+3y=-12$.

**17.** $2x+3y+4=0$ is equivalent to $y=-\frac{2}{3}x-\frac{4}{3}$, so the slope is $-\frac{2}{3}$. Point-slope form: $y-3=-\frac{2}{3}(x-(-5)) \Rightarrow y-3=-\frac{2}{3}x-\frac{10}{3} \Rightarrow$ Slope-intercept form: $y=-\frac{2}{3}x-\frac{1}{3}$. Multiply both sides by 3: $3y=-2x-1$ General form: $2x+3y=-1$.

**19.** $2x+3y+4=0$ is equivalent to $y=-\frac{2}{3}x-\frac{4}{3}$, so the slope is $\frac{3}{2}$. Point-slope form: $y-3=\frac{3}{2}(x-(-5)) \Rightarrow y-3=\frac{3}{2}x+\frac{15}{2} \Rightarrow$ Slope-intercept form: $y=\frac{3}{2}x+\frac{21}{2}$. Multiply both sides by 2: $2y=3x+21 \Rightarrow$ General form: $3x-2y=-21$.

**21.**
$$-9x-2y+18=0$$
$$-2y=9x-18$$
$$y=-\frac{9}{2}x+9$$

slope $=-\frac{9}{2}$; *y*-intercept: (0, 9)

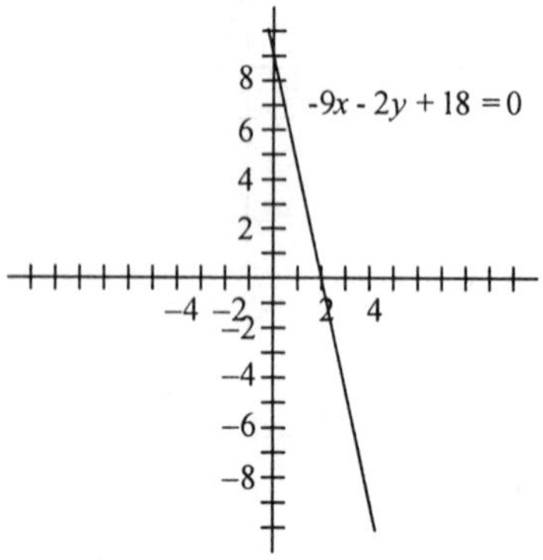


**23.**
$$4x+2y-9=0$$
$$2y=-4x+9$$
$$y=-2x+\frac{9}{2}$$

slope $=-2$; *y*-intercept: $\left(0, \frac{9}{2}\right)$

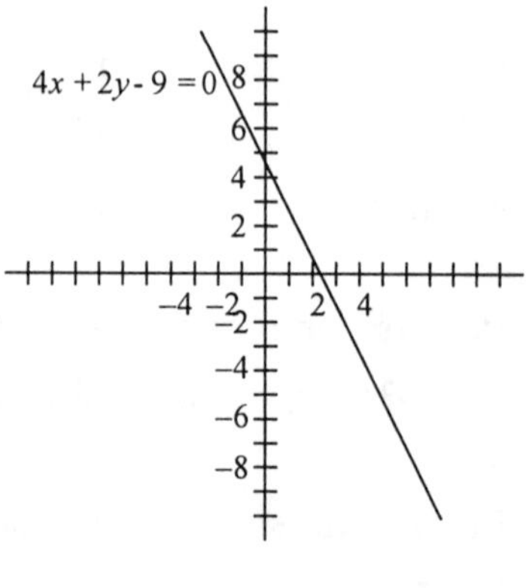

**25.** $3x-4y+12=0 \Rightarrow y=\frac{3}{4}x+3$; $6x-8y+9=0 \Rightarrow y=\frac{3}{4}x+\frac{9}{8}$: same slope, but different $y$-intercepts, so the lines are parallel.

**27.** $x-y+2=0 \Rightarrow y=x+2$; $3x-4y+12=0 \Rightarrow y=\frac{3}{4}+3$: different slopes $\left(1 \neq \frac{3}{4}\right)$, so the lines are intersecting.

**29.** $4x+6y+12=0 \Rightarrow y=-\frac{2}{3}x-2$; $2x+3y+6=0 \Rightarrow y=-\frac{2}{3}x-2$: same slopes and same $y$-intercepts, so the lines are coincident.

**31.** In slope-intercept form: $L$: $y=x-4$, $M$: $y=-\frac{1}{2}x+\frac{7}{2}$.

$$x_0-4=-\frac{1}{2}x_0+\frac{7}{2}$$
$$\frac{3}{2}x_0=\frac{15}{2}$$
$$x_0=5;\ y_0=5-4=1$$

Point of intersection: $(x_0, y_0)=(5, 1)$

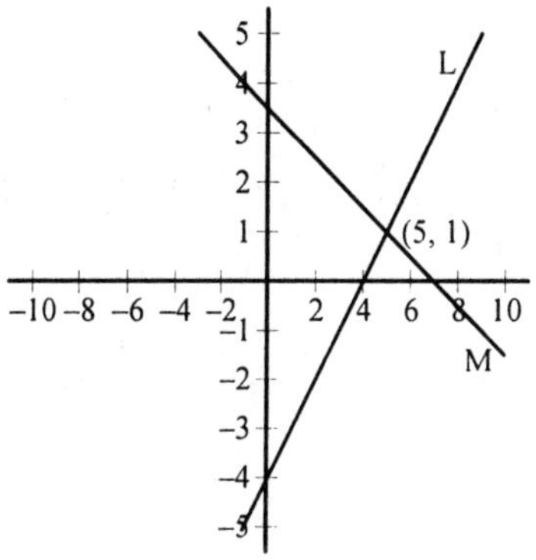


**33.** In slope-intercept form: $L$: $y=x+2$, $M$: $y=-\frac{1}{2}x+\frac{7}{2}$.

$$x_0+2=-\frac{1}{2}x_0+\frac{7}{2}$$
$$\frac{3}{2}x_0=\frac{3}{2}$$
$$x_0=1;\ y_0=1+2=3$$

Point of intersection: $(x_0, y_0)=(1, 3)$

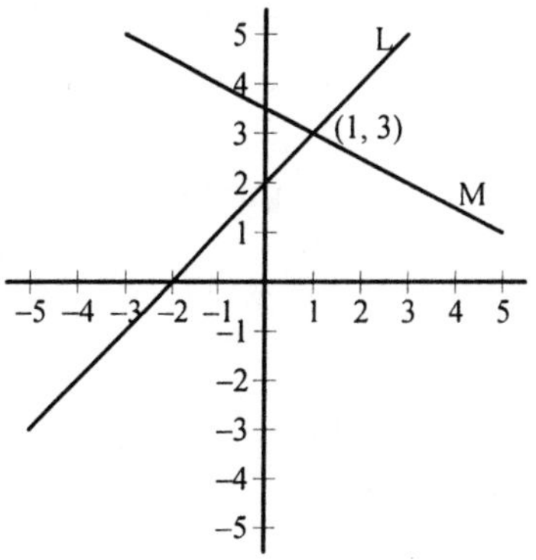


**35.** In slope-intercept form: $L$: $y=\frac{1}{2}x+2$, $M$: $y=-\frac{1}{2}x$.

$$\frac{1}{2}x_0+2=-\frac{1}{2}x_0$$
$$x_0=-2;\ y_0=-\frac{1}{2}\cdot(-2)=1$$

Point of intersection: $(x_0, y_0)=(-2, 1)$

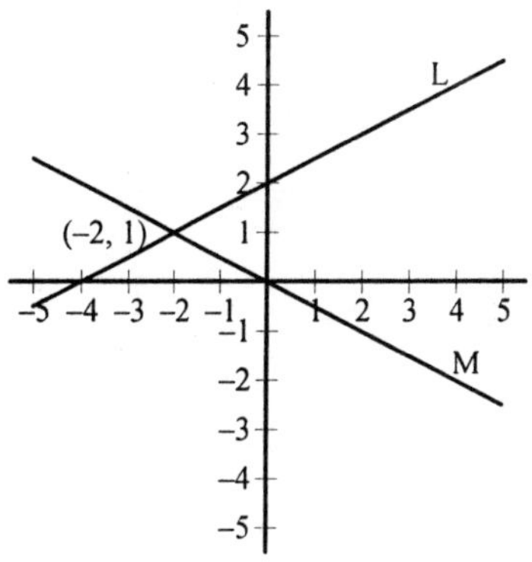

**37.** Let $x = \text{dollars invested in bonds}$ and $y = \text{dollars invested in the bank}$. Then $x + y = 90{,}000$ (total amount to invest) and $0.12x + 0.05y = 10{,}000$ (total annual interest income). The first equation is equivalent to $y = -x + 90{,}000$, and the second equation is equivalent to $y = -2.4x + 200{,}000$. Solve $-x + 90{,}000 = -2.4x + 200{,}000$; $1.4x = 110{,}000$; $x = 78{,}571.43$. $y = -78{,}571.43 + 90{,}000 = 11{,}428.57$. Invest \$78,571.43 in bonds and \$11,428.57 in the bank.

**39.** (a) & (b) See graph

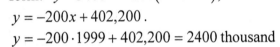

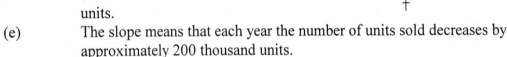

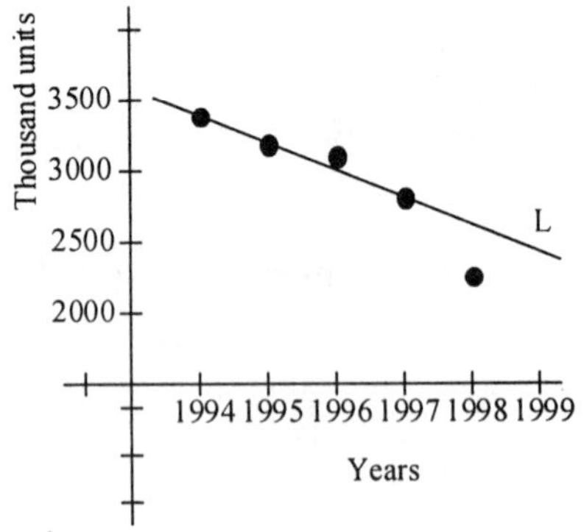


(c) *Note*: answers to (c) & (d) vary with your choice of *L*.
Using points (1994, 3400) and (1997, 2800), slope $= \dfrac{3400 - 2800}{1994 - 1997} = -200$; point-slope form: $y - 3400 = -200(x - 1994)$, or $y = -200x + 402{,}200$.

(d) $y = -200 \cdot 1999 + 402{,}200 = 2400$ thousand units.

(e) The slope means that each year the number of units sold decreases by approximately 200 thousand units.

## Professional Exam Questions

**1.** b: Let $x = \text{number of units}$, then $\text{costs} = 0.40(\$2)x + \$6000 = 0.8x + 6000$ dollars, and $\text{revenue} = 2x$ dollars. Solve $2x = 0.8x + 6000$: $1.2x = 6000$; $x = 5000$ units.

**3.** d: Let $x = \text{number of units}$, then $\text{costs} = (VC)x + FC$ and $\text{revenue} = (SP)x$. Solve $(VC)x + FC = (SP)x$: $FC = (SP - VC)x$; $x = \dfrac{FC}{SP - VC}$.

**5.** c: Straight-line depreciation is constant and sum-of-the-years'-digits depreciation is decreasing.

**7.** b: Total cost, i.e., total factory overhead.

# r 2

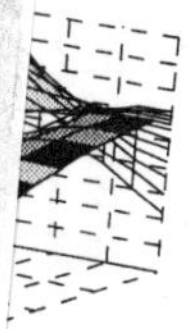

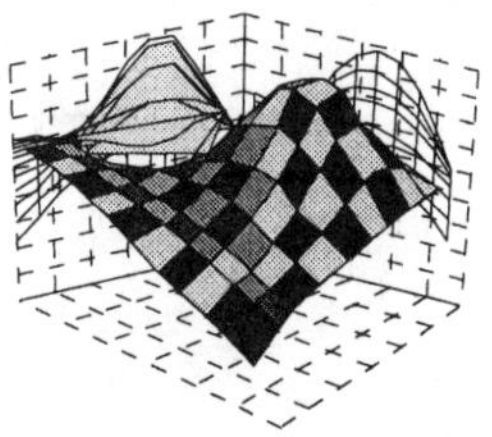

# Systems of Linear Equations; Matrices

## near Equations: Substitution; Elimination

$+1=5$

$10-2=8$

**3.** $\begin{cases} 3(2)-4\left(\frac{1}{2}\right)=6-2=4 \\ 2-3\left(\frac{1}{2}\right)=2-\frac{3}{2}=\frac{1}{2} \end{cases}$

$2(2)=3-3+4=4$

**5.** $1-1(-1)-2=1+1-2=0$

$2(-1)-3(2)=-2-6=-8$

**7.** $\begin{cases} x+y=9 \\ x-y=3 \end{cases}$ Add the two equations: $2x=12; x=6$. Back-substitute in the first equation: $6+y=9; y=3$. Solution: $x=6, y=3$.

**9.** $\begin{cases} 5x-\ \ y=13 \ (1) \\ 2x+3y=12 \ (2) \end{cases}$ $\begin{matrix} 3(1): \\ (2): \end{matrix} \begin{cases} 15x-3y=39 \\ 2x+3y=12 \end{cases}$ Add these two equations: $17x=51; x=3$. Back-substitute in (2): $2(3)+3y=12$; $6+3y=12$; $3y=6$; $y=2$. Solution: $x=3, y=2$.

**11.** $\begin{cases} 3x \quad\quad =24 \ (1) \\ x+2y=\ 0 \ (2) \end{cases}$ Solve (1) for $x$: $x=8$. Back-substitute in (2): $8+2y=0; 2y=-8$; $y=-4$. Solution: $x=8, y=-4$.

**13.** $\begin{cases} 3x-6y=24 & (1) \\ 5x+4y=12 & (2) \end{cases}$ $\quad \begin{matrix} 4(1): \\ 6(2): \end{matrix} \begin{cases} 12x-24y=96 \\ 30x+24y=72 \end{cases}$

Add these two equations: $42x = 168$; $x = 4$.
Back-substitute in (2): $5(4)+4y=12$;
$20+4y=12$; $4y=-8$; $y=-2$.
Solution: $x=4, y=-2$.

**15.** $\begin{cases} 2x+\ \ y=1 & (1) \\ 4x+2y=6 & (2) \end{cases}$ $\quad \begin{matrix} -2(1): \\ (2): \end{matrix} \begin{cases} -4x-2y=-2 \\ \ \ \ 4x+2y=\ \ 6 \end{cases}$

Add these two equations: $0 = 4$.
The system is inconsistent.

**17.** $\begin{cases} 2x-4y=-2 & (1) \\ 3x+2y=\ \ 3 & (2) \end{cases}$ $\quad \begin{matrix} (1): \\ 2(2): \end{matrix} \begin{cases} 2x-4y=-2 \\ 6x+4y=\ \ 6 \end{cases}$

Add these two equations: $8x = 4$; $x = \frac{1}{2}$.

Back-substitute in (2): $3\left(\frac{1}{2}\right)+2y=3$;

$\frac{3}{2}+2y=3$; $2y=\frac{3}{2}$; $y=\frac{3}{4}$.

Solution: $x=\frac{1}{2}$, $y=\frac{3}{4}$.

**19.** $\begin{cases} x+2y=4 & (1) \\ 2x+4y=8 & (2) \end{cases}$ $\quad \begin{matrix} -2(1): \\ (2): \end{matrix} \begin{cases} -2x-4y=-8 \\ \ \ \ 2x+4y=\ \ 8 \end{cases}$

Add these two equations: $0 = 0$; the system has infinite solutions.

Solution: Solve either equation for *y*: the solution is : $y = 2-\frac{1}{2}x$, $x$ is any real number; alternatively, solve either equation for *x*: the solution is: $x = 4-2y$, $y$ is any real number.

**21.** $\begin{cases} 2x-\ 3y=-1 & (1) \\ 10x+10y=\ \ 5 & (2) \end{cases}$ $\quad \begin{matrix} -5(1): \\ (2): \end{matrix} \begin{cases} -10x+15y=5 \\ \ \ \ 10x+10y=5 \end{cases}$

Add these two equations: $25y = 10$; $y = \frac{2}{5}$.

Back-substitute in (2): $10x+10\left(\frac{2}{5}\right)=5$;

$10x+4=5$; $10x=1$; $x=\frac{1}{10}$.

Solution: $x=\frac{1}{10}, y=\frac{2}{5}$.

**23.** $\begin{cases} 2x+3y=\ \ 6 & (1) \\ x-\ \ y=1/2 & (2) \end{cases}$ $\quad \begin{matrix} (1): \\ 3(2): \end{matrix} \begin{cases} 2x+3y=6 \\ 3x-3y=\frac{3}{2} \end{cases}$

Add these two equations: $5x = \frac{15}{2}$; $x = \frac{3}{2}$.

Back-substitute in (1): $2\left(\frac{3}{2}\right)+3y=6$;

$3+3y=6$; $3y=3$; $y=1$.

Solution: $x=\frac{3}{2}, y=1$.

**25.** $\begin{cases} 2x+3y=5 & (1) \\ 4x+6y=10 & (2) \end{cases}$ $\quad \begin{matrix} -2(1): \\ (2): \end{matrix} \begin{cases} -4x-6y=-10 \\ 4x+6y=10 \end{cases}$ Add these two equations: $0=0$; the system has infinite solutions.

Solution: Solve either equation for *y*: the solution is : $y=\frac{5}{3}-\frac{2}{3}x$, *x* is any real number;

alternatively, solve either equation for *x*: the solution is: $x=\frac{5}{2}-\frac{3}{2}y$, *y* is any real number.

**27.** $\begin{cases} 3x-5y=3 & (1) \\ 15x+5y=21 & (2) \end{cases}$ Add the two equations: $18x=24$; $x=\frac{4}{3}$.

Back-substitute in (1): $3\left(\frac{4}{3}\right)-5y=3$; $4-5y=3$; $-5y=-1$; $y=\frac{1}{5}$.

Solution: $x=\frac{4}{3}, y=\frac{1}{5}$.

**29.** $\begin{cases} x-y=6 & (1) \\ 2x-3z=16 & (2) \\ 2y+z=4 & (3) \end{cases} \Leftrightarrow \begin{cases} x-y=6 & (1) \\ 2y-3z=4 & (2) \leftarrow (2)-2(1) \\ 2y+z=4 & (3) \end{cases} \Leftrightarrow \begin{cases} x-y=6 & (1) \\ 2y-3z=4 & (2) \\ 4z=0 & (3) \leftarrow (3)-(2) \end{cases}$

$\Leftrightarrow \begin{cases} x-y=6 & (1) \\ 2y=4 & (2) \text{ :subst. } z=0 \\ z=0 & (3) \leftarrow 1/4\,(3) \end{cases} \Leftrightarrow \begin{cases} x-y=6 & (1) \\ y=2 & (2) \leftarrow 1/2\,(2) \\ z=0 & (3) \end{cases} \Leftrightarrow \begin{cases} x-2=6 & (1) \text{ :subst. } y=2 \\ y=2 & (2) \\ z=0 & (3) \end{cases}$

$\Leftrightarrow \begin{cases} x=8 & (1) \text{ :solve for } x \\ y=2 & (2) \\ z=0 & (3) \end{cases}$

**31.** $\begin{cases} x-2y+3z=7 & (1) \\ 2x+y+z=4 & (2) \\ -3x+2y-2z=-10 & (3) \end{cases} \Leftrightarrow \begin{cases} x-2y+3z=7 & (1) \\ 5y-5z=-10 & (2) \leftarrow (2)-2(1) \\ -4y+7z=11 & (3) \leftarrow (3)+3(1) \end{cases} \Leftrightarrow \begin{cases} x-2y+3z=7 & (1) \\ y-z=-2 & (2) \leftarrow 1/5\,(2) \\ -4y+7z=11 & (3) \end{cases}$

$\Leftrightarrow \begin{cases} x+z=3 & (1) \leftarrow (1)+2(2) \\ y-z=-2 & (2) \\ 3z=3 & (3) \leftarrow (3)+4(2) \end{cases} \Leftrightarrow \begin{cases} x+1=3 & (1) \text{ :subst. } z=1 \\ y-1=-2 & (2) \text{ :subst. } z=1 \\ z=1 & (3) \leftarrow 1/3\,(3) \end{cases} \Leftrightarrow \begin{cases} x=2 & (1) \\ y=-1 & (2) \\ z=1 & (3) \end{cases}$

**33.** $\begin{cases} x-y-z=1 & (1) \\ 2x+3y+z=2 & (2) \\ 3x+2y=0 & (3) \end{cases} \Leftrightarrow \begin{cases} x-y-z=1 & (1) \\ 5y+3z=0 & (2) \leftarrow (2)-2(1) \\ 5y+3z=-3 & (3) \leftarrow (3)-3(1) \end{cases} \Leftrightarrow \begin{cases} x-y-z=1 & (1) \\ 5y+3z=0 & (2) \\ 0=-3 & (3) \leftarrow (3)-(2) \end{cases}$

Inconsistent system.

**35.** $$\begin{cases} x-y-z=1 & (1) \\ -x+2y-3z=-4 & (2) \\ 3x-2y-7z=0 & (3) \end{cases} \Leftrightarrow \begin{cases} x-y-z=1 & (1) \\ y-4z=-3 & (2) \leftarrow (2)+(1) \\ y-4z=-3 & (3) \leftarrow (3)-3(1) \end{cases} \Leftrightarrow \begin{cases} x-5z=-2 & (1) \leftarrow (1)+(2) \\ y-4z=-3 & (2) \\ 0=0 & (3) \leftarrow (3)-(2) \end{cases}$$

Solution: $z$ is any real number, $x = 5z - 2$, and $y = 4z - 3$

**37.** $$\begin{cases} 2x-2y+3z=6 & (1) \\ 4x-3y+2z=0 & (2) \\ -2x+3y-7z=1 & (3) \end{cases} \Leftrightarrow \begin{cases} 2x-2y+3z=6 & (1) \\ y-4z=-12 & (2) \leftarrow (2)-2(1) \\ y-4z=7 & (3) \leftarrow (3)+(1) \end{cases} \Leftrightarrow \begin{cases} 2x-2y+3z=6 & (1) \\ y-4z=-12 & (2) \\ 0=19 & (3) \leftarrow (3)-(2) \end{cases}$$

Inconsistent system.

**39.** $$\begin{cases} x+y-z=6 & (1) \\ 3x-2y+z=-5 & (2) \\ x+3y-2z=14 & (3) \end{cases} \Leftrightarrow \begin{cases} x+y-z=6 & (1) \\ -5y+4z=-23 & (2) \leftarrow (2)-3(1) \\ 2y-z=8 & (3) \leftarrow (3)-(1) \end{cases} \Leftrightarrow \begin{cases} x+y-z=6 & (1) \\ -5y+4z=-23 & (2) \\ 3z=-6 & (3) \leftarrow 5(3)+2(2) \end{cases}$$

$$\Leftrightarrow \begin{cases} x+y+2=6 & (1) \text{ :subst. } z=-2 \\ -5y-8=-23 & (2) \text{ :subst. } z=-2 \\ z=-2 & (3) \leftarrow 1/3(3) \end{cases} \Leftrightarrow \begin{cases} x+3+2=6 & (1) \text{ :subst. } y=3 \\ y=3 & (2) \text{ :solve for } y \\ z=-2 & (3) \end{cases}$$

$$\Leftrightarrow \begin{cases} x=1 & (1) \text{ :solve for x} \\ y=3 & (2) \\ z=-2 & (3) \end{cases}$$

**41.** $$\begin{cases} x+2y-z=-3 & (1) \\ 2x-4y+z=-7 & (2) \\ -2x+2y-3z=4 & (3) \end{cases} \Leftrightarrow \begin{cases} x+2y-z=-3 & (1) \\ -8y+3z=-1 & (2) \leftarrow (2)-2(1) \\ 6y-5z=-2 & (3) \leftarrow (3)+2(1) \end{cases}$$

$$\Leftrightarrow \begin{cases} 4x-z=-13 & (1) \leftarrow 4(1)+(2) \\ -8y+3z=-1 & (2) \\ -11z=-11 & (3) \leftarrow 4(3)+3(2) \end{cases} \Leftrightarrow \begin{cases} 4x-1=-13 & (1) \text{ :subst. } z=1 \\ -8y+3=-1 & (2) \text{ :subst. } z=1 \\ z=1 & (3) \leftarrow -1/11(3) \end{cases}$$

$$\Leftrightarrow \begin{cases} x=-3 & (1) \text{ :solve for } x \\ y=1/2 & (2) \text{ :solve for } y \\ z=1 & (3) \end{cases}$$

**43.** Let $x$ represent one number and $y$ the other number. Then $x + y = 81$ and either $2x - 3y = 62$ or $3y - 2x = 62$. Thus, solve either $\begin{cases} x+y=81 \\ 2x-3y=62 \end{cases}$ or $\begin{cases} x+y=81 \\ -2x+3y=62 \end{cases}$

$$\begin{cases} x+y=81 & (1) \\ 2x-3y=62 & (2) \end{cases} \Leftrightarrow \begin{cases} x+y=81 & (1) \\ -5y=-100 & (2) \leftarrow (2)-2(1) \end{cases} \Leftrightarrow \begin{cases} x+20=81 & (1) \text{ : subst. } y=20 \\ y=20 & (2) \leftarrow -1/5(2) \end{cases} \Leftrightarrow \begin{cases} x=61 \\ y=20 \end{cases} \ldots$$

$$\text{or} \ldots \begin{cases} x + y = 81 & (1) \\ -2x + 3y = 62 & (2) \end{cases} \Leftrightarrow \begin{cases} x + y = 81 & (1) \\ 5y = 224 & (2) \leftarrow (2) + 2(1) \end{cases} \Leftrightarrow \begin{cases} x + 224/5 = 81 & (1) \text{ :subst. } y = 224/5 \\ y = 224/5 & (2) \leftarrow 1/5\,(2) \end{cases}$$

$$\Leftrightarrow \begin{cases} x = \frac{181}{5} \\ y = \frac{224}{5} \end{cases}$$ The numbers are 61 and 20 or $\frac{181}{5}$ and $\frac{224}{5}$.

**45.** Let $L$ represent the length of the room and $W$ the width of the room. Then $perimeter = 2L + 2W = 90$ feet and $L = 2W$. Using substitution, $2(2W) + 2W = 90$; $4W + 2W = 90$; $6W = 90; W = 15$. Since $L = 2W$; $L = 2 \cdot 15 = 30$. The floor is 30 feet long and 15 feet wide.

**47.** Let $x$ represent the price of a cheeseburger and $y$ the price of a chocolate shake. Then $4x + 2y = \$7.90$ and $2y - x = \$0.15$.

Solve: $$\begin{cases} 4x + 2y = 7.90 & (1) \\ -x + 2y = 0.15 & (2) \end{cases} \Leftrightarrow \begin{cases} 5x = 7.75 & (1) \leftarrow (1) - (2) \\ -x + 2y = 0.15 & (2) \end{cases}$$

$$\Leftrightarrow \begin{cases} x = 1.55 & (1) \leftarrow 1/5\,(1) \\ -1.55 + 2y = 0.15 & (2) \text{ :subst. } x = 1.55 \end{cases} \Leftrightarrow \begin{cases} x = 1.55 & (1) \\ y = 0.85 & (2) \text{ :solve for } y \end{cases}$$ . Cheeseburgers cost \$1.55 each, and chocolate shakes cost \$0.85 each.

**49.** Let $x$ represent the amount invested in AA bonds and $y$ the amount invested in Savings & Loan Certificates. Then $x + y = 50{,}000$ and $0.15x + 0.07y = 6000$:

$$\begin{aligned} 0.15x + 0.07(50{,}000 - x) &= 6000 \\ 0.15x + 3500 - 0.07x &= 6000 \\ 0.08x &= 2500 \\ x &= \$31{,}250 \\ y &= \$18{,}750 \end{aligned}$$

Invest \$31,250 in bonds and \$18,750 in Savings and Loan Certificates.

**51.** Let $x$ represent the number of nickels and $y$ the number of quarters. Then $x + y = 13$ and $5x + 25y = 165$; solving by substitution:

$$\begin{aligned} 5(13 - y) + 25y &= 165 \\ 65 - 5y + 25y &= 165 \\ 20y &= 100 \\ y &= 5 \\ x + 5 &= 13 \\ x &= 8 \end{aligned}$$

Yes, she has 8 nickels in her bank.

**53.** Let $x$ represent the cc of 30% acid solution and $y$ the cc of 10% acid solution. Solve

$$\begin{cases} x + y = 100 \\ 0.3x + 0.1y = 0.18 \cdot 100 \end{cases}$$

$$\begin{cases} x + y = 100_{(1)} \\ 0.3x + 0.1y = 18_{(2)} \end{cases} \Leftrightarrow \begin{cases} x + y = 100_{(1)} \\ -0.2y = -12_{(2) \leftarrow (2) - .3(1)} \end{cases} \Leftrightarrow \begin{cases} x + y = 100_{(1)} \\ y = 60_{(2) \leftarrow -5(2)} \end{cases} \Leftrightarrow \begin{cases} x = 40 \\ y = 60 \end{cases}$$

Combine 40 cc of the 30% acid solution with 60 cc of the 10% acid solution.

**55.** Let $x$ represent the number of adult patrons and $y$ the number of child patrons. Then $x + y = 5{,}200$ and $8x + 4y = 33{,}840$. Using substitution, $8x + 4(5{,}200 - x) = 33{,}840$; $8x + 20{,}800 - 4x = 33{,}840$; $4x + 20{,}800 = 33{,}840$; $4x = 13{,}040$; $x = 3{,}260$. $3260 + y = 5{,}200$; $y = 1940$.

There were 3,260 adult patrons (and 1,940 child patrons).

**57.** Let $x$ represent the dollars invested at 10% and $y$ the dollars invested at 12%. Then $x + y = 50{,}000$ and $0.10x + 0.12y = 5250$.

$$\text{Solve}: \begin{cases} x + y = 50000_{(1)} \\ 0.10x + 0.12y = 5250_{(2)} \end{cases} \Leftrightarrow \begin{cases} x + y = 50000_{(1)} \\ 0.02y = 250_{(2) \leftarrow (2) - 0.1(1)} \end{cases}$$

$$\Leftrightarrow \begin{cases} x + y = 50000_{(1)} \\ y = 12500_{(2) \leftarrow 50(2)} \end{cases} \Leftrightarrow \begin{cases} x = 37500 \\ y = 12500 \end{cases}$$

\$37,500 was invested at 10% and \$12,500 was invested at 12%.

**59.** Let $x$ represent the number of acres of corn and $y$ the number of acres of soybeans. Then $x + y = 1000$ and $62x + 44y = 45{,}800$. Using substitution:

$$62x + 44(1000 - x) = 45{,}800$$
$$62x + 44{,}000 - 44x = 45{,}800$$
$$18x = 1{,}800$$
$$x = 100$$
$$y = 1000 - x = 900$$

Mr. Smith should plant 100 acres of corn and 900 acres of sybeans.

**61.** Let $x$ represent the number of \$20 sets and $y$ the number of \$15 sets. Then $x + y = 200$ and $20x + 15y = 3200$. Using substitution:

$$20x + 15(200 - x) = 3200$$
$$20x + 3000 - 15x = 3200$$
$$5x + 3000 = 3200$$
$$5x = 200 \Rightarrow x = 40$$
$$y = 200 - x = 200 - 40 = 160$$

He should buy 40 of the \$20 sets and 160 of the \$15 sets.

## Technology Exercises

**1.** Intersection: $x = -2.54, y = 28.81$

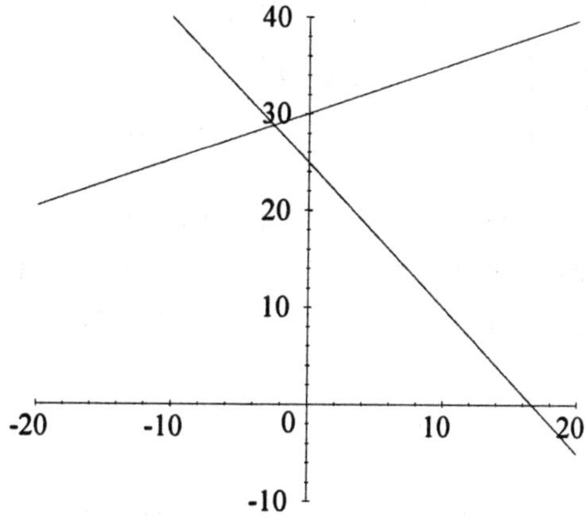


**3.** Intersection: $x = 3.07, y = -.22$

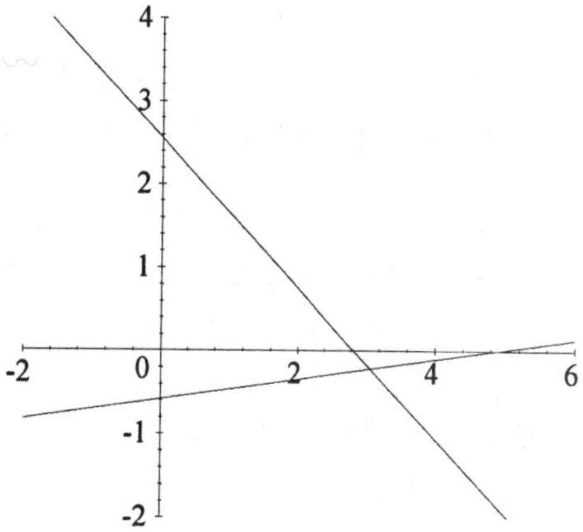


## 2.2 Systems of Linear Equations: Matrix Method

**1.** $\left[\begin{array}{cc|c} 2 & -3 & 5 \\ 1 & -1 & 3 \end{array}\right]$

**3.** $\left[\begin{array}{cc|c} 2 & 1 & -6 \\ 3 & 1 & -1 \end{array}\right]$

**5.** $\left[\begin{array}{ccc|c} 2 & -1 & -1 & 0 \\ 1 & -1 & -1 & 1 \\ 3 & -1 & 0 & 2 \end{array}\right]$

**7.** $\left[\begin{array}{ccc|c} 2 & -3 & 1 & 7 \\ 1 & 1 & -1 & 1 \\ 2 & 2 & -3 & -4 \end{array}\right]$

**9.** $\left[\begin{array}{cccc|c} 4 & -1 & 2 & -1 & 4 \\ 1 & 1 & 0 & 0 & -6 \\ 0 & 2 & -1 & 1 & 5 \end{array}\right]$

**11.** $\left[\begin{array}{cccc|c} 1 & -1 & 1 & -1 & 0 \\ 2 & 3 & -1 & 4 & 5 \end{array}\right]$

**13.** (a) $\left[\begin{array}{ccc|c} 1 & -3 & -5 & -2 \\ 0 & 1 & 6 & 9 \\ -3 & 5 & 4 & 6 \end{array}\right]$ (b) $\left[\begin{array}{ccc|c} 1 & -3 & -5 & -2 \\ 0 & 1 & 6 & 9 \\ 0 & -4 & -11 & 0 \end{array}\right]$ (c) $\left[\begin{array}{ccc|c} 1 & -3 & -5 & -2 \\ 0 & 1 & 6 & 9 \\ 0 & 0 & 13 & 36 \end{array}\right]$

**15.** (a) $\left[\begin{array}{ccc|c} 1 & -3 & 4 & 3 \\ 0 & 1 & -2 & 0 \\ -3 & 3 & 4 & 6 \end{array}\right]$ (b) $\left[\begin{array}{ccc|c} 1 & -3 & 4 & 3 \\ 0 & 1 & -2 & 0 \\ 0 & -6 & 16 & 15 \end{array}\right]$ (c) $\left[\begin{array}{ccc|c} 1 & -3 & 4 & 3 \\ 0 & 1 & -2 & 0 \\ 0 & 0 & 4 & 15 \end{array}\right]$

**17.** (a) $\left[\begin{array}{ccc|c} 1 & -3 & 2 & -6 \\ 0 & 1 & -1 & 8 \\ -3 & -6 & 4 & 6 \end{array}\right]$ (b) $\left[\begin{array}{ccc|c} 1 & -3 & 2 & -6 \\ 0 & 1 & -1 & 8 \\ 0 & -15 & 10 & -12 \end{array}\right]$ (c) $\left[\begin{array}{ccc|c} 1 & -3 & 2 & -6 \\ 0 & 1 & -1 & 8 \\ 0 & 0 & -5 & 108 \end{array}\right]$

**19.** (a) $\left[\begin{array}{rrr|r} 1 & -3 & 1 & -2 \\ 0 & 1 & 4 & 2 \\ -3 & 1 & 4 & 6 \end{array}\right]$ (b) $\left[\begin{array}{rrr|r} 1 & -3 & 1 & -2 \\ 0 & 1 & 4 & 2 \\ 0 & -8 & 7 & 0 \end{array}\right]$ (c) $\left[\begin{array}{rrr|r} 1 & -3 & 1 & -2 \\ 0 & 1 & 4 & 2 \\ 0 & 0 & 39 & 16 \end{array}\right]$

**21.** $\begin{cases} x+2y = 5 \text{ (1)} \\ y=-1 \text{ (2)} \end{cases} \Leftrightarrow \begin{cases} x+2(-1) = 5 \text{ (1) : subst. } y=-1 \\ y=-1 \text{ (2)} \end{cases} \Leftrightarrow \begin{cases} x = 7 \text{ (1) : solve for } x \\ y=-1 \text{ (2)} \end{cases}$

Consistent

**23.** $\begin{cases} x+2y+3z = 1 \text{ (1)} \\ y+4z = 2 \text{ (2)} \\ 0x+0y+0z = 3 \text{ (3)} \end{cases}$

Inconsistent

**25.** $\begin{cases} x + 2z = -1 \text{ (1)} \\ y-4z = -2 \text{ (2)} \\ 0x+0y+0z = 0 \text{ (3)} \end{cases}$ $\quad$ $\begin{array}{l} x = -2z-1 \text{ (1) : solve for } x \\ y = 4z-2 \text{ (2) : solve for } y \\ z \text{ is any real number} \end{array}$

Consistent

**27.** $\begin{cases} x_1+2x_2-x_3+x_4 = 1 \text{ (1)} \\ x_2+4x_3+x_4 = 2 \text{ (2)} \\ x_3+2x_4 = 3 \text{ (3)} \end{cases}$ $\quad$ $\begin{array}{l} x_1 = 1-2x_2+x_3-x_4 \text{ (1) : solve for } x_1 \\ x_2 = 2-4x_3-x_4 \text{ (2) : solve for } x_2 \\ x_3 = 3-2x_4 \text{ (3) : solve for } x_3 \end{array}$

Consistent

Express $x_1$ and $x_2$ in terms of $x_4$:

$x_2 = 2-4x_3-x_4 = 2-4(3-2x_4)-x_4 = 2-12+8x_4-x_4 = -10+7x_4$

$x_1 = 1-2x_2+x_3-x_4 = 1-2(-10+7x_4)+(3-2x_4)-x_4 = 1+20-14x_4+3-2x_4-x_4$

$= 24-17x_4$

Solution: $\begin{cases} x_1 = 24-17x_4 \\ x_2 = -10+7x_4 \\ x_3 = 3-2x_4 \\ x_4 \text{ is any real number} \end{cases}$

**29.** $\begin{cases} x_1+2x_2+4x_3 = 2 \text{ (1)} \\ x_2+x_3+3x_4 = 3 \text{ (2)} \\ 0x_1+0x_2+0x_3+0x_4 = 0 \text{ (3)} \end{cases}$ $\quad$ $\begin{array}{l} x_1 = 2-2x_2-4x_4 \text{ (1) : solve for } x_1 \\ x_2 = 3-x_3-3x_4 \text{ (2) : solve for } x_2 \end{array}$

Consistent

Express $x_1$ in terms of $x_3$ and $x_4$:

$$x_1 = 2 - 2(3 - x_3 - 3x_4) - 4x_4 = 2 - 6 + 2x_3 + 6x_4 - 4x_4 = -4 + 2x_3 + 2x_4$$

$$\text{Solution: } \begin{cases} x_1 = -4 + 2x_3 + 2x_4 \\ x_2 = 3 - x_3 - 3x_4 \\ x_3 \text{ is any real number} \\ x_4 \text{ is any real number} \end{cases}$$

**31.** $$\begin{cases} x_1 - 2x_2 + x_4 = -2 \ (1) \\ x_2 - 3x_3 + 2x_4 = 2 \ (2) \\ x_3 - x_4 = 0 \ (3) \\ 0x_1 + 0x_2 + 0x_3 + 0x_4 = 0 \ (4) \end{cases} \qquad \begin{aligned} x_1 &= -2 + 2x_2 - x_4 \ \text{(1) : solve for } x_1 \\ x_2 &= 2 + 3x_3 - 2x_4 \ \text{(2) : solve for } x_2 \\ x_3 &= x_4 \ \text{(3) : solve for } x_3 \end{aligned}$$

Consistent

Express $x_1$ and $x_2$ in terms of $x_4$:

$x_2 = 2 + 3x_4 - 2x_4 = 2 + x_4$; $x_1 = -2 + 2(2 + x_4) - x_4 = -2 + 4 + 2x_4 - x_4 = 2 + x_4$.

$$\text{Solution: } \begin{cases} x_1 = 2 + x_4 \\ x_2 = 2 + x_4 \\ x_3 = x_4 \\ x_4 \text{ is any real number} \end{cases}$$

**33.** $$\left[\begin{array}{cc|c} 1 & 1 & 6 \\ 2 & -1 & 0 \end{array}\right] \xrightarrow[R_2 = r_2 - 2r_1]{} \left[\begin{array}{cc|c} 1 & 1 & 6 \\ 0 & -3 & -12 \end{array}\right] \xrightarrow[R_2 = -\frac{1}{3}r_2]{} \left[\begin{array}{cc|c} 1 & 1 & 6 \\ 0 & 1 & 4 \end{array}\right]$$

$$\xrightarrow[R_1 = r_1 - r_2]{} \left[\begin{array}{cc|c} 1 & 0 & 2 \\ 0 & 1 & 4 \end{array}\right]. \text{ Solution: } x = 2,\ y = 4$$

**35.** $$\left[\begin{array}{cc|c} 2 & 1 & 5 \\ 1 & -1 & 1 \end{array}\right] \xrightarrow[R_1 = r_2,\ R_2 = r_1]{} \left[\begin{array}{cc|c} 1 & -1 & 1 \\ 2 & 1 & 5 \end{array}\right] \xrightarrow[R_2 = r_2 - 2r_1]{} \left[\begin{array}{cc|c} 1 & -1 & 1 \\ 0 & 3 & 3 \end{array}\right] \xrightarrow[R_2 = \frac{1}{3}r_2]{} \left[\begin{array}{cc|c} 1 & -1 & 1 \\ 0 & 1 & 1 \end{array}\right]$$

$$\xrightarrow[R_1 = r_1 + r_2]{} \left[\begin{array}{cc|c} 1 & 0 & 2 \\ 0 & 1 & 1 \end{array}\right]. \text{ Solution: } x = 2,\ y = 1$$

**37.** $$\left[\begin{array}{cc|c} 2 & 3 & 7 \\ 3 & -1 & 5 \end{array}\right] \xrightarrow[R_1 = r_2,\ R_2 = r_1]{} \left[\begin{array}{cc|c} 3 & -1 & 5 \\ 2 & 3 & 7 \end{array}\right] \xrightarrow[R_1 = r_1 - r_2]{} \left[\begin{array}{cc|c} 1 & -4 & -2 \\ 2 & 3 & 7 \end{array}\right] \xrightarrow[R_2 = r_2 - 2r_1]{} \left[\begin{array}{cc|c} 1 & -4 & -2 \\ 0 & 11 & 11 \end{array}\right]$$

$$\xrightarrow[R_2 = \frac{1}{11}r_2]{} \left[\begin{array}{cc|c} 1 & -4 & -2 \\ 0 & 1 & 1 \end{array}\right] \xrightarrow[R_1 = r_1 + 4r_2]{} \left[\begin{array}{cc|c} 1 & 0 & 2 \\ 0 & 1 & 1 \end{array}\right]. \text{ Solution: } x = 2,\ y = 1$$

**39.**
$$\left[\begin{array}{rr|r} 5 & -7 & 31 \\ 3 & 2 & 0 \end{array}\right] \xrightarrow[R_1=r_1-2r_2]{} \left[\begin{array}{rr|r} -1 & -11 & 31 \\ 3 & 2 & 0 \end{array}\right] \xrightarrow[R_1=-r_1]{} \left[\begin{array}{rr|r} 1 & 11 & -31 \\ 3 & 2 & 0 \end{array}\right]$$
$$\xrightarrow[R_2=r_2-3r_1]{} \left[\begin{array}{rr|r} 1 & 11 & -31 \\ 0 & -31 & 93 \end{array}\right] \xrightarrow[R_2=-\frac{1}{31}r_2]{} \left[\begin{array}{rr|r} 1 & 11 & -31 \\ 0 & 1 & -3 \end{array}\right] \xrightarrow[R_1=r_1-11r_2]{} \left[\begin{array}{rr|r} 1 & 0 & 2 \\ 0 & 1 & -3 \end{array}\right].$$

Solution: $x=2,\ y=-3$

**41.**
$$\left[\begin{array}{rr|r} 2 & -3 & 0 \\ 4 & 9 & 5 \end{array}\right] \xrightarrow[R_1=\frac{1}{2}r_1]{} \left[\begin{array}{rr|r} 1 & -3/2 & 0 \\ 4 & 9 & 5 \end{array}\right] \xrightarrow[R_2=r_2-4r_1]{} \left[\begin{array}{rr|r} 1 & -3/2 & 0 \\ 0 & 15 & 5 \end{array}\right]$$
$$\xrightarrow[R_2=\frac{1}{15}r_2]{} \left[\begin{array}{rr|r} 1 & -3/2 & 0 \\ 0 & 1 & 1/3 \end{array}\right] \xrightarrow[R_1=r_1+\frac{3}{2}r_2]{} \left[\begin{array}{rr|r} 1 & 0 & 1/2 \\ 0 & 1 & 1/3 \end{array}\right].$$

Solution: $x=\frac{1}{2},\ y=\frac{1}{3}$

**43.**
$$\left[\begin{array}{rr|r} 4 & -3 & 4 \\ 2 & 6 & 7 \end{array}\right] \xrightarrow[R_1=\frac{1}{4}r_1]{} \left[\begin{array}{rr|r} 1 & -3/4 & 1 \\ 2 & 6 & 7 \end{array}\right] \xrightarrow[R_2=r_2-2r_1]{} \left[\begin{array}{rr|r} 1 & -3/4 & 1 \\ 0 & 15/2 & 5 \end{array}\right]$$
$$\xrightarrow[R_2=\frac{2}{15}r_2]{} \left[\begin{array}{rr|r} 1 & -3/4 & 1 \\ 0 & 1 & 2/3 \end{array}\right] \xrightarrow[R_1=r_1+\frac{3}{4}r_2]{} \left[\begin{array}{rr|r} 1 & 0 & 3/2 \\ 0 & 1 & 2/3 \end{array}\right].$$

Solution: $x=\frac{3}{2},\ y=\frac{2}{3}$

**45.**
$$\left[\begin{array}{rr|r} 1/2 & 1/3 & 2 \\ 1 & 1 & 5 \end{array}\right] \xrightarrow[\substack{R_1=r_2 \\ R_2=r_1}]{} \left[\begin{array}{rr|r} 1 & 1 & 5 \\ 1/2 & 1/3 & 2 \end{array}\right] \xrightarrow[R_2=r_2-\frac{1}{2}r_1]{} \left[\begin{array}{rr|r} 1 & 1 & 5 \\ 0 & -1/6 & -1/2 \end{array}\right]$$
$$\xrightarrow[R_2=-6r_2]{} \left[\begin{array}{rr|r} 1 & 1 & 5 \\ 0 & 1 & 3 \end{array}\right] \xrightarrow[R_1=r_1-r_2]{} \left[\begin{array}{rr|r} 1 & 0 & 2 \\ 0 & 1 & 3 \end{array}\right].$$

Solution: $x=2,\ y=3$

**47.**
$$\left[\begin{array}{rr|r} 1 & 1 & 1 \\ 3 & -2 & 4/3 \end{array}\right] \xrightarrow[R_2=r_2-3r_1]{} \left[\begin{array}{rr|r} 1 & 1 & 1 \\ 0 & -5 & -5/3 \end{array}\right] \xrightarrow[R_2=-\frac{1}{5}r_2]{} \left[\begin{array}{rr|r} 1 & 1 & 1 \\ 0 & 1 & 1/3 \end{array}\right]$$
$$\xrightarrow[R_1=r_1-r_2]{} \left[\begin{array}{rr|r} 1 & 0 & 2/3 \\ 0 & 1 & 1/3 \end{array}\right].$$

Solution: $x=\frac{2}{3},\ y=\frac{1}{3}$

**49.**
$$\left[\begin{array}{rrr|r} 2 & 1 & 1 & 6 \\ 1 & -1 & -1 & -3 \\ 3 & 1 & 2 & 7 \end{array}\right] \xrightarrow[\substack{R_1=r_2 \\ R_2=r_1}]{} \left[\begin{array}{rrr|r} 1 & -1 & -1 & -3 \\ 2 & 1 & 1 & 6 \\ 3 & 1 & 2 & 7 \end{array}\right] \xrightarrow[\substack{R_2=r_2-2r_1 \\ R_1=r_3-3r_1}]{} \left[\begin{array}{rrr|r} 1 & -1 & -1 & -3 \\ 0 & 3 & 3 & 12 \\ 0 & 4 & 5 & 16 \end{array}\right]$$
$$\xrightarrow[R_2=\frac{1}{3}r_2]{} \left[\begin{array}{rrr|r} 1 & -1 & -1 & -3 \\ 0 & 1 & 1 & 4 \\ 0 & 4 & 5 & 16 \end{array}\right] \xrightarrow[\substack{R_1=r_1+r_2 \\ R_3=r_3-4r_2}]{} \left[\begin{array}{rrr|r} 1 & 0 & 0 & 1 \\ 0 & 1 & 1 & 4 \\ 0 & 0 & 1 & 0 \end{array}\right] \xrightarrow[R_2=r_2-r_3]{} \left[\begin{array}{rrr|r} 1 & 0 & 0 & 1 \\ 0 & 1 & 0 & 4 \\ 0 & 0 & 1 & 0 \end{array}\right].$$

Solution: $x=1,\ y=4,\ z=0$

**51.** $\left[\begin{array}{ccc|c} 1 & 1 & -1 & -2 \\ 3 & 1 & 1 & 0 \\ 2 & -1 & 2 & 1 \end{array}\right] \xrightarrow[\substack{R_2=r_2-3r_1 \\ R_3=r_3-2r_1}]{} \left[\begin{array}{ccc|c} 1 & 1 & -1 & -2 \\ 0 & -2 & 4 & 6 \\ 0 & -3 & 4 & 5 \end{array}\right] \xrightarrow[R_2=-\frac{1}{2}r_2]{} \left[\begin{array}{ccc|c} 1 & 1 & -1 & -2 \\ 0 & 1 & -2 & -3 \\ 0 & -3 & 4 & 5 \end{array}\right]$

$\xrightarrow[\substack{R_1=r_1-r_2 \\ R_3=r_3+3r_2}]{} \left[\begin{array}{ccc|c} 1 & 0 & 1 & 1 \\ 0 & 1 & -2 & -3 \\ 0 & 0 & -2 & -4 \end{array}\right] \xrightarrow[R_3=-\frac{1}{2}r_3]{} \left[\begin{array}{ccc|c} 1 & 0 & 1 & 1 \\ 0 & 1 & -2 & -3 \\ 0 & 0 & 1 & 2 \end{array}\right]$

$\xrightarrow[\substack{R_1=r_1-r_3 \\ R_2=r_2+2r_3}]{} \left[\begin{array}{ccc|c} 1 & 0 & 0 & -1 \\ 0 & 1 & 0 & 1 \\ 0 & 0 & 1 & 2 \end{array}\right]$. Solution: $x=-1,\ y=1,\ z=2$

**53.** $\left[\begin{array}{ccc|c} 2 & 1 & -1 & 2 \\ 1 & 3 & 2 & 1 \\ 1 & 1 & 1 & 2 \end{array}\right] \xrightarrow[\substack{R_1=r_2 \\ R_2=r_1}]{} \left[\begin{array}{ccc|c} 1 & 3 & 2 & 1 \\ 2 & 1 & -1 & 2 \\ 1 & 1 & 1 & 2 \end{array}\right] \xrightarrow[\substack{R_2=r_2-2r_1 \\ R_3=r_3-r_1}]{} \left[\begin{array}{ccc|c} 1 & 3 & 2 & 1 \\ 0 & -5 & -5 & 0 \\ 0 & -2 & -1 & 1 \end{array}\right]$

$\xrightarrow[R_2=-\frac{1}{5}r_2]{} \left[\begin{array}{ccc|c} 1 & 3 & 2 & 1 \\ 0 & 1 & 1 & 0 \\ 0 & -2 & -1 & 1 \end{array}\right] \xrightarrow[\substack{R_1=r_1-3r_2 \\ R_3=r_3+2r_2}]{} \left[\begin{array}{ccc|c} 1 & 0 & -1 & 1 \\ 0 & 1 & 1 & 0 \\ 0 & 0 & 1 & 1 \end{array}\right]$

$\xrightarrow[\substack{R_1=r_1+r_3 \\ R_2=r_2-r_3}]{} \left[\begin{array}{ccc|c} 1 & 0 & 0 & 2 \\ 0 & 1 & 0 & -1 \\ 0 & 0 & 1 & 1 \end{array}\right]$. Solution: $x=2,\ y=-1,\ z=1$

**55.** $\left[\begin{array}{ccc|c} 1 & 1 & -1 & 0 \\ 2 & 4 & -4 & -1 \\ 2 & 1 & 1 & 2 \end{array}\right] \xrightarrow[\substack{R_2=r_2-2r_1 \\ R_3=r_3-2r_1}]{} \left[\begin{array}{ccc|c} 1 & 1 & -1 & 0 \\ 0 & 2 & -2 & -1 \\ 0 & -1 & 3 & 2 \end{array}\right] \xrightarrow[\substack{R_2=r_3 \\ R_3=r_2}]{} \left[\begin{array}{ccc|c} 1 & 1 & -1 & 0 \\ 0 & -1 & 3 & 2 \\ 0 & 2 & -2 & -1 \end{array}\right]$

$\xrightarrow[\substack{R_1=r_1+r_2 \\ R_3=r_3+2r_2}]{} \left[\begin{array}{ccc|c} 1 & 0 & 2 & 2 \\ 0 & -1 & 3 & 2 \\ 0 & 0 & 4 & 3 \end{array}\right] \xrightarrow[\substack{R_2=-r_2 \\ R_3=\frac{1}{4}r_3}]{} \left[\begin{array}{ccc|c} 1 & 0 & 2 & 2 \\ 0 & 1 & -3 & -2 \\ 0 & 0 & 1 & 3/4 \end{array}\right]$

$\xrightarrow[\substack{R_1=r_1-2r_3 \\ R_2=r_2+3r_3}]{} \left[\begin{array}{ccc|c} 1 & 0 & 0 & 1/2 \\ 0 & 1 & 0 & 1/4 \\ 0 & 0 & 1 & 3/4 \end{array}\right]$. Solution: $x=\frac{1}{2},\ y=\frac{1}{4},\ z=\frac{3}{4}$.

**57.** $\left[\begin{array}{ccc|c} 3 & 1 & -1 & 2/3 \\ 2 & -1 & 1 & 1 \\ 4 & 2 & 0 & 8/3 \end{array}\right] \xrightarrow[R_1=r_1-r_2]{} \left[\begin{array}{ccc|c} 1 & 2 & -2 & -1/3 \\ 2 & -1 & 1 & 1 \\ 4 & 2 & 0 & 8/3 \end{array}\right] \xrightarrow[\substack{R_2=r_2-2r_1 \\ R_3=r_3-4r_1}]{} \left[\begin{array}{ccc|c} 1 & 2 & -2 & -1/3 \\ 0 & -5 & 5 & 5/3 \\ 0 & -6 & 8 & 4 \end{array}\right]$

$$\xrightarrow[R_2=-\frac{1}{5}r_2]{}\left[\begin{array}{ccc|c}1&2&-2&-1/3\\0&1&-1&-1/3\\0&-6&8&4\end{array}\right]\xrightarrow[\substack{R_1=r_1-2r_2\\R_3=r_3+6r_2}]{}\left[\begin{array}{ccc|c}1&0&0&1/3\\0&1&-1&-1/3\\0&0&2&2\end{array}\right]$$

$$\xrightarrow[R_3=\frac{1}{2}r_3]{}\left[\begin{array}{ccc|c}1&0&0&1/3\\0&1&-1&-1/3\\0&0&1&1\end{array}\right]\xrightarrow[R_2=r_2+r_3]{}\left[\begin{array}{ccc|c}1&0&0&1/3\\0&1&0&2/3\\0&0&1&1\end{array}\right].\ \text{Solution: } x=\frac{1}{3},\ y=\frac{2}{3},\ z=1$$

**59.** $\left[\begin{array}{cc|c}1&-1&5\\2&-2&6\end{array}\right]\xrightarrow[R_2=r_2-2r_1]{}\left[\begin{array}{cc|c}1&-1&5\\0&0&-4\end{array}\right]$. Inconsistent; No solution

**61.** $\left[\begin{array}{cc|c}2&-3&6\\4&-6&12\end{array}\right]\xrightarrow[R_2=r_2-2r_1]{}\left[\begin{array}{cc|c}2&-3&6\\0&0&0\end{array}\right]$. Consistent; Infinitely many solutions

**63.** $\left[\begin{array}{cc|c}5&-6&1\\-10&12&0\end{array}\right]\xrightarrow[R_2=r_2+2r_1]{}\left[\begin{array}{cc|c}5&-6&1\\0&0&2\end{array}\right]$. Inconsistent; No solution

**65.** $\left[\begin{array}{cc|c}2&3&5\\4&4&8\end{array}\right]\xrightarrow[R_2=r_2-2r_1]{}\left[\begin{array}{cc|c}2&3&5\\0&-2&-2\end{array}\right]$. Consistent; Unique solution

**67.** $\left[\begin{array}{cc|c}1&2&6\\2&1&6\end{array}\right]\xrightarrow[R_2=r_2-2r_1]{}\left[\begin{array}{cc|c}1&2&6\\0&-3&-6\end{array}\right]$. Consistent; Unique solution

**69.** $\left[\begin{array}{cc|c}1&-1&3\\-2&2&4\end{array}\right]\xrightarrow[R_2=r_2+2r_1]{}\left[\begin{array}{cc|c}1&-1&3\\0&0&10\end{array}\right]$. Inconsistent; No solution

**71.** Let $x$ represent the number of pounds of cashews.
Then $\underbrace{\$5x}_{\text{Revenue from cashews}}+\underbrace{\$1.50(30)}_{\text{Revenue from peanuts}}=\underbrace{\$3(x+30)}_{\text{Revenue from mixture}}$; $5x+45=3x+90$; $2x=45$; $x=22.5$.

Mix 22.5 lbs. of cashews with the peanuts.

**73.** Let $x$ represent the number of two-student workstations and $y$ the number of three-student workstations. Then $2x+3y=38$ students and $x+y=16$ workstations. Solve $\begin{cases}2x+3y=38\\x+y=16\end{cases}$:

$$\left[\begin{array}{cc|c}2&3&38\\1&1&16\end{array}\right]\xrightarrow[R_1=r_1-3r_2]{}\left[\begin{array}{cc|c}-1&0&-10\\1&1&16\end{array}\right]\xrightarrow[R_2=r_2+r_1]{}\left[\begin{array}{cc|c}-1&0&-10\\0&1&6\end{array}\right]$$

$$\xrightarrow[R_1=-r_1]{}\left[\begin{array}{cc|c}1&0&10\\0&1&6\end{array}\right];\ x=10, y=6.$$

There are 10 two-student workstations and 6 three-student workstations.

**75.** Let $x$ represent the price per pound of bacon and $y$ the price per carton of eggs. Then we seek $2x+2y$, where $x$ and $y$ satisfy $3x+2y=7.45$ and $2x+3y=6.45$.

Solve $\begin{cases}3x+2y=7.45\\2x+3y=6.45\end{cases}$ for $x$ and $y$, and evaluate $2x+2y$.

$$\left[\begin{array}{cc|c}3&2&7.45\\2&3&6.45\end{array}\right]\xrightarrow[R_1=r_1-r_2]{}\left[\begin{array}{cc|c}1&-1&1\\2&3&6.45\end{array}\right]\xrightarrow[R_2=r_2-2r_1]{}\left[\begin{array}{cc|c}1&-1&1\\0&5&4.45\end{array}\right]$$

$$\xrightarrow[R_2=\frac{1}{5}r_2]{}\left[\begin{array}{cc|c}1&-1&1\\0&1&.89\end{array}\right]\xrightarrow[R_1=r_1+r_2]{}\left[\begin{array}{cc|c}1&0&1.89\\0&1&.89\end{array}\right];\ x=\$1.89, y=\$0.89;$$

$2x+2y=\$3.78+\$1.78=\$5.56$. The store should refund \$5.56.

**77.** Let $x$, $y$, and $z$ represent the number of orchestra, main and balcony seats in the theater, respectively. Then $x+y+z=500$ seats, $75x+50y+35z=23{,}000$ dollars and

$75\left(\frac{x}{2}\right)+50y+35z=21{,}500$ dollars. Solve $\begin{cases}x+y+z=500\\75x+50y+35z=23{,}000\\37.5x+50y+35z=21{,}500\end{cases}$:

$$\left[\begin{array}{ccc|c}1&1&1&500\\75&50&35&23000\\37.5&50&35&21500\end{array}\right]\xrightarrow[\substack{R_2=r_2-75r_1\\R_3=r_3-37.5r_1}]{}\left[\begin{array}{ccc|c}1&1&1&500\\0&-25&-40&-14500\\0&12.5&-2.5&2750\end{array}\right]$$

$$\xrightarrow[R_2=-\frac{1}{25}r_2]{}\left[\begin{array}{ccc|c}1&1&1&500\\0&1&1.6&580\\0&12.5&-2.5&2750\end{array}\right]\xrightarrow[\substack{R_1=r_1-r_2\\R_3=r_3-12.5r_2}]{}\left[\begin{array}{ccc|c}1&0&-.6&-80\\0&1&1.6&580\\0&0&-22.5&-4500\end{array}\right]$$

$$\xrightarrow[R_3=-\frac{1}{22.5}r_3]{}\left[\begin{array}{ccc|c}1&0&-.6&-80\\0&1&1.6&580\\0&0&1&200\end{array}\right]\xrightarrow[\substack{R_1=r_1+0.6r_3\\R_2=r_2-1.6r_3}]{}\left[\begin{array}{ccc|c}1&0&0&40\\0&1&0&260\\0&0&1&200\end{array}\right];\ x=40,\ y=260,\ z=200.$$

There are 40 orchestra, 260 main, and 200 balcony seats.

**79.** Let $x$, $y$ and $z$ represent the amounts invested at 6%, 8% and 9%, respectively. Then $x+y+z=\$6500$, $0.06x+0.08y+0.09z=\$480$ and $0.09z=0.08y+60$. Solve

$$\begin{cases}x+y+z=6500\\0.06x+0.08y+0.09z=480\\-0.08y+0.09z=60\end{cases}:$$

$$\left[\begin{array}{ccc|c} 1 & 1 & 1 & 6500 \\ 0.06 & 0.08 & 0.09 & 480 \\ 0 & -0.08 & 0.09 & 60 \end{array}\right] \xrightarrow[\substack{R_2=100r_2 \\ R_3=100r_3}]{} \left[\begin{array}{ccc|c} 1 & 1 & 1 & 6500 \\ 6 & 8 & 9 & 48000 \\ 0 & -8 & 9 & 6000 \end{array}\right]$$

$$\xrightarrow[R_2=r_2-6r_1]{} \left[\begin{array}{ccc|c} 1 & 1 & 1 & 6500 \\ 0 & 2 & 3 & 9000 \\ 0 & -8 & 9 & 6000 \end{array}\right] \xrightarrow[\substack{R_1=r_1-\frac{1}{2}r_2 \\ R_3=r_3+4r_2}]{} \left[\begin{array}{ccc|c} 1 & 0 & -1/2 & 2000 \\ 0 & 2 & 3 & 9000 \\ 0 & 0 & 21 & 42000 \end{array}\right]$$

$$\xrightarrow[R_3=\frac{1}{21}r_3]{} \left[\begin{array}{ccc|c} 1 & 0 & -1/2 & 2000 \\ 0 & 2 & 3 & 9000 \\ 0 & 0 & 1 & 2000 \end{array}\right] \xrightarrow[\substack{R_1=r_1+\frac{1}{2}r_3 \\ R_2=r_2-3r_3}]{} \left[\begin{array}{ccc|c} 1 & 0 & 0 & 3000 \\ 0 & 2 & 0 & 3000 \\ 0 & 0 & 1 & 2000 \end{array}\right]$$

$$\xrightarrow[R_2=\frac{1}{2}r_2]{} \left[\begin{array}{ccc|c} 1 & 0 & 0 & 3000 \\ 0 & 1 & 0 & 1500 \\ 0 & 0 & 1 & 2000 \end{array}\right] : \ x=3000,\ y=1500,\ z=2000.$$

\$3000 was invested at 6%, \$1500 invested at 8% and \$2000 at 9%.

**81.** Let $x$, $y$, and $z$ represent the number of cases of orange juice, tomato juice, and pineapple juice, respectively. Then $10x+12y+9z=398$ minutes cleaning, $4x+4y+6z=164$ minutes filling, and $2x+y+z=58$ minutes labeling.

$$\text{Solve: } \begin{cases} 10x+12y+9z = 398 \\ 4x+4y+6z = 164 \\ 2x+y+z = 58 \end{cases} \quad \left[\begin{array}{ccc|c} 10 & 12 & 9 & 398 \\ 4 & 4 & 6 & 164 \\ 2 & 1 & 1 & 58 \end{array}\right] \xrightarrow[\substack{R_1=r_2 \\ R_2=r_1}]{} \left[\begin{array}{ccc|c} 4 & 4 & 6 & 164 \\ 10 & 12 & 9 & 398 \\ 2 & 1 & 1 & 58 \end{array}\right]$$

$$\xrightarrow[\substack{R_1=\frac{1}{4}r_1 \\ R_2=r_2-5r_3}]{} \left[\begin{array}{ccc|c} 1 & 1 & 3/2 & 41 \\ 0 & 7 & 4 & 108 \\ 2 & 1 & 1 & 58 \end{array}\right] \xrightarrow[R_3=r_3-2r_1]{} \left[\begin{array}{ccc|c} 1 & 1 & 3/2 & 41 \\ 0 & 7 & 4 & 108 \\ 0 & -1 & -2 & -24 \end{array}\right]$$

$$\xrightarrow[\substack{R_2=-r_3 \\ R_3=r_2}]{} \left[\begin{array}{ccc|c} 1 & 1 & 3/2 & 41 \\ 0 & 1 & 2 & 24 \\ 0 & 7 & 4 & 108 \end{array}\right] \xrightarrow[\substack{R_1=r_1-r_2 \\ R_3=r_3-7r_2}]{} \left[\begin{array}{ccc|c} 1 & 0 & -1/2 & 17 \\ 0 & 1 & 2 & 24 \\ 0 & 0 & -10 & -60 \end{array}\right]$$

$$\xrightarrow[R_3=-\frac{1}{10}r_3]{} \left[\begin{array}{ccc|c} 1 & 0 & -1/2 & 17 \\ 0 & 1 & 2 & 24 \\ 0 & 0 & 1 & 6 \end{array}\right] \xrightarrow[\substack{R_1=r_1+\frac{1}{2}r_3 \\ R_2=r_2-2r_3}]{} \left[\begin{array}{ccc|c} 1 & 0 & 0 & 20 \\ 0 & 1 & 0 & 12 \\ 0 & 0 & 1 & 6 \end{array}\right]$$

$x=20, y=12, z=6$. 20 cases of orange juice, 12 cases of tomato juice and 6 cases of pineapple juice are prepared.

**83.** Let $x$, $y$ and $z$ represent the number of the first, second and third types of packages ordered, respectively. Then $20x+40z=200$ sheets of white paper; $15x+3y+30z=180$ sheets of blue paper; and $x+y=12$ sheets of red paper. Solve $\begin{cases} 20x \quad +40z=200 \\ 15x+3y+30z=180 \\ x+y \quad =12 \end{cases}$:

$$\left[\begin{array}{ccc|c} 20 & 0 & 40 & 200 \\ 15 & 3 & 30 & 180 \\ 1 & 1 & 0 & 12 \end{array}\right] \xrightarrow[R_2=\frac{1}{3}r_2]{R_1=\frac{1}{20}r_1} \left[\begin{array}{ccc|c} 1 & 0 & 2 & 10 \\ 5 & 1 & 10 & 60 \\ 1 & 1 & 0 & 12 \end{array}\right] \xrightarrow[R_3=r_3-r_1]{R_2=r_2-5r_1} \left[\begin{array}{ccc|c} 1 & 0 & 2 & 10 \\ 0 & 1 & 0 & 10 \\ 0 & 1 & -2 & 2 \end{array}\right]$$

$$\xrightarrow{R_3=r_3-r_2} \left[\begin{array}{ccc|c} 1 & 0 & 2 & 10 \\ 0 & 1 & 0 & 10 \\ 0 & 0 & -2 & -8 \end{array}\right] \xrightarrow{R_1=r_1+r_3} \left[\begin{array}{ccc|c} 1 & 0 & 0 & 2 \\ 0 & 1 & 0 & 10 \\ 0 & 0 & -2 & -8 \end{array}\right] \xrightarrow{R_3=-\frac{1}{2}r_3} \left[\begin{array}{ccc|c} 1 & 0 & 0 & 2 \\ 0 & 1 & 0 & 10 \\ 0 & 0 & 1 & 4 \end{array}\right]; x=2,$$

$y=10$, $z=4$.

Order 2 packages of the first type, 10 packages of the second type, and 4 packages of the third type.

**85.** Let $x$, $y$ and $z$ represent the number of units of foods I, II and III, respectively, needed to create a meal. Then $10x+5y+15z=100$ units of protein; $3x+6y+3z=50$ units of carbohydrates; and $4x+4y+6z=50$ units of iron. Solve $\begin{cases} 10x+5y+15z=100 \\ 3x+6y+3z=50 \\ 4x+4y+6z=50 \end{cases}$:

$$\left[\begin{array}{ccc|c} 10 & 5 & 15 & 100 \\ 3 & 6 & 3 & 50 \\ 4 & 4 & 6 & 50 \end{array}\right] \xrightarrow[R_3=\frac{1}{2}r_3]{R_1=\frac{1}{5}r_1} \left[\begin{array}{ccc|c} 2 & 1 & 3 & 20 \\ 3 & 6 & 3 & 50 \\ 2 & 2 & 3 & 25 \end{array}\right] \xrightarrow[R_2=r_2-r_3]{R_1=r_1-r_3} \left[\begin{array}{ccc|c} 0 & -1 & 0 & -5 \\ 1 & 4 & 0 & 25 \\ 2 & 2 & 3 & 25 \end{array}\right]$$

$$\xrightarrow[R_2=-r_1]{R_1=r_2} \left[\begin{array}{ccc|c} 1 & 4 & 0 & 25 \\ 0 & 1 & 0 & 5 \\ 2 & 2 & 3 & 25 \end{array}\right] \xrightarrow{R_3=r_3-2r_1} \left[\begin{array}{ccc|c} 1 & 4 & 0 & 25 \\ 0 & 1 & 0 & 5 \\ 0 & -6 & 3 & -25 \end{array}\right] \xrightarrow[R_3=r_3+6r_2]{R_1=r_1-4r_2} \left[\begin{array}{ccc|c} 1 & 0 & 0 & 5 \\ 0 & 1 & 0 & 5 \\ 0 & 0 & 3 & 5 \end{array}\right]$$

$$\xrightarrow{R_3=\frac{1}{3}r_3} \left[\begin{array}{ccc|c} 1 & 0 & 0 & 5 \\ 0 & 1 & 0 & 5 \\ 0 & 0 & 1 & 5/3 \end{array}\right] \quad x=5, y=5, z=\frac{5}{3}.$$

Use 5 units of food I, 5 units of food II and $\frac{5}{3}$ units of food III.

**87.** Let $x$, $y$ and $z$ represent the number of assorted, mixed and single, respectively, cartons ordered. Then $2x+4y=40$ rock CDs, $4x+2y=32$ western CDs and $x+2z=14$ blues CDs.

Solve $\begin{cases} 2x+4y \quad\quad =40 \\ 4x+2y \quad\quad =32 \\ x \quad\quad +2z=14 \end{cases}$: $\left[\begin{array}{ccc|c} 2 & 4 & 0 & 40 \\ 4 & 2 & 0 & 32 \\ 1 & 0 & 2 & 14 \end{array}\right] \xrightarrow[R_2=r_2-2r_1,\ R_3=r_3-\frac{1}{2}r_1]{} \left[\begin{array}{ccc|c} 2 & 4 & 0 & 40 \\ 0 & -6 & 0 & -48 \\ 0 & -2 & 2 & -6 \end{array}\right]$

$\xrightarrow[R_1=\frac{1}{2}r_1,\ R_2=-\frac{1}{6}r_2,\ R_3=\frac{1}{2}r_3]{} \left[\begin{array}{ccc|c} 1 & 2 & 0 & 20 \\ 0 & 1 & 0 & 8 \\ 0 & -1 & 1 & -3 \end{array}\right] \xrightarrow[R_1=r_1-2r_2,\ R_3=r_3+r_2]{} \left[\begin{array}{ccc|c} 1 & 0 & 0 & 4 \\ 0 & 1 & 0 & 8 \\ 0 & 0 & 1 & 5 \end{array}\right]$: $x=4, y=8, z=5$.

Order 4 assorted cartons, 8 mixed cartons and 5 single cartons.

**89.** Let $x$, $y$ and $z$ represent the number of large, mammoth and giant, respectively, cans used to fill the order. Then $y+z=6$ pounds of walnuts; $2x+6y+4z=34$ pounds of peanuts; and $x+2y+2z=15$ pounds of cashews. Solve $\begin{cases} y+z=6 \\ 2x+6y+4z=34 \\ x+2y+2z=15 \end{cases}$:

$\left[\begin{array}{ccc|c} 0 & 1 & 1 & 6 \\ 2 & 6 & 4 & 34 \\ 1 & 2 & 2 & 15 \end{array}\right] \xrightarrow[R_1=r_3,\ R_3=r_1]{} \left[\begin{array}{ccc|c} 1 & 2 & 2 & 15 \\ 2 & 6 & 4 & 34 \\ 0 & 1 & 1 & 6 \end{array}\right] \xrightarrow[R_r=r_2-2r_1]{} \left[\begin{array}{ccc|c} 1 & 2 & 2 & 15 \\ 0 & 2 & 0 & 4 \\ 0 & 1 & 1 & 6 \end{array}\right]$

$\xrightarrow[R_2=\frac{1}{2}r_2]{} \left[\begin{array}{ccc|c} 1 & 2 & 2 & 15 \\ 0 & 1 & 0 & 2 \\ 0 & 1 & 1 & 6 \end{array}\right] \xrightarrow[R_1=r_1-2r_2,\ R_3=r_3-r_2]{} \left[\begin{array}{ccc|c} 1 & 0 & 2 & 11 \\ 0 & 1 & 0 & 2 \\ 0 & 0 & 1 & 4 \end{array}\right] \xrightarrow[R_1=r_1-2r_3]{} \left[\begin{array}{ccc|c} 1 & 0 & 0 & 3 \\ 0 & 1 & 0 & 2 \\ 0 & 0 & 1 & 4 \end{array}\right]$

$x=3, y=2, z=4$. Fill the order with 3 large cans, 2 mammoth cans and 4 giant cans.

## Technology Exercises

**1.** $\left[\begin{array}{ccc|c} -2 & 0 & -8 & 0 \\ -2 & 1 & -4 & 1 \\ 3 & -1 & 0 & 1 \end{array}\right]$

**3.** $\left[\begin{array}{ccc|c} 1 & 0 & 4 & 0 \\ 1 & 1 & 8 & 1 \\ 3 & -1 & 0 & 1 \end{array}\right]$

**5.** REF: $\left[\begin{array}{ccc|c} 1 & -1 & 0.5 & 1 \\ 0 & 1 & 3 & 0 \\ 0 & 0 & 1 & 0.22 \end{array}\right]$ or $\left[\begin{array}{ccc|c} 1 & -1 & 1/2 & 1 \\ 0 & 1 & 3 & 0 \\ 0 & 0 & 1 & 2/9 \end{array}\right]$;

RREF: $\left[\begin{array}{ccc|c} 1 & 0 & 0 & 0.22 \\ 0 & 1 & 0 & -0.67 \\ 0 & 0 & 1 & 0.22 \end{array}\right]$ or $\left[\begin{array}{ccc|c} 1 & 0 & 0 & 2/9 \\ 0 & 1 & 0 & -2/3 \\ 0 & 0 & 1 & 2/9 \end{array}\right]$

$x=.22, y=-.67, z=.22$ or $x=\frac{2}{9}, y=-\frac{2}{3}, z=\frac{2}{9}$.

**7.** REF: $\left[\begin{array}{ccc|c} 1 & 1 & 1 & 4 \\ 0 & 1 & 1 & 2 \\ 0 & 0 & 1 & 3 \end{array}\right]$; RREF: $\left[\begin{array}{ccc|c} 1 & 0 & 0 & 2 \\ 0 & 1 & 0 & -1 \\ 0 & 0 & 1 & 3 \end{array}\right]$

$x = 2, y = -1, z = 3.$

**9.** REF: $\left[\begin{array}{cccc|c} 1 & 1 & 1 & 1 & 20 \\ 0 & 1 & 1 & 1 & 0 \\ 0 & 0 & 1 & 1 & 13 \\ 0 & 0 & 0 & 1 & -4 \end{array}\right]$; RREF: $\left[\begin{array}{cccc|c} 1 & 0 & 0 & 0 & 20 \\ 0 & 1 & 0 & 0 & -13 \\ 0 & 0 & 1 & 0 & 17 \\ 0 & 0 & 0 & 1 & -4 \end{array}\right]$

$x_1 = 20, x_2 = -13, x_3 = 17, x_4 = -4$.

## 2.3 Systems of *m* Linear Equations Containing *n* Variables

**1.** Not in reduced row-echelon form (the 2nd row contains all 0's while the third row does not).

**3.** Not in reduced row-echelon form (there is no zero above the 1 in the second row.)

**5.** Not in reduced row-echelon form (the leftmost 1 in the 3rd row is not to the right of the leftmost 1 in a previous row).

**7.** Reduced row-echelon form

**9.** Reduced row-echelon form

**11.** Reduced row-echelon form

**13.** Infinitely many solutions

**15.** One solution

**17.** Infinitely many solutions

**19.** Infinitely many solutions

**21.** Infinitely many solutions

**23.** One solution

**25.** Infinitely many solutions

**27.** Infinitely many solutions

**29.** $$\left[\begin{array}{cc|c} 1 & 1 & 3 \\ 2 & -1 & 3 \end{array}\right] \xrightarrow[R_2 = r_2 - 2r_1]{} \left[\begin{array}{cc|c} 1 & 1 & 3 \\ 0 & -3 & -3 \end{array}\right] \xrightarrow[R_2 = -\frac{1}{3}r_2]{} \left[\begin{array}{cc|c} 1 & 1 & 3 \\ 0 & 1 & 1 \end{array}\right] \xrightarrow[R_1 = r_1 - r_2]{} \left[\begin{array}{cc|c} 1 & 0 & 2 \\ 0 & 1 & 1 \end{array}\right]$$

$x = 2, y = 1.$

**31.** $$\left[\begin{array}{cc|c} 3 & -3 & 12 \\ 3 & 2 & -3 \\ 2 & 1 & 4 \end{array}\right] \xrightarrow[R_1 = \frac{1}{3}r_1]{} \left[\begin{array}{cc|c} 1 & -1 & 4 \\ 3 & 2 & -3 \\ 2 & 1 & 4 \end{array}\right] \xrightarrow[\substack{R_2 = r_2 - 3r_1 \\ R_3 = r_3 - 2r_1}]{} \left[\begin{array}{cc|c} 1 & -1 & 4 \\ 0 & 5 & -15 \\ 0 & 3 & -4 \end{array}\right]$$

$$\xrightarrow[R_2=\frac{1}{5}r_2]{}\left[\begin{array}{rr|r}1 & -1 & 4\\ 0 & 1 & -3\\ 0 & 3 & -4\end{array}\right]\xrightarrow[\substack{R_1=r_1+r_2\\ R_3=r_3-3r_2}]{}\left[\begin{array}{rr|r}1 & 0 & 1\\ 0 & 1 & -3\\ 0 & 0 & 5\end{array}\right]\xrightarrow[R_3=\frac{1}{5}r_3]{}\left[\begin{array}{rr|r}1 & 0 & 1\\ 0 & 1 & -3\\ 0 & 0 & 1\end{array}\right]$$

$$\xrightarrow[\substack{R_1=r_1-r_3\\ R_2=r_2+3r_3}]{}\left[\begin{array}{rr|r}1 & 0 & 0\\ 0 & 1 & 0\\ 0 & 0 & 1\end{array}\right]; \text{ Inconsistent system}$$

**33.** $$\left[\begin{array}{rr|r}2 & -4 & 8\\ 1 & -2 & 4\\ -1 & 2 & -4\end{array}\right]\xrightarrow[R_1=\frac{1}{2}r_1]{}\left[\begin{array}{rr|r}1 & -2 & 4\\ 1 & -2 & 4\\ -1 & 2 & -4\end{array}\right]\xrightarrow[\substack{R_2=r_2-r_1\\ R_3=r_3+r_1}]{}\left[\begin{array}{rr|r}1 & -2 & 4\\ 0 & 0 & 0\\ 0 & 0 & 0\end{array}\right]$$

Solution: $x = 4 + 2y$, $y$ is any real number.

**35.** $$\left[\begin{array}{rrr|r}2 & 1 & 3 & -1\\ -1 & 1 & 3 & 8\\ 2 & -2 & -6 & -16\end{array}\right]\xrightarrow[\substack{R_1=r_2\\ R_2=r_1}]{}\left[\begin{array}{rrr|r}-1 & 1 & 3 & 8\\ 2 & 1 & 3 & -1\\ 2 & -2 & -6 & -16\end{array}\right]\xrightarrow[\substack{R_2=r_2+2r_1\\ R_3=r_3+2r_1}]{}\left[\begin{array}{rrr|r}-1 & 1 & 3 & 8\\ 0 & 3 & 9 & 15\\ 0 & 0 & 0 & 0\end{array}\right]$$

$$\xrightarrow[\substack{R_1=-r_1\\ R_2=\frac{1}{3}r_2}]{}\left[\begin{array}{rrr|r}1 & -1 & -3 & -8\\ 0 & 1 & 3 & 5\\ 0 & 0 & 0 & 0\end{array}\right]\xrightarrow[R_1=r_1+r_2]{}\left[\begin{array}{rrr|r}1 & 0 & 0 & -3\\ 0 & 1 & 3 & 5\\ 0 & 0 & 0 & 0\end{array}\right]$$

Solution: $x = -3, y = 5 - 3z$, $z$ is any real number.

**37.** $$\left[\begin{array}{rrr|r}1 & -1 & 0 & 1\\ 0 & 1 & -1 & 6\\ 1 & 0 & 1 & -1\end{array}\right]\xrightarrow[R_3=r_3-r_1]{}\left[\begin{array}{rrr|r}1 & -1 & 0 & 1\\ 0 & 1 & -1 & 6\\ 0 & 1 & 1 & -2\end{array}\right]\xrightarrow[\substack{R_1=r_1+r_2\\ R_3=r_3-r_2}]{}\left[\begin{array}{rrr|r}1 & 0 & -1 & 7\\ 0 & 1 & -1 & 6\\ 0 & 0 & 2 & -8\end{array}\right]$$

$$\xrightarrow[R_3=\frac{1}{2}r_3]{}\left[\begin{array}{rrr|r}1 & 0 & -1 & 7\\ 0 & 1 & -1 & 6\\ 0 & 0 & 1 & -4\end{array}\right]\xrightarrow[\substack{R_1=r_1+r_3\\ R_2=r_2+r_3}]{}\left[\begin{array}{rrr|r}1 & 0 & 0 & 3\\ 0 & 1 & 0 & 2\\ 0 & 0 & 1 & -4\end{array}\right]$$

Solution: $x = 3, y = 2, z = -4$

**39.** $$\left[\begin{array}{rrrr|r}1 & 1 & 0 & 0 & 7\\ 0 & 1 & -1 & 1 & 5\\ 1 & -1 & 1 & 1 & 6\\ 0 & 1 & 0 & -1 & 10\end{array}\right]\xrightarrow[R_3=r_3-r_1]{}\left[\begin{array}{rrrr|r}1 & 1 & 0 & 0 & 7\\ 0 & 1 & -1 & 1 & 5\\ 0 & -2 & 1 & 1 & -1\\ 0 & 1 & 0 & -1 & 10\end{array}\right]$$

$$\xrightarrow[\substack{R_1=r_1-r_2\\ R_3=r_3+2r_2\\ R_4=r_4-r_2}]{}\left[\begin{array}{rrrr|r}1 & 0 & 1 & -1 & 2\\ 0 & 1 & -1 & 1 & 5\\ 0 & 0 & -1 & 3 & 9\\ 0 & 0 & 1 & -2 & 5\end{array}\right]\xrightarrow[\substack{R_1=r_1+r_3\\ R_2=r_2-r_3\\ R_4=r_4+r_3}]{}\left[\begin{array}{rrrr|r}1 & 0 & 0 & 2 & 11\\ 0 & 1 & 0 & -2 & -4\\ 0 & 0 & -1 & 3 & 9\\ 0 & 0 & 0 & 1 & 14\end{array}\right]$$

$$\xrightarrow[\substack{R_1=r_1-2r_4\\ R_2=r_2+2r_4\\ R_3=r_3-3r_4}]{}\left[\begin{array}{cccc|c}1&0&0&0&-17\\0&1&0&0&24\\0&0&-1&0&-33\\0&0&0&1&14\end{array}\right]\xrightarrow[R_3=-r_3]{}\left[\begin{array}{cccc|c}1&0&0&0&-17\\0&1&0&0&24\\0&0&1&0&33\\0&0&0&1&14\end{array}\right]$$

Solution: $x_1=-17,\ x_2=24,\ x_3=33,\ x_4=14$

**41.** $$\left[\begin{array}{cccc|c}1&2&3&-1&0\\3&0&0&-1&4\\0&1&-1&-1&2\end{array}\right]\xrightarrow[R_2=r_2-3r_1]{}\left[\begin{array}{cccc|c}1&2&3&-1&0\\0&-6&-9&2&4\\0&1&-1&-1&2\end{array}\right]$$

$$\xrightarrow[\substack{R_2=r_3\\ R_3=r_2}]{}\left[\begin{array}{cccc|c}1&2&3&-1&0\\0&1&-1&-1&2\\0&-6&-9&2&4\end{array}\right]\xrightarrow[\substack{R_1=r_1-2r_2\\ R_3=r_3+6r_2}]{}\left[\begin{array}{cccc|c}1&0&5&1&-4\\0&1&-1&-1&2\\0&0&-15&-4&16\end{array}\right]$$

$$\xrightarrow[R_3=-\frac{1}{15}r_3]{}\left[\begin{array}{cccc|c}1&0&5&1&-4\\0&1&-1&-1&2\\0&0&1&4/15&-16/15\end{array}\right]\xrightarrow[\substack{R_1=r_1-5r_3\\ R_2=r_2+r_3}]{}\left[\begin{array}{cccc|c}1&0&0&-1/3&4/3\\0&1&0&-11/15&14/15\\0&0&1&4/15&-16/15\end{array}\right]$$

Solution: $x_1=\frac{4}{3}+\frac{1}{3}x_4, x_2=\frac{14}{15}+\frac{11}{15}x_4, x_3=-\frac{16}{15}-\frac{4}{15}x_4$, $x_4$ is any real number.

**43.** $$\left[\begin{array}{ccc|c}1&-1&1&5\\2&-2&2&8\end{array}\right]\xrightarrow[R_2=r_2-2r_1]{}\left[\begin{array}{ccc|c}1&-1&1&5\\0&0&0&-2\end{array}\right]\xrightarrow[R_2=-\frac{1}{2}r_2]{}\left[\begin{array}{ccc|c}1&-1&1&5\\0&0&0&1\end{array}\right]$$

$$\xrightarrow[R_1=r_1-5r_2]{}\left[\begin{array}{ccc|c}1&-1&1&0\\0&0&0&1\end{array}\right]$$: Inconsistent system.

**45.** $$\left[\begin{array}{ccc|c}3&-1&2&3\\3&3&1&3\\3&-5&3&12\end{array}\right]\xrightarrow[\substack{R_2=r_2-r_1\\ R_3=r_3-r_1}]{}\left[\begin{array}{ccc|c}3&-1&2&3\\0&4&-1&0\\0&-4&1&9\end{array}\right]\xrightarrow[\substack{R_1=r_1+\frac{1}{4}r_2\\ R_3=r_3+r_2}]{}\left[\begin{array}{ccc|c}3&0&7/4&3\\0&4&-1&0\\0&0&0&9\end{array}\right]$$

Inconsistent system.

**47.** $$\left[\begin{array}{cccc|c}1&1&1&1&4\\2&-1&1&0&0\\3&2&1&-1&6\\1&-2&-2&2&-1\end{array}\right]\xrightarrow[\substack{R_2=r_2-2r_1\\ R_3=r_3-3r_1\\ R_4=r_4-r_1}]{}\left[\begin{array}{cccc|c}1&1&1&1&4\\0&-3&-1&-2&-8\\0&-1&-2&-4&-6\\0&-3&-3&1&-5\end{array}\right]$$

$$\xrightarrow[\substack{R_2=-r_3\\ R_3=-r_2}]{}\left[\begin{array}{cccc|c}1&1&1&1&4\\0&1&2&4&6\\0&3&1&2&8\\0&-3&-3&1&-5\end{array}\right]\xrightarrow[\substack{R_1=r_1-r_2\\ R_3=r_3-3r_2\\ R_4=r_4+3r_2}]{}\left[\begin{array}{cccc|c}1&0&-1&-3&-2\\0&1&2&4&6\\0&0&-5&-10&-10\\0&0&3&13&13\end{array}\right]$$

$$\xrightarrow[R_3=-\frac{1}{5}r_3]{}\left[\begin{array}{cccc|c}1&0&-1&-3&-2\\0&1&2&4&6\\0&0&1&2&2\\0&0&3&13&13\end{array}\right]\xrightarrow[\substack{R_1=r_1+r_3\\R_2=r_2-2r_3\\R_4=r_4-3r_3}]{}\left[\begin{array}{cccc|c}1&0&0&-1&0\\0&1&0&0&2\\0&0&1&2&2\\0&0&0&7&7\end{array}\right]$$

$$\xrightarrow[R_4=\frac{1}{7}r_4]{}\left[\begin{array}{cccc|c}1&0&0&-1&0\\0&1&0&0&2\\0&0&1&2&2\\0&0&0&1&1\end{array}\right]\xrightarrow[\substack{R_1=r_1+r_4\\R_3=r_3-2r_4}]{}\left[\begin{array}{cccc|c}1&0&0&0&1\\0&1&0&0&2\\0&0&1&0&0\\0&0&0&1&1\end{array}\right]$$

Solution: $x_1=1,\ x_2=2,\ x_3=0,\ x_4=1$

**49.** $$\left[\begin{array}{ccc|c}2&-1&-1&0\\1&-1&-1&1\\3&-1&-1&2\end{array}\right]\xrightarrow[\substack{R_1=r_2\\R_2=r_1}]{}\left[\begin{array}{ccc|c}1&-1&-1&1\\2&-1&-1&0\\3&-1&-1&2\end{array}\right]\xrightarrow[\substack{R_2=r_2-2r_1\\R_3=r_3-3r_1}]{}\left[\begin{array}{ccc|c}1&-1&-1&1\\0&1&1&-2\\0&2&2&-1\end{array}\right]$$

$$\xrightarrow[\substack{R_1=r_1+r_2\\R_3=r_3-2r_2}]{}\left[\begin{array}{ccc|c}1&0&0&-1\\0&1&1&-2\\0&0&0&3\end{array}\right]$$ : Inconsistent system.

**51.** $$\left[\begin{array}{ccc|c}2&-1&1&6\\3&-1&1&6\\4&-2&2&12\end{array}\right]\xrightarrow[\substack{R_1=r_1-r_2\\R_3=r_3-2r_1}]{}\left[\begin{array}{ccc|c}-1&0&0&0\\3&-1&1&6\\0&0&0&0\end{array}\right]\xrightarrow[R_2=r_2+3r_1]{}\left[\begin{array}{ccc|c}-1&0&0&0\\0&-1&1&6\\0&0&0&0\end{array}\right]$$

$$\xrightarrow[\substack{R_1=-r_1\\R_2=-r_2}]{}\left[\begin{array}{ccc|c}1&0&0&0\\0&1&-1&-6\\0&0&0&0\end{array}\right]$$ : Solution: $x=0, y=-6+z$, $z$ is any real number.

**53.** Let $x$, $y$ and $z$ represent the amounts of the 10%, 30% and 50% acid solutions, respectively, in the mixture. Then $x+y+z=100$ liters and $0.10x+0.30y+0.50z=0.25(100)$. (Note $x\ge 0$, $y\ge 0$, $z\ge 0$.) Solve $\begin{cases}x+y+z=100\\0.1x+0.3y+0.5z=25\end{cases}$:

$$\left[\begin{array}{ccc|c}1&1&1&100\\0.1&0.3&0.5&25\end{array}\right]\xrightarrow[R_2=r_2-0.1r_1]{}\left[\begin{array}{ccc|c}1&1&1&100\\0&0.2&0.4&15\end{array}\right]\xrightarrow[R_2=5r_2]{}\left[\begin{array}{ccc|c}1&1&1&100\\0&1&2&75\end{array}\right]$$

$$\xrightarrow[R_2=r_1-r_2]{}\left[\begin{array}{ccc|c}1&0&-1&25\\0&1&2&75\end{array}\right]$$ : $x=25+z,\ y=75-2z,\ x\ge 0,\ y\ge 0,\ z\ge 0$

| | | | | | | | | |
|---|---|---|---|---|---|---|---|---|
| $25+z$: liters of 10% acid solution | 25 | 30 | 35 | 40 | 45 | 50 | 55 | 60 |
| $75-2z$: liters of 30% acid solution | 75 | 65 | 55 | 45 | 35 | 25 | 15 | 5 |
| $z$: liters of 50% acid solution | 0 | 5 | 10 | 15 | 20 | 25 | 30 | 35 |

**55.** Let $x$, $y$ and $z$ represent the prices for a hamburger, an order of large fries and a large cola, respectively. Then $8x+6y+6z=\$26.10$, $10x+6y+8z=\$31.60$, $\$1.75 \le x \le \$2.25$, $\$0.75 \le y \le \$1.00$ and $\$0.60 \le z \le \$0.90$. Solve $\begin{cases} 8x+6y+6z=26.1 \\ 10x+6y+8z=31.6 \end{cases}$:

$$\left[\begin{array}{ccc|c} 8 & 6 & 6 & 26.1 \\ 10 & 6 & 8 & 31.6 \end{array}\right] \xrightarrow[R_1=r_1-r_2]{} \left[\begin{array}{ccc|c} -2 & 0 & -2 & -5.5 \\ 10 & 6 & 8 & 31.6 \end{array}\right] \xrightarrow[R_2=r_2+5r_1]{} \left[\begin{array}{ccc|c} -2 & 0 & -2 & -5.5 \\ 0 & 6 & -2 & 4.1 \end{array}\right]$$

$$\xrightarrow[R_1=-\frac{1}{2}r_1,\ R_2=\frac{1}{6}r_2]{} \left[\begin{array}{ccc|c} 1 & 0 & 1 & 2.75 \\ 0 & 1 & -1/3 & 41/60 \end{array}\right]: \ x=2.75-z,\ y=\frac{41}{60}+\frac{1}{3}z$$

No, there is not sufficient information to determine the price of each food item.

| | | | | | | | | | | |
|---|---|---|---|---|---|---|---|---|---|---|
| $2.75-z$: hamburger price | \$2.13 | \$2.10 | \$2.07 | \$2.04 | \$2.01 | \$1.98 | \$1.95 | \$1.92 | \$1.89 | \$1.86 |
| $\frac{41}{60}+\frac{1}{3}z$: order of fries price | \$0.89 | \$0.90 | \$0.91 | \$0.92 | \$0.93 | \$0.94 | \$0.95 | \$0.96 | \$0.97 | \$0.98 |
| $z$: large cola price | \$0.62 | \$0.65 | \$0.68 | \$0.71 | \$0.74 | \$0.77 | \$0.80 | \$0.83 | \$0.86 | \$0.89 |

**57.** Let $x$, $y$ and $z$ represent the amounts the couple invests in Treasury bills, corporate bonds and junk bonds, respectively. Then $x+y+z=\$25{,}000$ and $0.07x+0.09y+0.11z=\$2000$, with $x>0$, $y>0$ and $z>0$. Solve $\begin{cases} x+y+z=25{,}000 \\ 0.07x+0.09y+0.11z=2{,}000 \end{cases}$:

$$\left[\begin{array}{ccc|c} 1 & 1 & 1 & 25000 \\ 0.07 & 0.09 & 0.11 & 2000 \end{array}\right] \xrightarrow[R_2=r_2-0.07r_1]{} \left[\begin{array}{ccc|c} 1 & 1 & 1 & 25000 \\ 0 & 0.02 & 0.04 & 250 \end{array}\right]$$

$$\xrightarrow[R_2=50r_2]{} \left[\begin{array}{ccc|c} 1 & 1 & 1 & 25000 \\ 0 & 1 & 2 & 12500 \end{array}\right] \xrightarrow[R_1=r_1-r_2]{} \left[\begin{array}{ccc|c} 1 & 0 & -1 & 12500 \\ 0 & 1 & 2 & 12500 \end{array}\right]:$$

$x=12{,}500+z$, $y=12{,}500-2z$. $x>0, y>0, z>0$. Possibilities include:

| | | | | | | |
|---|---|---|---|---|---|---|
| $12{,}500+z$: Treasury bills | \$13,500 | \$14,500 | \$15,500 | \$16,500 | \$17,500 | \$18,500 |
| $12500-2z$: corporate bonds | \$10,500 | \$8,500 | \$6,500 | \$4,500 | \$2,500 | \$500 |
| $z$: junk bonds | \$1,000 | \$2,000 | \$3,000 | \$4,000 | \$5,000 | \$6,000 |

**59.** (a) Let $x$, $y$ and $z$ represent the amounts the couple invests in Treasury bills, corporate bonds and junk bonds, respectively. Then $x+y+z=\$25{,}000$ and $0.07x+0.09y+0.11z=\$1500$, with $x>0$, $y>0$ and $z>0$. Solve $\begin{cases} x+y+z=25{,}000 \\ 0.07x+0.09y+0.11z=1{,}500 \end{cases}$:

$$\left[\begin{array}{ccc|c} 1 & 1 & 1 & 25000 \\ 0.07 & 0.09 & 0.11 & 1500 \end{array}\right] \xrightarrow[R_2=r_2-0.07r_1]{} \left[\begin{array}{ccc|c} 1 & 1 & 1 & 25000 \\ 0 & 0.02 & 0.04 & -250 \end{array}\right]$$

$$\xrightarrow[R_2=50r_2]{} \left[\begin{array}{ccc|c} 1 & 1 & 1 & 25000 \\ 0 & 1 & 2 & -12500 \end{array}\right] \xrightarrow[R_1=r_1-r_2]{} \left[\begin{array}{ccc|c} 1 & 0 & -1 & 37500 \\ 0 & 1 & 2 & -12500 \end{array}\right]:$$

$x=37{,}500+z$, $y=-12{,}500-2z$. $x>0, y>0, z>0$. Note that, for $z>0$, this solution implies that $x > 37{,}500$ and $y < -12{,}500 < 0$, which is clearly not possible. The couple could invest all \$25,000 in Treasury bills, yielding an annual return of \$1750, already more than the \$1500 they require.

(b) $x+y+z = \$25{,}000$ and $0.07x+0.09y+0.11z = \$2500$, with $x>0$, $y>0$ and $z>0$.

Solve $\begin{cases} x+ \quad y+ \quad z=25{,}000 \\ 0.07x+0.09y+0.11z= 2{,}500 \end{cases}$:

$$\left[\begin{array}{ccc|c} 1 & 1 & 1 & 25000 \\ 0.07 & 0.09 & 0.11 & 2500 \end{array}\right] \xrightarrow[R_2=r_2-0.07r_1]{} \left[\begin{array}{ccc|c} 1 & 1 & 1 & 25000 \\ 0 & 0.02 & 0.04 & 750 \end{array}\right]$$

$$\xrightarrow[R_2=50r_2]{} \left[\begin{array}{ccc|c} 1 & 1 & 1 & 25000 \\ 0 & 1 & 2 & 37500 \end{array}\right] \xrightarrow[R_1=r_1-r_2]{} \left[\begin{array}{ccc|c} 1 & 0 & -1 & -12500 \\ 0 & 1 & 2 & 37500 \end{array}\right]:$$

$x = -12{,}500 + z$, $y = 37{,}500 - 2z$. $x>0, y>0, z>0$. Possibilities include:

| | | | | | | |
|---|---|---|---|---|---|---|
| $-12{,}500+z$: Treasury bills | \$500 | \$1,500 | \$2,500 | \$3,500 | \$4,500 | \$5,500 |
| $37{,}500-2z$: corporate bonds | \$11,500 | \$9,500 | \$7,500 | \$5,500 | \$3,500 | \$1,500 |
| $z$: junk bonds | \$13,000 | \$14,000 | \$15,000 | \$16,000 | \$17,000 | \$18,000 |

(c) As the required return increases, the amount they can invest in Treasury bills diminishes, and the amount they must invest in junk bonds increases (thus increasing the risk of their investment).

## Technology Exercises

**1.** REF: $\left[\begin{array}{cccc|c} 1 & 2 & 1 & 3 & 4 \\ 0 & 1 & .33 & 1.67 & .67 \\ 0 & 0 & 1 & 2 & 8 \\ 0 & 0 & 0 & 1 & 3 \end{array}\right]$ or $\left[\begin{array}{cccc|c} 1 & 2 & 1 & 3 & 4 \\ 0 & 1 & 1/3 & 5/3 & 2/3 \\ 0 & 0 & 1 & 2 & 8 \\ 0 & 0 & 0 & 1 & 3 \end{array}\right]$;

RREF: $\left[\begin{array}{cccc|c} 1 & 0 & 0 & 0 & 3 \\ 0 & 1 & 0 & 0 & -5 \\ 0 & 0 & 1 & 0 & 2 \\ 0 & 0 & 0 & 1 & 3 \end{array}\right]$: $x_1 = 3, x_2 = -5,$ $x_3 = 2, x_4 = 3$

**3.** REF: $\left[\begin{array}{cccc|c} 1 & 1 & -1 & -1 & 6 \\ 0 & 1 & -3 & -1 & 12 \\ 0 & 0 & 1 & 1 & -10 \\ 0 & 0 & 0 & 1 & -19 \end{array}\right]$ RREF: $\left[\begin{array}{cccc|c} 1 & 0 & 0 & 0 & -24 \\ 0 & 1 & 0 & 0 & 20 \\ 0 & 0 & 1 & 0 & 9 \\ 0 & 0 & 0 & 1 & -19 \end{array}\right]$: $x_1 = -24, x_2 = 20,$ $x_3 = 9, x_4 = -19$

**5.** REF: $\left[\begin{array}{cccc|c} 1 & -2 & -5 & 0 & -25 \\ 0 & 1 & 1 & 0 & 15 \\ 0 & 0 & 1 & 4 & -15 \\ 0 & 0 & 0 & 1 & -10 \end{array}\right]$ RREF: $\left[\begin{array}{cccc|c} 1 & 0 & 0 & 0 & 80 \\ 0 & 1 & 0 & 0 & -10 \\ 0 & 0 & 1 & 0 & 25 \\ 0 & 0 & 0 & 1 & -10 \end{array}\right]$: $x_1 = 80, x_2 = -10,$ $x_3 = 25, x_4 = -10$

## 2.4 Matrix Algebra

**1.** $2\times 2$ **3.** $2\times 3$ **5.** $3\times 2$

**7.** $3\times 2$ **9.** $2\times 1$ **11.** $1\times 1$

**13.** False: equal matrices must have equal dimensions. This compares a $2\times 1$ matrix with a $1\times 2$ matrix.

**15.** True **17.** True **19.** True **21.** True

**23.** False: The sum of two $2\times 1$ matrices is another $2\times 1$ matrix.

**25.** $\begin{bmatrix} 3+-2 & -1+2 \\ 4+2 & 2+5 \end{bmatrix}=\begin{bmatrix} 1 & 1 \\ 6 & 7 \end{bmatrix}$

**27.** $\begin{bmatrix} 3\cdot 2 & 3\cdot 6 & 3\cdot 0 \\ 3\cdot 4 & 3\cdot(-2) & 3\cdot 1 \end{bmatrix}=\begin{bmatrix} 6 & 18 & 0 \\ 12 & -6 & 3 \end{bmatrix}$

**29.** $\begin{bmatrix} 2-0 & -2+6 & 16-24 \\ 4-3 & 8-12 & 2-3 \end{bmatrix}=\begin{bmatrix} 2 & 4 & -8 \\ 1 & -4 & -1 \end{bmatrix}$

**31.** $\begin{bmatrix} 3a+10a & 24+30 \\ 3b-5b & 3-10 \\ 3c-5c & -6+0 \end{bmatrix}=\begin{bmatrix} 13a & 54 \\ -2b & -7 \\ -2c & -6 \end{bmatrix}$

**33.** $A+B=\begin{bmatrix} 2+1 & -3+(-2) & 4+0 \\ 0+5 & 2+1 & 1+2 \end{bmatrix}=\begin{bmatrix} 3 & -5 & 4 \\ 5 & 3 & 3 \end{bmatrix}$

**35.** $2A-3C=\begin{bmatrix} 4 & -6 & 8 \\ 0 & 4 & 2 \end{bmatrix}-\begin{bmatrix} -9 & 0 & 15 \\ 6 & 3 & 9 \end{bmatrix}=\begin{bmatrix} 4+9 & -6-0 & 8-15 \\ 0-6 & 4-3 & 2-9 \end{bmatrix}=\begin{bmatrix} 13 & -6 & -7 \\ -6 & 1 & -7 \end{bmatrix}$

**37.** $(A+B)-2C=\begin{bmatrix} 3 & -5 & 4 \\ 5 & 3 & 3 \end{bmatrix}-\begin{bmatrix} -6 & 0 & 10 \\ 4 & 2 & 6 \end{bmatrix}=\begin{bmatrix} 3+6 & -5-0 & 4-10 \\ 5-4 & 3-2 & 3-6 \end{bmatrix}=\begin{bmatrix} 9 & -5 & -6 \\ 1 & 1 & -3 \end{bmatrix}$

**39.** $3A+4(B+C)=\begin{bmatrix} 6 & -9 & 12 \\ 0 & 6 & 3 \end{bmatrix}+4\begin{bmatrix} -2 & -2 & 5 \\ 7 & 2 & 5 \end{bmatrix}=\begin{bmatrix} 6-8 & -9-8 & 12+20 \\ 0+28 & 6+8 & 3+20 \end{bmatrix}$

$=\begin{bmatrix} -2 & -17 & 32 \\ 28 & 14 & 23 \end{bmatrix}$

**41.** $2(A-B)-C=2\begin{bmatrix}2-1 & -3-(-2) & 4-0\\ 0-5 & 2-1 & 1-2\end{bmatrix}-\begin{bmatrix}-3 & 0 & 5\\ 2 & 1 & 3\end{bmatrix}$

$=2\begin{bmatrix}1 & -1 & 4\\ -5 & 1 & -1\end{bmatrix}-\begin{bmatrix}-3 & 0 & 5\\ 2 & 1 & 3\end{bmatrix}=\begin{bmatrix}2+3 & -2-0 & 8-5\\ -10-2 & 2-1 & -2-3\end{bmatrix}=\begin{bmatrix}5 & -2 & 3\\ -12 & 1 & -5\end{bmatrix}$

**43.** $3A-B-6C=\begin{bmatrix}6 & -9 & 12\\ 0 & 6 & 3\end{bmatrix}-\begin{bmatrix}1 & -2 & 0\\ 5 & 1 & 2\end{bmatrix}-\begin{bmatrix}-18 & 0 & 30\\ 12 & 6 & 18\end{bmatrix}$

$=\begin{bmatrix}6-1+18 & -9+2-0 & 12-0-30\\ 0-5-12 & 6-1-6 & 3-2-18\end{bmatrix}=\begin{bmatrix}23 & -7 & -18\\ -17 & -1 & -17\end{bmatrix}$

**45.** $A+B=\begin{bmatrix}2+1 & -3-2 & 4+0\\ 0+5 & 2+1 & 1+2\end{bmatrix}=\begin{bmatrix}3 & -5 & 4\\ 5 & 3 & 3\end{bmatrix}$;

$B+A=\begin{bmatrix}1+2 & -2-3 & 0+4\\ 5+0 & 1+2 & 2+1\end{bmatrix}=\begin{bmatrix}3 & -5 & 4\\ 5 & 3 & 3\end{bmatrix}=A+B$

**47.** $A+(-A)=\begin{bmatrix}2 & -3 & 4\\ 0 & 2 & 1\end{bmatrix}+\begin{bmatrix}-2 & 3 & -4\\ 0 & -2 & -1\end{bmatrix}=\begin{bmatrix}2-2 & -3+3 & 4-4\\ 0+0 & 2-2 & 1-1\end{bmatrix}=\begin{bmatrix}0 & 0 & 0\\ 0 & 0 & 0\end{bmatrix}$

**49.** $2B+3B=2\begin{bmatrix}1 & -2 & 0\\ 5 & 1 & 2\end{bmatrix}+3\begin{bmatrix}1 & -2 & 0\\ 5 & 1 & 2\end{bmatrix}=\begin{bmatrix}2 & -4 & 0\\ 10 & 2 & 4\end{bmatrix}+\begin{bmatrix}3 & -6 & 0\\ 15 & 3 & 6\end{bmatrix}$

$=\begin{bmatrix}2+3 & -4-6 & 0+0\\ 10+15 & 2+3 & 4+6\end{bmatrix}=\begin{bmatrix}5 & -10 & 0\\ 25 & 5 & 10\end{bmatrix}=5\begin{bmatrix}1 & -2 & 0\\ 5 & 1 & 2\end{bmatrix}=5B$

**51.** $x=-4,\ z=4$

**53.** Equating corresponding entries, $\begin{cases}x-2y=3\\ 0=0\\ -2=-2\\ 6=x+y\end{cases}$, so solve $\begin{cases}x-2y=3\\ x+\ y=6\end{cases}$:

$\left[\begin{array}{cc|c}1 & -2 & 3\\ 1 & 1 & 6\end{array}\right]\xrightarrow[R_2=r_2-r_1]{}\left[\begin{array}{cc|c}1 & -2 & 3\\ 0 & 3 & 3\end{array}\right]\xrightarrow[R_2=\frac{1}{3}r_2]{}\left[\begin{array}{cc|c}1 & -2 & 3\\ 0 & 1 & 1\end{array}\right]\xrightarrow[R_1=r_1+2r_2]{}\left[\begin{array}{cc|c}1 & 0 & 5\\ 0 & 1 & 1\end{array}\right]$

$x=5,\ y=1$

**55.** $[2+x \quad 3+2y \quad -4+z]=[6 \quad -9 \quad 2]$. Thus, $\begin{cases}2+\ x=6\\ 3+2y=-9\\ -4+\ z=2\end{cases}$, so $\begin{cases}x=4\\ y=-6\\ z=6\end{cases}$.

**57.**

| | 1/2-inch | 1-inch | 2-inch |
|---|---|---|---|
| steel | 25 | 45 | 35 |
| aluminum | 13 | 20 | 23 |

as a $2 \times 3$ matrix: $\begin{bmatrix} 25 & 45 & 35 \\ 13 & 20 & 23 \end{bmatrix}$ or

| | steel | aluminum |
|---|---|---|
| 1/2-inch | 25 | 13 |
| 1-inch | 45 | 20 |
| 2-inch | 35 | 23 |

as a $3 \times 2$ matrix: $\begin{bmatrix} 25 & 13 \\ 45 & 20 \\ 35 & 23 \end{bmatrix}$

**59.**

| | Democrats | Republicans | Independents |
|---|---|---|---|
| < \$25,000 | 351 | 271 | 73 |
| ≥ \$25,000 | 203 | 215 | 55 |

as a $2 \times 3$ matrix: $\begin{matrix} & \text{Dem.} & \text{Rep.} & \text{Ind.} \\ <\$25{,}000 & \\ \geq\$25{,}000 & \end{matrix} \begin{bmatrix} 351 & 271 & 73 \\ 203 & 215 & 55 \end{bmatrix}$ or

| | < \$25,000 | ≥ \$25,000 |
|---|---|---|
| Democrats | 351 | 203 |
| Republicans | 271 | 215 |
| Independents | 73 | 55 |

as a $3 \times 2$ matrix: $\begin{matrix} & <\$25{,}000 & \$\geq 25{,}000 \\ \text{Dem.} \\ \text{Rep.} \\ \text{Ind.} \end{matrix} \begin{bmatrix} 351 & 203 \\ 271 & 215 \\ 73 & 55 \end{bmatrix}$

**61.**

| | LAS | ENG | EDUC |
|---|---|---|---|
| male | $500 - 250 = 250$ | $0.75 \cdot 300 = 225$ | $(1000 - 800) - 120 = 80$ |
| female | $0.5 \cdot 500 = 250$ | $300 - 225 = 75$ | $0.60 \cdot (1000 - 800) = 120$ |

as a $2 \times 3$ matrix: $\begin{matrix} & \text{LAS} & \text{ENG} & \text{EDUC} \\ \text{male} \\ \text{female} \end{matrix} \begin{bmatrix} 250 & 225 & 80 \\ 250 & 75 & 120 \end{bmatrix}$

## Technology Exercises

1. $\begin{bmatrix} -2 & 1 & 7 & 5 \\ 4 & 6 & 7 & 5 \\ -3.5 & 8 & -4 & 13 \\ 12 & -1 & 7 & 6 \end{bmatrix}$

3. $\begin{bmatrix} 19 & -11 & -14 & -15 \\ -12 & -13 & -21 & -17 \\ 15.5 & -24 & 19 & -39 \\ -29 & 10 & -14 & -11 \end{bmatrix}$

5. $\begin{bmatrix} 37 & -17 & 30 & 15 \\ 4 & 9 & 13 & 1 \\ 20.5 & 16 & -3 & 31 \\ 29 & 18 & 22 & 43 \end{bmatrix}$

## 2.5 Multiplication of Matrices

1. $[1\cdot 2+3\cdot 4]=[14]$

3. $[1\cdot 0+(-2)\cdot 1+3\cdot 2]=[4]$

5. $[1\cdot 2+4\cdot 4 \quad 1\cdot 0+4\cdot(-2)]=[18 \quad -8]$

7. $\begin{bmatrix} 2\cdot 2+0\cdot 3 & 2\cdot 1+0\cdot(-2) \\ 4\cdot 2+(-2)\cdot 3 & 4\cdot 1+(-2)(-2) \end{bmatrix}=\begin{bmatrix} 4 & 2 \\ 2 & 8 \end{bmatrix}$

9. $[1\cdot 0+(-2)\cdot 1+3\cdot 2 \quad 1\cdot 1+(-2)\cdot 2+3\cdot 3]=[4 \quad 6]$

11. $\begin{bmatrix} 1\cdot 0+(-2)\cdot 1+3\cdot 2 & 1\cdot(-2)+(-2)\cdot 0+3\cdot(-4) \\ 4\cdot 0+0\cdot 1+6\cdot 2 & 4\cdot(-2)+0\cdot 0+6\cdot(-4) \end{bmatrix}=\begin{bmatrix} 4 & -14 \\ 12 & -32 \end{bmatrix}$

13. $\begin{bmatrix} 2\cdot 2+0\cdot 3 & 2\cdot 1+0\cdot(-2) \\ 4\cdot 2+(-2)\cdot 3 & 4\cdot 1+(-2)(-2) \\ 6\cdot 2+(-1)\cdot 3 & 6\cdot 1+(-1)(-2) \end{bmatrix}=\begin{bmatrix} 4 & 2 \\ 2 & 8 \\ 9 & 8 \end{bmatrix}$

15. $\begin{bmatrix} 1\cdot 3+(-1)\cdot 0+6\cdot 1 & 1\cdot 2+(-1)\cdot 1+6\cdot 0 \\ 2\cdot 3+0\cdot 0+(-1)\cdot 1 & 2\cdot 2+0\cdot 1+(-1)\cdot 0 \\ 3\cdot 3+1\cdot 0+2\cdot 1 & 3\cdot 2+1\cdot 1+2\cdot 0 \end{bmatrix}=\begin{bmatrix} 9 & 1 \\ 5 & 4 \\ 11 & 7 \end{bmatrix}$

17. $BA$ is defined with dimensions $3\times 4$.

19. $AB$ is not defined.

21. $(BA)C$ is not defined ($BA$ is $3\times 4$ and $C$ is $2\times 3$)

23. $BA + A$ is defined with dimensions $3\times 4$.

25. $DC + B$ is defined with dimension $3\times 3$.

**27.** $AB=\begin{bmatrix}1-2 & 2+8 & 3-4\\ 0-4 & 0+16 & 0-8\end{bmatrix}=\begin{bmatrix}-1 & 10 & -1\\ -4 & 16 & -8\end{bmatrix}$

**29.** $BC=\begin{bmatrix}3+8+0 & 1-2+6\\ -3+16+0 & -1-4-4\end{bmatrix}=\begin{bmatrix}11 & 5\\ 13 & -9\end{bmatrix}$

**31.** $(D+I_3)C=\begin{bmatrix}2 & 0 & 4\\ 0 & 2 & 2\\ 0 & -1 & 2\end{bmatrix}\begin{bmatrix}3 & 1\\ 4 & -1\\ 0 & 2\end{bmatrix}=\begin{bmatrix}6+0+0 & 2+0+8\\ 0+8+0 & 0-2+4\\ 0-4+0 & 0+1+4\end{bmatrix}=\begin{bmatrix}6 & 10\\ 8 & 2\\ -4 & 5\end{bmatrix}$

**33.** $EI_2=\begin{bmatrix}3\cdot1+-1\cdot0 & 3\cdot0+-1\cdot1\\ 4\cdot1+2\cdot0 & 4\cdot0+2\cdot1\end{bmatrix}=\begin{bmatrix}3 & -1\\ 4 & 2\end{bmatrix}$

**35.** $(2E)B=\begin{bmatrix}6 & -2\\ 8 & 4\end{bmatrix}\cdot B=\begin{bmatrix}6\cdot1+-2\cdot-1 & 6\cdot2+-2\cdot4 & 6\cdot3+-2\cdot-2\\ 8\cdot1+4\cdot-1 & 8\cdot2+4\cdot4 & 8\cdot3+4\cdot-2\end{bmatrix}=\begin{bmatrix}8 & 4 & 22\\ 4 & 32 & 16\end{bmatrix}$

**37.** $-5E+A=\begin{bmatrix}-15 & 5\\ -20 & -10\end{bmatrix}+\begin{bmatrix}1 & 2\\ 0 & 4\end{bmatrix}=\begin{bmatrix}-14 & 7\\ -20 & -6\end{bmatrix}$

**39.** $3CB+4D=3\begin{bmatrix}3\cdot1+1\cdot-1 & 3\cdot2+1\cdot4 & 3\cdot3+1\cdot-2\\ 4\cdot1+-1\cdot-1 & 4\cdot2+-1\cdot4 & 4\cdot3+-1\cdot-2\\ 0\cdot1+2\cdot-1 & 0\cdot2+2\cdot4 & 0\cdot3+2\cdot-2\end{bmatrix}+\begin{bmatrix}4 & 0 & 16\\ 0 & 4 & 8\\ 0 & -4 & 4\end{bmatrix}$

$$=\begin{bmatrix}6+4 & 30+0 & 21+16\\ 15+0 & 12+4 & 42+8\\ -6+0 & 24-4 & -12+4\end{bmatrix}=\begin{bmatrix}10 & 30 & 37\\ 15 & 16 & 50\\ -6 & 20 & -8\end{bmatrix}.$$

**41.** $CB=\begin{bmatrix}3\cdot1+1\cdot-1 & 3\cdot2+1\cdot4 & 3\cdot3+1\cdot-2\\ 4\cdot1+-1\cdot-1 & 4\cdot2-1\cdot4 & 4\cdot3+-1\cdot-2\\ 0\cdot1+2\cdot-1 & 0\cdot2+2\cdot4 & 0\cdot3+2\cdot-2\end{bmatrix}=\begin{bmatrix}2 & 10 & 7\\ 5 & 4 & 14\\ -2 & 8 & -4\end{bmatrix};$

$$D(CB)=\begin{bmatrix}1\cdot2+0\cdot5+4\cdot-2 & 1\cdot10+0\cdot4+4\cdot8 & 1\cdot7+0\cdot14+4\cdot-4\\ 0\cdot2+1\cdot5+2\cdot-2 & 0\cdot10+1\cdot4+2\cdot8 & 0\cdot7+1\cdot14+2\cdot-4\\ 0\cdot2+-1\cdot5+1\cdot-2 & 0\cdot10+-1\cdot4+1\cdot8 & 0\cdot7+-1\cdot14+1\cdot-4\end{bmatrix}=\begin{bmatrix}-6 & 42 & -9\\ 1 & 20 & 6\\ -7 & 4 & -18\end{bmatrix}$$

$$DC = \begin{bmatrix} 1\cdot3+0\cdot4+4\cdot0 & 1\cdot1+0\cdot-1+4\cdot2 \\ 0\cdot3+1\cdot4+2\cdot0 & 0\cdot1+1\cdot-1+2\cdot2 \\ 0\cdot3+-1\cdot4+1\cdot0 & 0\cdot1+-1\cdot-1+1\cdot2 \end{bmatrix} = \begin{bmatrix} 3 & 9 \\ 4 & 3 \\ -4 & 3 \end{bmatrix};$$

$$(DC)B = \begin{bmatrix} 3\cdot1+9\cdot-1 & 3\cdot2+9\cdot4 & 3\cdot3+9\cdot-2 \\ 4\cdot1+3\cdot-1 & 4\cdot2+3\cdot4 & 4\cdot3+3\cdot-2 \\ -4\cdot1+3\cdot-1 & -4\cdot2+3\cdot4 & -4\cdot3+3\cdot-2 \end{bmatrix} = \begin{bmatrix} -6 & 42 & -9 \\ 1 & 20 & 6 \\ -7 & 4 & -18 \end{bmatrix}.$$

**43.** $AB = \begin{bmatrix} 3+2 & 2-4 \\ 6+0 & 4+0 \end{bmatrix} = \begin{bmatrix} 5 & -2 \\ 6 & 4 \end{bmatrix}$; $BA = \begin{bmatrix} 3+4 & -3+0 \\ -2+8 & 2+0 \end{bmatrix} = \begin{bmatrix} 7 & -3 \\ 6 & 2 \end{bmatrix}$

**45.** $\begin{bmatrix} a & b \\ c & d \end{bmatrix}\begin{bmatrix} 0 & 1 \\ 2 & -1 \end{bmatrix} = \begin{bmatrix} 2b & a-b \\ 2d & c-d \end{bmatrix} = \begin{bmatrix} 2 & 1 \\ -1 & 0 \end{bmatrix}$: $2b = 2$, so $b = 1$; $a - b = a - 1 = 1$, so $a = 2$.

$2d = -1$, so $d = -\frac{1}{2}$; $c - d = c + \frac{1}{2} = 0$, so $c = -\frac{1}{2}$; $A = \begin{bmatrix} 2 & 1 \\ -1/2 & -1/2 \end{bmatrix}$

**47.** $A^2 = AA = \begin{bmatrix} 1+4-5 & 2+8-10 & 5+20-25 \\ 2+8-10 & 4+16-20 & 10+40-50 \\ -1-4+5 & -2-8+10 & -5-20+25 \end{bmatrix} = \begin{bmatrix} 0 & 0 & 0 \\ 0 & 0 & 0 \\ 0 & 0 & 0 \end{bmatrix}$

**49.** $A^2 + A = \begin{bmatrix} a^2+b^2 & ab+ba \\ ba+ab & b^2+a^2 \end{bmatrix} + \begin{bmatrix} a & b \\ b & a \end{bmatrix} = \begin{bmatrix} a^2+a+b^2 & 2ab+b \\ 2ab+b & a^2+a+b^2 \end{bmatrix} = \begin{bmatrix} 0 & 0 \\ 0 & 0 \end{bmatrix}$:

$2ab + b = (2a+1)b = 0$, so either $2a + 1 = 0$ or $b = 0$. If $2a + 1 = 0$, then $a = -\frac{1}{2}$, so $a^2 + a + b^2 = -\frac{1}{4} + b^2 = 0$, so $b^2 = \frac{1}{4}$, and $b = \pm\frac{1}{2}$. If $b = 0$, then $a^2 + a + b = a^2 + a = a(a+1) = 0$, so either $a = 0$ or $a + 1 = 0$, i.e., $a = -1$. Solutions: $a = -\frac{1}{2}$, $b = -\frac{1}{2}$; $a = -\frac{1}{2}$, $b = \frac{1}{2}$; $a = 0$, $b = 0$; $a = -1$, $b = 0$.

**51.** $\begin{bmatrix} \frac{x_1}{2} + \frac{x_2}{4} & \frac{x_1}{2} + \frac{3x_2}{4} \end{bmatrix} = \begin{bmatrix} x_1 & x_2 \end{bmatrix}$: $\begin{cases} \frac{x_1}{2} + \frac{x_2}{4} = x_1 & \text{or} \quad \frac{-x_1}{2} + \frac{x_2}{4} = 0; \\ \frac{x_1}{2} + \frac{3x_2}{4} = x_2 & \text{or} \quad \frac{x_1}{2} - \frac{x_2}{4} = 0; \\ & \text{and} \quad x_1 + x_2 = 1 \end{cases}$

$$\left[\begin{array}{cc|c} -1/2 & 1/4 & 0 \\ 1/2 & -1/4 & 0 \\ 1 & 1 & 1 \end{array}\right] \underset{\substack{R_1=r_3 \\ R_2=4r_2 \\ R_3=4r_1}}{\longrightarrow} \left[\begin{array}{cc|c} 1 & 1 & 1 \\ 2 & -1 & 0 \\ -2 & 1 & 0 \end{array}\right] \underset{\substack{R_2=r_2-2r_1 \\ R_3=r_3+2r_1}}{\longrightarrow} \left[\begin{array}{cc|c} 1 & 1 & 1 \\ 0 & -3 & -2 \\ 0 & 3 & 2 \end{array}\right]$$

$$\underset{\substack{R_1=r_1+\frac{1}{3}r_2 \\ R_3=r_3+r_2}}{\longrightarrow} \left[\begin{array}{cc|c} 1 & 0 & 1/3 \\ 0 & -3 & -2 \\ 0 & 0 & 0 \end{array}\right] \underset{R_2=-\frac{1}{3}r_2}{\longrightarrow} \left[\begin{array}{cc|c} 1 & 0 & 1/3 \\ 0 & 1 & 2/3 \\ 0 & 0 & 0 \end{array}\right] : \begin{bmatrix} x_1 & x_2 \end{bmatrix} = \begin{bmatrix} \frac{1}{3} & \frac{2}{3} \end{bmatrix}.$$

**53.** (a) $PQ = [14+84+40 \quad 21+108+60 \quad 7+60+30 \quad 84+240+75]$ $= [138 \quad 189 \quad 97 \quad 399]$. The entries in $PQ$ are the amounts of raw materials $M_1$, $M_2$, $M_3$, $M_4$ needed to fill the order.

(b) $QC = \begin{bmatrix} 20+36+15+240 \\ 70+108+75+400 \\ 80+144+90+300 \end{bmatrix} = \begin{bmatrix} 311 \\ 653 \\ 614 \end{bmatrix}$. The entries in $QC$ are the costs for producing each of the products $P_1$, $P_2$, $P_3$.

(c) $PQC = (PQ)C = [1380+2268+1455+7980] = [13{,}083]$. The entry in $PQC$ is the total cost to produce the order.

## Technology Exercises

**1.** $\begin{bmatrix} 0.5 & 16 & -30 & 25 \\ 21 & 14 & 28 & 64 \\ 19.5 & 8 & -23 & 33 \\ -9.5 & 45 & -9 & 83 \end{bmatrix}$

**3.** $\begin{bmatrix} 31.5 & 251 & -31.5 & 143 \\ 861 & 350 & 791 & 420 \\ 369.5 & 115 & 206.5 & 215 \\ 412.5 & 882 & 451.5 & 491 \end{bmatrix}$

**5.** $\begin{bmatrix} 66 & 74 & 94 & 38 \\ 71 & 13 & 106 & 28 \\ 165 & 124.5 & 79 & 52 \\ 158 & -3 & 152 & 46 \end{bmatrix}$

**7.** $\begin{bmatrix} -5 & 23 & -102 & 44 \\ -108 & -56 & -70 & 122 \\ 108 & -152 & -67 & 36 \\ -346 & 279 & -249 & 187 \end{bmatrix}$

**9.** (a) $A^2 = \begin{bmatrix} -2.95 & -0.5 & -1.21 & 1.6 & -0.49 \\ -2.8 & 2.56 & 2.54 & 1.4 & 1.44 \\ 5.2 & 8.6 & 1.11 & -2.6 & -0.26 \\ -1.5 & -1 & 0.3 & 1 & 0.3 \\ 3.6 & -1.8 & -0.63 & -1.8 & -0.18 \end{bmatrix}$

$$A^{10} = \begin{bmatrix} 433.0971 & -1583.5617 & -369.8169 & -216.5481 & -141.6383 \\ 1207.6949 & 5998.4376 & 2563.2274 & -603.8474 & 1045.7047 \\ -1423.0228 & 8065.0704 & 4271.7984 & 711.5114 & 2023.3641 \\ 479.4764 & -246.5381 & -231.5431 & -239.7372 & -140.5573 \\ -899.3405 & -3064.3130 & -1627.5015 & 449.6703 & -698.6086 \end{bmatrix}$$

(Rounded to four decimal places.)

$$A^{15} = \begin{bmatrix} -2247.8448 & -118449.3176 & -69122.3644 & 1123.9224 & -32134.3035 \\ -11094.8095 & 542433.5133 & 249928.4018 & 5547.4047 & 110820.2098 \\ 100366.7830 & 831934.5635 & 386083.7906 & -50183.3915 & 165597.1358 \\ -18782.2970 & -38432.0936 & -18654.0856 & 9391.1485 & -7395.9599 \\ -1633.6655 & -306291.0222 & -130154.5685 & 816.8327 & -56015.4536 \end{bmatrix}$$

(Rounded to four decimal places.)

(b) $$A^2 = \begin{bmatrix} 1.16 & 1.42 & -0.408 & 0.29 \\ 1.25 & -1.26 & 0.506 & 0.91 \\ 13.9 & -5.29 & 0.632 & 7.8 \\ -2.74 & -0.53 & 0.6 & -0.788 \end{bmatrix}$$

$$A^{10} = \begin{bmatrix} -31.1428 & 14.0580 & 1.0561 & -16.1157 \\ 28.4909 & -0.3398 & -3.2689 & 14.6394 \\ -50.8954 & 149.5593 & -30.8932 & -34.3845 \\ 52.3686 & -47.5375 & 5.6205 & 29.5703 \end{bmatrix}$$

(Rounded to four decimal places.)

$$A^{15} = \begin{bmatrix} 111.7124 & -169.5588 & 31.0626 & 68.5252 \\ -155.7116 & 145.0638 & -15.6641 & -86.5465 \\ -621.4964 & -218.2318 & 157.5564 & -283.4745 \\ -1.3722 & 252.7427 & -69.4323 & -20.2157 \end{bmatrix}$$

(Rounded to four decimal places.)

(c) $$A^2 = \begin{bmatrix} 1 & 0 & 1 \\ 1 & 2 & 1 \\ 2 & 2 & 2 \end{bmatrix}$$

$$A^{10} = \begin{bmatrix} 171 & 170 & 171 \\ 341 & 342 & 341 \\ 512 & 512 & 512 \end{bmatrix}$$

$$A^{15} = \begin{bmatrix} 5461 & 5462 & 5461 \\ 10923 & 10922 & 10923 \\ 16384 & 16384 & 16384 \end{bmatrix}$$

## 2.6 Inverse of a Matrix

**1.** $\begin{bmatrix}1 & 2\\ 2 & 3\end{bmatrix}\begin{bmatrix}-3 & 2\\ 2 & -1\end{bmatrix}=\begin{bmatrix}-3+4 & 2-2\\ -6+6 & 4-3\end{bmatrix}=\begin{bmatrix}1 & 0\\ 0 & 1\end{bmatrix}=I_2$

**3.** $\begin{bmatrix}-1 & -2\\ 3 & 4\end{bmatrix}\begin{bmatrix}2 & 1\\ -3/2 & -1/2\end{bmatrix}=\begin{bmatrix}-2+3 & -1+1\\ 6-6 & 3-2\end{bmatrix}=\begin{bmatrix}1 & 0\\ 0 & 1\end{bmatrix}=I_2$

**5.** $\begin{bmatrix}1 & 2 & 3\\ 2 & 3 & 4\\ 1 & 2 & 1\end{bmatrix}\begin{bmatrix}-5/2 & 2 & -1/2\\ 1 & -1 & 1\\ 1/2 & 0 & -1/2\end{bmatrix}=\begin{bmatrix}-5/2+4/2+3/2 & 2-2+0 & -1/2+4/2-3/2\\ -5+3+2 & 4-3+0 & -1+3-2\\ -5/2+4/2+1/2 & 2-2+0 & -1/2+4/2-1/2\end{bmatrix}$

$=\begin{bmatrix}1 & 0 & 0\\ 0 & 1 & 0\\ 0 & 0 & 1\end{bmatrix}=I_3$

**7.** $\left[\begin{array}{cc|cc}3 & 7 & 1 & 0\\ 2 & 5 & 0 & 1\end{array}\right]\xrightarrow[R_1=r_1-r_2]{}\left[\begin{array}{cc|cc}1 & 2 & 1 & -1\\ 2 & 5 & 0 & 1\end{array}\right]\xrightarrow[R_2=r_2-2r_1]{}\left[\begin{array}{cc|cc}1 & 2 & 1 & -1\\ 0 & 1 & -2 & 3\end{array}\right]$

$\xrightarrow[R_1=r_1-2r_2]{}\left[\begin{array}{cc|cc}1 & 0 & 5 & -7\\ 0 & 1 & -2 & 3\end{array}\right];\ \begin{bmatrix}3 & 7\\ 2 & 5\end{bmatrix}^{-1}=\begin{bmatrix}5 & -7\\ -2 & 3\end{bmatrix}$

**9.** $\left[\begin{array}{cc|cc}1 & -1 & 1 & 0\\ 3 & -4 & 0 & 1\end{array}\right]\xrightarrow[R_2=r_2-3r_1]{}\left[\begin{array}{cc|cc}1 & -1 & 1 & 0\\ 0 & -1 & -3 & 1\end{array}\right]\xrightarrow[R_1=r_1-r_2]{}\left[\begin{array}{cc|cc}1 & 0 & 4 & -1\\ 0 & -1 & -3 & 1\end{array}\right]$

$\xrightarrow[R_2=-r_2]{}\left[\begin{array}{cc|cc}1 & 0 & 4 & -1\\ 0 & 1 & 3 & -1\end{array}\right];\ \begin{bmatrix}1 & -1\\ 3 & -4\end{bmatrix}^{-1}=\begin{bmatrix}4 & -1\\ 3 & -1\end{bmatrix}$

**11.** $\left[\begin{array}{cc|cc}2 & 1 & 1 & 0\\ 4 & 3 & 0 & 1\end{array}\right]\xrightarrow[R_2=r_2-2r_1]{}\left[\begin{array}{cc|cc}2 & 1 & 1 & 0\\ 0 & 1 & -2 & 1\end{array}\right]\xrightarrow[R_1=r_1-r_2]{}\left[\begin{array}{cc|cc}2 & 0 & 3 & -1\\ 0 & 1 & -2 & 1\end{array}\right]$

$\xrightarrow[R_1=\frac{1}{2}r_1]{}\left[\begin{array}{cc|cc}1 & 0 & 3/2 & -1/2\\ 0 & 1 & -2 & 1\end{array}\right];\ \begin{bmatrix}2 & 1\\ 4 & 3\end{bmatrix}^{-1}=\begin{bmatrix}3/2 & -1/2\\ -2 & 1\end{bmatrix}$

**13.** $\left[\begin{array}{ccc|ccc}0 & 0 & 1 & 1 & 0 & 0\\ 0 & 1 & 0 & 0 & 1 & 0\\ 1 & 0 & 0 & 0 & 0 & 1\end{array}\right]\xrightarrow[\substack{R_1=r_3\\ R_3=r_1}]{}\left[\begin{array}{ccc|ccc}1 & 0 & 0 & 0 & 0 & 1\\ 0 & 1 & 0 & 0 & 1 & 0\\ 0 & 0 & 1 & 1 & 0 & 0\end{array}\right];\ \begin{bmatrix}0 & 0 & 1\\ 0 & 1 & 0\\ 1 & 0 & 0\end{bmatrix}^{-1}=\begin{bmatrix}0 & 0 & 1\\ 0 & 1 & 0\\ 1 & 0 & 0\end{bmatrix}$

**15.**
$$\left[\begin{array}{ccc|ccc} 1 & 1 & -1 & 1 & 0 & 0 \\ 3 & -1 & 0 & 0 & 1 & 0 \\ 2 & -3 & 4 & 0 & 0 & 1 \end{array}\right] \xrightarrow[\substack{R_2 = r_2 - 3r_1 \\ R_3 = r_3 - 2r_1}]{} \left[\begin{array}{ccc|ccc} 1 & 1 & -1 & 1 & 0 & 0 \\ 0 & -4 & 3 & -3 & 1 & 0 \\ 0 & -5 & 6 & -2 & 0 & 1 \end{array}\right]$$

$$\xrightarrow[R_2 = -\frac{1}{4} r_2]{} \left[\begin{array}{ccc|ccc} 1 & 1 & -1 & 1 & 0 & 0 \\ 0 & 1 & -3/4 & 3/4 & -1/4 & 0 \\ 0 & -5 & 6 & -2 & 0 & 1 \end{array}\right]$$

$$\xrightarrow[\substack{R_1 = r_1 - r_2 \\ R_3 = r_3 + 5r_2}]{} \left[\begin{array}{ccc|ccc} 1 & 0 & -1/4 & 1/4 & 1/4 & 0 \\ 0 & 1 & -3/4 & 3/4 & -1/4 & 0 \\ 0 & 0 & 9/4 & 7/4 & -5/4 & 1 \end{array}\right]$$

$$\xrightarrow[R_3 = \frac{4}{9} r_3]{} \left[\begin{array}{ccc|ccc} 1 & 0 & -1/4 & 1/4 & 1/4 & 0 \\ 0 & 1 & -3/4 & 3/4 & -1/4 & 0 \\ 0 & 0 & 1 & 7/9 & -5/9 & 4/9 \end{array}\right]$$

$$\xrightarrow[\substack{R_1 = r_1 + \frac{1}{4} r_3 \\ R_2 = r_2 + \frac{3}{4} r_3}]{} \left[\begin{array}{ccc|ccc} 1 & 0 & 0 & 4/9 & 1/9 & 1/9 \\ 0 & 1 & 0 & 4/3 & -2/3 & 1/3 \\ 0 & 0 & 1 & 7/9 & -5/9 & 4/9 \end{array}\right]; \begin{bmatrix} 1 & 1 & -1 \\ 3 & -1 & 0 \\ 2 & -3 & 4 \end{bmatrix}^{-1} = \begin{bmatrix} 4/9 & 1/9 & 1/9 \\ 4/3 & -2/3 & 1/3 \\ 7/9 & -5/9 & 4/9 \end{bmatrix}$$

**17.**
$$\left[\begin{array}{ccc|ccc} 1 & 1 & -1 & 1 & 0 & 0 \\ 2 & 1 & 1 & 0 & 1 & 0 \\ 1 & 0 & 1 & 0 & 0 & 1 \end{array}\right] \xrightarrow[\substack{R_2 = R_2 - 2r_1 \\ R_3 = R_3 - r_1}]{} \left[\begin{array}{ccc|ccc} 1 & 1 & -1 & 1 & 0 & 0 \\ 0 & -1 & 3 & -2 & 1 & 0 \\ 0 & -1 & 2 & -1 & 0 & 1 \end{array}\right]$$

$$\xrightarrow[\substack{R_1 = r_1 + r_2 \\ R_3 = r_3 - r_2}]{} \left[\begin{array}{ccc|ccc} 1 & 0 & 2 & -1 & 1 & 0 \\ 0 & -1 & 3 & -2 & 1 & 0 \\ 0 & 0 & -1 & 1 & -1 & 1 \end{array}\right] \xrightarrow[\substack{R_1 = r_1 + 2r_3 \\ R_2 = r_2 + 3r_3}]{} \left[\begin{array}{ccc|ccc} 1 & 0 & 0 & 1 & -1 & 2 \\ 0 & -1 & 0 & 1 & -2 & 3 \\ 0 & 0 & -1 & 1 & -1 & 1 \end{array}\right]$$

$$\xrightarrow[\substack{R_2 = -r_2 \\ R_3 = -r_3}]{} \left[\begin{array}{ccc|ccc} 1 & 0 & 0 & 1 & -1 & 2 \\ 0 & 1 & 0 & -1 & 2 & -3 \\ 0 & 0 & 1 & -1 & 1 & -1 \end{array}\right]; \begin{bmatrix} 1 & 1 & -1 \\ 2 & 1 & 1 \\ 1 & 0 & 1 \end{bmatrix}^{-1} = \begin{bmatrix} 1 & -1 & 2 \\ -1 & 2 & -3 \\ -1 & 1 & -1 \end{bmatrix}$$

**19.**
$$\left[\begin{array}{cccc|cccc} 1 & 1 & 0 & 0 & 1 & 0 & 0 & 0 \\ 0 & 1 & -1 & 1 & 0 & 1 & 0 & 0 \\ 1 & -1 & 1 & 1 & 0 & 0 & 1 & 0 \\ 0 & 1 & 0 & -1 & 0 & 0 & 0 & 1 \end{array}\right] \xrightarrow[R_3 = r_3 - r_1]{} \left[\begin{array}{cccc|cccc} 1 & 1 & 0 & 0 & 1 & 0 & 0 & 0 \\ 0 & 1 & -1 & 1 & 0 & 1 & 0 & 0 \\ 0 & -2 & 1 & 1 & -1 & 0 & 1 & 0 \\ 0 & 1 & 0 & -1 & 0 & 0 & 0 & 1 \end{array}\right]$$

$$\xrightarrow[\substack{R_1=r_1-r_2\\R_3=r_3+2r_2\\R_4=r_4-r_2}]{}\left[\begin{array}{cccc|cccc}1&0&1&-1&1&-1&0&0\\0&1&-1&1&0&1&0&0\\0&0&-1&3&-1&2&1&0\\0&0&1&-2&0&-1&0&1\end{array}\right]$$

$$\xrightarrow[\substack{R_1=r_1+r_3\\R_2=r_2-r_3\\R_4=r_4+r_3}]{}\left[\begin{array}{cccc|cccc}1&0&0&2&0&1&1&0\\0&1&0&-2&1&-1&-1&0\\0&0&-1&3&-1&2&1&0\\0&0&0&1&-1&1&1&1\end{array}\right]$$

$$\xrightarrow[\substack{R_1=r_1-2r_4\\R_2=r_2+2r_4\\R_3=r_3-3r_4}]{}\left[\begin{array}{cccc|cccc}1&0&0&0&2&-1&-1&-2\\0&1&0&0&-1&1&1&2\\0&0&-1&0&2&-1&-2&-3\\0&0&0&1&-1&1&1&1\end{array}\right]$$

$$\xrightarrow[R_3=-r_3]{}\left[\begin{array}{cccc|cccc}1&0&0&0&2&-1&-1&-2\\0&1&0&0&-1&1&1&2\\0&0&1&0&-2&1&2&3\\0&0&0&1&-1&1&1&1\end{array}\right];\ \begin{bmatrix}1&1&0&0\\0&1&-1&1\\1&-1&1&1\\0&1&0&-1\end{bmatrix}^{-1}=\begin{bmatrix}2&-1&-1&-2\\-1&1&1&2\\-2&1&2&3\\-1&1&1&1\end{bmatrix}$$

**21.** $\left[\begin{array}{cc|cc}4&6&1&0\\2&3&0&1\end{array}\right]\xrightarrow[R_1=r_1-2r_2]{}\left[\begin{array}{cc|cc}\mathbf{0}&\mathbf{0}&1&-2\\2&3&0&1\end{array}\right]$: row of zeros $\Rightarrow$ no inverse

**23.** $\left[\begin{array}{cc|cc}-8&4&1&0\\-4&2&0&1\end{array}\right]\xrightarrow[R_1=r_1-2r_2]{}\left[\begin{array}{cc|cc}\mathbf{0}&\mathbf{0}&1&-2\\-4&2&0&1\end{array}\right]$: row of zeros $\Rightarrow$ no inverse

**25.** $\left[\begin{array}{ccc|ccc}1&1&1&1&0&0\\3&-4&2&0&1&0\\\mathbf{0}&\mathbf{0}&\mathbf{0}&0&0&1\end{array}\right]$: row of zeros $\Rightarrow$ no inverse

**27.** $\left[\begin{array}{cc|cc}1&1&1&0\\1&2&0&1\end{array}\right]\xrightarrow[R_2=r_2-r_1]{}\left[\begin{array}{cc|cc}1&1&1&0\\0&1&-1&1\end{array}\right]\xrightarrow[R_1=r_1-r_2]{}\left[\begin{array}{cc|cc}1&0&2&-1\\0&1&-1&1\end{array}\right];$

$\begin{bmatrix}1&1\\1&2\end{bmatrix}^{-1}=\begin{bmatrix}2&-1\\-1&1\end{bmatrix}$

**29.** $\left[\begin{array}{cc|cc} 3 & -2 & 1 & 0 \\ 0 & 4 & 0 & 1 \end{array}\right] \xrightarrow[R_2=\frac{1}{4}r_2]{} \left[\begin{array}{cc|cc} 3 & -2 & 1 & 0 \\ 0 & 1 & 0 & 1/4 \end{array}\right] \xrightarrow[R_1=r_1+2r_2]{} \left[\begin{array}{cc|cc} 3 & 0 & 1 & 1/2 \\ 0 & 1 & 0 & 1/4 \end{array}\right]$

$\xrightarrow[R_1=\frac{1}{3}r_1]{} \left[\begin{array}{cc|cc} 1 & 0 & 1/3 & 1/6 \\ 0 & 1 & 0 & 1/4 \end{array}\right]; \begin{bmatrix} 3 & -2 \\ 0 & 4 \end{bmatrix}^{-1} = \begin{bmatrix} 1/3 & 1/6 \\ 0 & 1/4 \end{bmatrix}$

**31.** $\left[\begin{array}{cc|cc} 3 & 2 & 1 & 0 \\ 6 & 4 & 0 & 1 \end{array}\right] \xrightarrow[R_2=r_2-2r_1]{} \left[\begin{array}{cc|cc} 3 & 2 & 1 & 0 \\ \mathbf{0} & \mathbf{0} & -2 & 1 \end{array}\right]$: row of zeros $\Rightarrow$ no inverse

**33.** $\left[\begin{array}{ccc|ccc} 1 & -2 & -1 & 1 & 0 & 0 \\ -2 & 5 & 4 & 0 & 1 & 0 \\ 3 & -8 & -5 & 0 & 0 & 1 \end{array}\right] \xrightarrow[\substack{R_2=r_2+2r_1 \\ R_3=r_3-3r_1}]{} \left[\begin{array}{ccc|ccc} 1 & -2 & -1 & 1 & 0 & 0 \\ 0 & 1 & 2 & 2 & 1 & 0 \\ 0 & -2 & -2 & -3 & 0 & 1 \end{array}\right]$

$\xrightarrow[\substack{R_1=r_1+2r_2 \\ R_3=r_3+2r_2}]{} \left[\begin{array}{ccc|ccc} 1 & 0 & 3 & 5 & 2 & 0 \\ 0 & 1 & 2 & 2 & 1 & 0 \\ 0 & 0 & 2 & 1 & 2 & 1 \end{array}\right] \xrightarrow[R_3=\frac{1}{2}r_3]{} \left[\begin{array}{ccc|ccc} 1 & 0 & 3 & 5 & 2 & 0 \\ 0 & 1 & 2 & 2 & 1 & 0 \\ 0 & 0 & 1 & 1/2 & 1 & 1/2 \end{array}\right]$

$\xrightarrow[\substack{R_1=r_1-3r_3 \\ R_2=r_2-2r_3}]{} \left[\begin{array}{ccc|ccc} 1 & 0 & 0 & 7/2 & -1 & -3/2 \\ 0 & 1 & 0 & 1 & -1 & -1 \\ 0 & 0 & 1 & 1/2 & 1 & 1/2 \end{array}\right]; \begin{bmatrix} 1 & -2 & -1 \\ -2 & 5 & 4 \\ 3 & -8 & -5 \end{bmatrix}^{-1} = \begin{bmatrix} 7/2 & -1 & -3/2 \\ 1 & -1 & -1 \\ 1/2 & 1 & 1/2 \end{bmatrix}$

**35.** $\left[\begin{array}{cc|cc} 1 & 2 & 1 & 0 \\ 2 & -1 & 0 & 1 \end{array}\right] \xrightarrow[R_2=r_2-2r_1]{} \left[\begin{array}{cc|cc} 1 & 2 & 1 & 0 \\ 0 & -5 & -2 & 1 \end{array}\right] \xrightarrow[R_2=-\frac{1}{5}r_2]{} \left[\begin{array}{cc|cc} 1 & 2 & 1 & 0 \\ 0 & 1 & 2/5 & -1/5 \end{array}\right]$

$\xrightarrow[R_1=r_1-2r_2]{} \left[\begin{array}{cc|cc} 1 & 0 & 1/5 & 2/5 \\ 0 & 1 & 2/5 & -1/5 \end{array}\right]; A^{-1} = \begin{bmatrix} 1/5 & 2/5 \\ 2/5 & -1/5 \end{bmatrix}$

$\left[\begin{array}{cc|cc} 1 & 3 & 1 & 0 \\ 2 & 1 & 0 & 1 \end{array}\right] \xrightarrow[R_2=r_2-2r_1]{} \left[\begin{array}{cc|cc} 1 & 3 & 1 & 0 \\ 0 & -5 & -2 & 1 \end{array}\right] \xrightarrow[R_2=-\frac{1}{5}r_2]{} \left[\begin{array}{cc|cc} 1 & 3 & 1 & 0 \\ 0 & 1 & 2/5 & -1/5 \end{array}\right]$

$\xrightarrow[R_1=r_1-3r_2]{} \left[\begin{array}{cc|cc} 1 & 0 & -1/5 & 3/5 \\ 0 & 1 & 2/5 & -1/5 \end{array}\right]; B^{-1} = \begin{bmatrix} -1/5 & 3/5 \\ 2/5 & -1/5 \end{bmatrix}; A^{-1} - B^{-1} = \begin{bmatrix} 2/5 & -1/5 \\ 0 & 0 \end{bmatrix}$

**37.** $A = \begin{bmatrix} 1 & 3 & 2 \\ 2 & 7 & 3 \\ 1 & 0 & 6 \end{bmatrix}; \left[\begin{array}{ccc|ccc} 1 & 3 & 2 & 1 & 0 & 0 \\ 2 & 7 & 3 & 0 & 1 & 0 \\ 1 & 0 & 6 & 0 & 0 & 1 \end{array}\right] \xrightarrow[\substack{R_2=r_2-2r_1 \\ R_3=r_3-r_1}]{} \left[\begin{array}{ccc|ccc} 1 & 3 & 2 & 1 & 0 & 0 \\ 0 & 1 & -1 & -2 & 1 & 0 \\ 0 & -3 & 4 & -1 & 0 & 1 \end{array}\right]$

$\xrightarrow[\substack{R_1=r_1-3r_2 \\ R_3=r_3+3r_2}]{} \left[\begin{array}{ccc|ccc} 1 & 0 & 5 & 7 & -3 & 0 \\ 0 & 1 & -1 & -2 & 1 & 0 \\ 0 & 0 & 1 & -7 & 3 & 1 \end{array}\right] \xrightarrow[\substack{R_1=r_1-5r_3 \\ R_2=r_2+r_3}]{} \left[\begin{array}{ccc|ccc} 1 & 0 & 0 & 42 & -18 & -5 \\ 0 & 1 & 0 & -9 & 4 & 1 \\ 0 & 0 & 1 & -7 & 3 & 1 \end{array}\right];$

$A^{-1}=\begin{bmatrix}42 & -18 & -5\\ -9 & 4 & 1\\ -7 & 3 & 1\end{bmatrix}$. The solution to this system is given by $A^{-1}B=\begin{bmatrix}42 & -18 & -5\\ -9 & 4 & 1\\ -7 & 3 & 1\end{bmatrix}\begin{bmatrix}2\\ 1\\ 3\end{bmatrix}$.

**39.** $A=\begin{bmatrix}3 & 7\\ 2 & 5\end{bmatrix}, X=\begin{bmatrix}x\\ y\end{bmatrix}, B=\begin{bmatrix}10\\ 7\end{bmatrix}$. From problem 7, $A^{-1}=\begin{bmatrix}5 & -7\\ -2 & 3\end{bmatrix}$,

so $X=A^{-1}B\Rightarrow\begin{bmatrix}x\\ y\end{bmatrix}=\begin{bmatrix}5 & -7\\ -2 & 3\end{bmatrix}\begin{bmatrix}10\\ 2\end{bmatrix}=\begin{bmatrix}36\\ -14\end{bmatrix}$; $x=36, y=-14$

**41.** $\begin{bmatrix}x\\ y\end{bmatrix}=\begin{bmatrix}3 & 7\\ 2 & 5\end{bmatrix}^{-1}\begin{bmatrix}13\\ 9\end{bmatrix}=\begin{bmatrix}5 & -7\\ -2 & 3\end{bmatrix}\begin{bmatrix}13\\ 9\end{bmatrix}=\begin{bmatrix}2\\ 1\end{bmatrix}$; $x=2,\ y=1$

**43.** $\begin{bmatrix}x\\ y\end{bmatrix}=\begin{bmatrix}5 & -7\\ -2 & 3\end{bmatrix}\begin{bmatrix}12\\ -4\end{bmatrix}=\begin{bmatrix}88\\ -36\end{bmatrix}$; $x=88, y=-36$.

**45.** $A=\begin{bmatrix}1 & 1 & -1\\ 3 & -1 & 0\\ 2 & -3 & 4\end{bmatrix},\ X=\begin{bmatrix}x\\ y\\ z\end{bmatrix},\ B=\begin{bmatrix}3\\ -4\\ 6\end{bmatrix}$. From problem 15, $A^{-1}=\begin{bmatrix}4/9 & 1/9 & 1/9\\ 4/3 & -2/3 & 1/3\\ 7/9 & -5/9 & 4/9\end{bmatrix}$,

so $X=A^{-1}B\Rightarrow\begin{bmatrix}x\\ y\\ z\end{bmatrix}=\begin{bmatrix}4/9 & 1/9 & 1/9\\ 4/3 & -2/3 & 1/3\\ 7/9 & -5/9 & 4/9\end{bmatrix}\begin{bmatrix}3\\ -4\\ 6\end{bmatrix}=\begin{bmatrix}14/9\\ 26/3\\ 65/9\end{bmatrix}$; $x=\frac{14}{9}, y=\frac{26}{3}, z=\frac{65}{9}$.

**47.** $\begin{bmatrix}x\\ y\\ z\end{bmatrix}=\begin{bmatrix}4/9 & 1/9 & 1/9\\ 4/3 & -2/3 & 1/3\\ 7/9 & -5/9 & 4/9\end{bmatrix}\begin{bmatrix}12\\ -4\\ 16\end{bmatrix}=\begin{bmatrix}20/3\\ 24\\ 56/3\end{bmatrix}$; $x=\frac{20}{3}, y=24, z=\frac{56}{3}$.

**49.** $\begin{bmatrix}x\\ y\\ z\end{bmatrix}=\begin{bmatrix}4/9 & 1/9 & 1/9\\ 4/3 & -2/3 & 1/3\\ 7/9 & -5/9 & 4/9\end{bmatrix}\begin{bmatrix}0\\ -8\\ -6\end{bmatrix}=\begin{bmatrix}-14/9\\ 10/3\\ 16/9\end{bmatrix}$; $x=-\frac{14}{9}, y=\frac{10}{3}, z=\frac{16}{9}$.

**51.** $\begin{bmatrix}a & b\\ c & d\end{bmatrix}\begin{bmatrix}d/\Delta & -b/\Delta\\ -c/\Delta & a/\Delta\end{bmatrix}=\begin{bmatrix}(ad-bc)/\Delta & (-ab+ba)/\Delta\\ (cd-dc)/\Delta & (-cb+da)/\Delta\end{bmatrix}=\begin{bmatrix}\frac{ad-bc}{ad-bc} & 0\\ 0 & \frac{ad-bc}{ad-bc}\end{bmatrix}=\begin{bmatrix}1 & 0\\ 0 & 1\end{bmatrix}$, so

$\begin{bmatrix}a & b\\ c & d\end{bmatrix}^{-1}=\begin{bmatrix}d/\Delta & -b/\Delta\\ -c/\Delta & a/\Delta\end{bmatrix}$

**53.** $\Delta = 1\cdot 0 - 5\cdot 2 = -10;\ \begin{bmatrix} 1 & 5 \\ 2 & 0 \end{bmatrix}^{-1} = \begin{bmatrix} 0/(-10) & -5/(-10) \\ -2/(-10) & 1/(-10) \end{bmatrix} = \begin{bmatrix} 0 & 1/2 \\ 1/5 & -1/10 \end{bmatrix}$

**55.** $\Delta = 1\cdot 15 - 2\cdot 8 = -1;\ \begin{bmatrix} 1 & 2 \\ 8 & 15 \end{bmatrix}^{-1} = \begin{bmatrix} 15/(-1) & (-2)/(-1) \\ (-8)/(-1) & 1/(-1) \end{bmatrix} = \begin{bmatrix} -15 & 2 \\ 8 & -1 \end{bmatrix}$

## Technology Exercises

**1.** $\begin{bmatrix} 0.0054 & 0.0509 & -0.0066 \\ 0.0104 & -0.0186 & 0.0095 \\ -0.0193 & 0.0116 & 0.0344 \end{bmatrix}$

(rounded to four decimal places)

**3.** $\begin{bmatrix} 0.0249 & -0.0360 & -0.0057 & 0.0059 \\ -0.0171 & 0.0521 & 0.0292 & -0.0305 \\ 0.0206 & 0.0081 & -0.0421 & 0.0005 \\ -0.0175 & 0.0570 & 0.0657 & 0.0619 \end{bmatrix}$

(rounded to four decimal places)

**5.** $A = \begin{bmatrix} 25 & 61 & -12 \\ 18 & -12 & 7 \\ 3 & 4 & -1 \end{bmatrix}; A^{-1} = \begin{bmatrix} -0.023426 & 0.019034 & 0.41435 \\ 0.057101 & 0.016105 & -0.57247 \\ 0.158126 & 0.121523 & -2.04685 \end{bmatrix}; \begin{bmatrix} x \\ y \\ z \end{bmatrix} = A^{-1}\begin{bmatrix} 10 \\ -9 \\ 12 \end{bmatrix};$

$x = 4.5666, y = -6.4436, z = -24.0747$.

**7.** $\begin{bmatrix} x \\ y \\ z \end{bmatrix} = A^{-1}\begin{bmatrix} 21 \\ 7 \\ -2 \end{bmatrix}$ (see problem 5); $x = -1.1874, y = 2.4568, z = 8.2650$.

**9.** $A^{-1} = \begin{bmatrix} .25 & -.0625 & -.28125 & .3125 & .09375 \\ -1.5 & .125 & 3.0625 & -2.625 & .3125 \\ 1.75 & -.1875 & -2.84375 & 2.9375 & -.71875 \\ -.5 & .375 & .6875 & -.875 & .4375 \\ 1.25 & .3125 & 2.40625 & -2.5625 & .53125 \end{bmatrix};\ I_5 = \begin{bmatrix} 1 & 0 & 0 & 0 & 0 \\ 0 & 1 & 0 & 0 & 0 \\ 0 & 0 & 1 & 0 & 0 \\ 0 & 0 & 0 & 1 & 0 \\ 0 & 0 & 0 & 0 & 1 \end{bmatrix}$

## 2.7 Applications: Leontief Model; Cryptography; Accounting; The Method of Least Squares

### Application 1:

**1.** Let $x$ = $A$'s wages, $y$ = $B$'s wages, and $z$ = $C$'s wages. Solve

$$\left(I_3 - \begin{bmatrix} 1/2 & 1/3 & 1/4 \\ 1/4 & 1/3 & 1/4 \\ 1/4 & 1/3 & 1/2 \end{bmatrix}\right)\begin{bmatrix} x \\ y \\ z \end{bmatrix} = \begin{bmatrix} 0 \\ 0 \\ 0 \end{bmatrix}: \left[\begin{array}{ccc|c} 1/2 & -1/3 & -1/4 & 0 \\ -1/4 & 2/3 & -1/4 & 0 \\ -1/4 & -1/3 & 1/2 & 0 \end{array}\right]$$

$$\xrightarrow[\substack{R_2 = r_2 + \frac{1}{2}r_1 \\ R_3 = r_3 + \frac{1}{2}r_1}]{} \left[\begin{array}{ccc|c} 1/2 & -1/3 & -1/4 & 0 \\ 0 & 1/2 & -3/8 & 0 \\ 0 & -1/2 & 3/8 & 0 \end{array}\right] \xrightarrow[R_3 = r_3 + r_2]{} \left[\begin{array}{ccc|c} 1/2 & -1/3 & -1/4 & 0 \\ 0 & 1/2 & -3/8 & 0 \\ 0 & 0 & 0 & 0 \end{array}\right]$$

$$\xrightarrow[\substack{R_1 = 2r_1 \\ R_2 = 2r_2}]{} \left[\begin{array}{ccc|c} 1 & -2/3 & -1/2 & 0 \\ 0 & 1 & -3/4 & 0 \\ 0 & 0 & 0 & 0 \end{array}\right] \xrightarrow[R_1 = r_1 + \frac{2}{3}r_2]{} \left[\begin{array}{ccc|c} 1 & 0 & -1 & 0 \\ 0 & 1 & -3/4 & 0 \\ 0 & 0 & 0 & 0 \end{array}\right]:$$

$\begin{cases} x - z = 0 \\ y - \frac{3}{4}z = 0 \end{cases}$ $\begin{matrix} x = z \\ y = \frac{3}{4}z \end{matrix}$. When $z = \$30{,}000$, $x = \$30{,}000$ and $y = \frac{3}{4} \cdot \$30{,}000 = \$22{,}500$.

*Answer*: $A$'s wages are \$30,000, $B$'s wages are \$22,500, and $C$'s wages are \$30,000.

**3.** Let $x$ = $A$'s wages, $y$ = $B$'s wages, and $z$ = $C$'s wages. Solve

$$\left(I_3 - \begin{bmatrix} 0.2 & 0.3 & 0.1 \\ 0.6 & 0.4 & 0.2 \\ 0.2 & 0.3 & 0.7 \end{bmatrix}\right)\begin{bmatrix} x \\ y \\ z \end{bmatrix} = \begin{bmatrix} 0 \\ 0 \\ 0 \end{bmatrix}: \left[\begin{array}{ccc|c} 0.8 & -0.3 & -0.1 & 0 \\ -0.6 & 0.6 & -0.2 & 0 \\ -0.2 & -0.3 & 0.3 & 0 \end{array}\right]$$

$$\xrightarrow[\substack{R_1 = 10r_1 \\ R_2 = 10r_2 \\ R_3 = 10r_3}]{} \left[\begin{array}{ccc|c} 8 & -3 & -1 & 0 \\ -6 & 6 & -2 & 0 \\ -2 & -3 & 3 & 0 \end{array}\right] \xrightarrow[\substack{R_1 = r_1 + 4r_3 \\ R_2 = r_2 - 3r_3}]{} \left[\begin{array}{ccc|c} 0 & -15 & 11 & 0 \\ 0 & 15 & -11 & 0 \\ -2 & -3 & 3 & 0 \end{array}\right]$$

$$\xrightarrow[\substack{R_1 = r_3 \\ R_3 = r_1}]{} \left[\begin{array}{ccc|c} -2 & -3 & 3 & 0 \\ 0 & 15 & -11 & 0 \\ 0 & -15 & 11 & 0 \end{array}\right] \xrightarrow[\substack{R_1 = -\frac{1}{2}r_1 \\ R_3 = r_3 + r_2}]{} \left[\begin{array}{ccc|c} 1 & 3/2 & -3/2 & 0 \\ 0 & 15 & -11 & 0 \\ 0 & 0 & 0 & 0 \end{array}\right]$$

$$\xrightarrow[R_2 = \frac{1}{15}r_2]{} \left[\begin{array}{ccc|c} 1 & 3/2 & -3/2 & 0 \\ 0 & 1 & -11/15 & 0 \\ 0 & 0 & 0 & 0 \end{array}\right] \xrightarrow[R_1 = r_1 - \frac{3}{2}r_2]{} \left[\begin{array}{ccc|c} 1 & 0 & -2/5 & 0 \\ 0 & 1 & -11/15 & 0 \\ 0 & 0 & 0 & 0 \end{array}\right]:$$

$\begin{cases} x-\frac{2}{5}z=0, & x=\frac{2}{5}z \\ y-\frac{11}{15}z=0 & y=\frac{11}{15}z \end{cases}$. When $z=\$30{,}000$, $x=\frac{2}{5}\cdot\$30{,}000=\$12{,}000$, and

$y=\frac{11}{15}\cdot\$30{,}000=\$22{,}000$.

*Answer*: *A*'s wages are \$12,000, *B*'s wages are \$22,000, and *C*'s wages are \$30,000.

**5.** Solve $AX+D_2=X$ where $A=\begin{bmatrix} 50/180 & 20/160 & 40/120 \\ 20/180 & 30/160 & 20/120 \\ 30/180 & 20/160 & 20/120 \end{bmatrix}$, $X=\begin{bmatrix} x \\ y \\ z \end{bmatrix}$, and $D_2=\begin{bmatrix} 80 \\ 90 \\ 60 \end{bmatrix}$. The solution is $X=(I_3-A)^{-1}D_2$. Using the text's approximation for $(I-A)^{-1}$,

$$X=\begin{bmatrix} 1.6048 & 0.3568 & 0.7131 \\ 0.2946 & 1.3363 & 0.3857 \\ 0.3660 & 0.2721 & 1.4013 \end{bmatrix}\begin{bmatrix} 80 \\ 90 \\ 60 \end{bmatrix}=\begin{bmatrix} 203.28 \\ 166.98 \\ 137.85 \end{bmatrix}.$$

**7.** Let $f=$farmer's wages, $b=$builder's wages, $t=$tailor's wages, and $r=$rancher's wages.

Solve $X=AX$, or $(I_4-A)X=\mathbf{0}$, where $X=\begin{bmatrix} f \\ b \\ t \\ r \end{bmatrix}$ and $A=\begin{bmatrix} 0.3 & 0.3 & 0.3 & 0.2 \\ 0.2 & 0.3 & 0.3 & 0.2 \\ 0.2 & 0.1 & 0.1 & 0.2 \\ 0.3 & 0.3 & 0.3 & 0.4 \end{bmatrix}$.

$$\left[\begin{array}{cccc|c} 0.7 & -0.3 & -0.3 & -0.2 & 0 \\ -0.2 & 0.7 & -0.3 & -0.2 & 0 \\ -0.2 & -0.1 & 0.9 & -0.2 & 0 \\ -0.3 & -0.3 & -0.3 & 0.6 & 0 \end{array}\right] \to \left[\begin{array}{cccc|c} 1 & -3/7 & -3/7 & -2/7 & 0 \\ 0 & 43/70 & -27/70 & -9/35 & 0 \\ 0 & -13/70 & 57/70 & -9/35 & 0 \\ 0 & -3/7 & -3/7 & 18/35 & 0 \end{array}\right]$$

$$\left[\begin{array}{cccc|c} 1 & 0 & -30/43 & -20/43 & 0 \\ 0 & 1 & -27/43 & -18/43 & 0 \\ 0 & 0 & 30/43 & -72/215 & 0 \\ 0 & 0 & -30/43 & 72/215 & 0 \end{array}\right] \to \left[\begin{array}{cccc|c} 1 & 0 & 0 & -4/5 & 0 \\ 0 & 1 & 0 & -18/25 & 0 \\ 0 & 0 & 1 & -12/25 & 0 \\ 0 & 0 & 0 & 0 & 0 \end{array}\right] : \begin{cases} f-\frac{4}{5}r=0, & f=\frac{4}{5}r \\ b-\frac{18}{25}r=0 & b=\frac{18}{25}r \\ t-\frac{12}{25}r=0 & t=\frac{12}{25}r \end{cases}$$

If $r=\$25{,}000$, $f=\frac{4}{5}\cdot\$25{,}000=\$20{,}000$, $b=\frac{18}{25}\cdot\$25{,}000=\$18{,}000$, and $t=\frac{12}{25}\cdot\$25{,}000=\$12{,}000$. *Answer*: the farmer's wages are \$20,000, the builder's wages are \$18,000, and the tailor's wages are \$12,000, when the rancher's wages are \$25,000.

**9.** Solve $AX + D_2 = X$, where $A = \begin{bmatrix} 30/130 & 40/70 \\ 20/130 & 10/70 \end{bmatrix}$, $X = \begin{bmatrix} x \\ y \end{bmatrix}$, and $D_2 = \begin{bmatrix} 80 \\ 40 \end{bmatrix}$. The solution is $X = (I_2 - A)^{-1} D_2$. $I_2 - A = \begin{bmatrix} 10/13 & -4/7 \\ -2/13 & 6/7 \end{bmatrix}$. Using the formula from Problem 51, Section 2.6 for this matrix, $\Delta = \frac{52}{13 \cdot 7} = \frac{4}{7}$, and $(I_2 - A)^{-1} = \begin{bmatrix} \frac{6}{7} \cdot \frac{7}{4} & \frac{4}{7} \cdot \frac{7}{4} \\ \frac{2}{13} \cdot \frac{7}{4} & \frac{10}{13} \cdot \frac{7}{4} \end{bmatrix} = \begin{bmatrix} \frac{3}{2} & 1 \\ \frac{7}{26} & \frac{35}{26} \end{bmatrix}$. Thus,

$$X = \begin{bmatrix} \frac{3}{2} & 1 \\ \frac{7}{26} & \frac{35}{26} \end{bmatrix} \begin{bmatrix} 80 \\ 40 \end{bmatrix} = \begin{bmatrix} 160 \\ 980/13 \end{bmatrix} \approx \begin{bmatrix} 160 \\ 75.38 \end{bmatrix}.$$

## Technology Exercises

**1.** $$(I_4 - A)^{-1} = \begin{bmatrix} 1.22391 & 0.502846 & 0.44592 & 0.341556 \\ 0.602467 & 1.79791 & 0.839658 & 0.749526 \\ 0.666509 & 1.35911 & 1.67694 & 0.592979 \\ 0.303605 & 0.512334 & 0.265655 & 1.48008 \end{bmatrix}; \; X = (I_4 - A)^{-1} D = \begin{bmatrix} 7.58065 \\ 9.27419 \\ 16.00806 \\ 4.51613 \end{bmatrix}$$

**3.** $$(I_4 - A)^{-1} = \begin{bmatrix} 1.05672 & 0.26418 & 0.283605 & 0.493395 \\ 0.0815851 & 1.2704 & 0.407925 & 0.0932401 \\ 0.21756 & 0.0543901 & 1.0878 & 0.24864 \\ 0.0621601 & 0.01554 & 0.3108 & 1.49961 \end{bmatrix}; \; X = (I_4 - A)^{-1} D = \begin{bmatrix} 7.70008 \\ 10.11655 \\ 3.64413 \\ 1.04118 \end{bmatrix}$$

**5.** $$A = \begin{bmatrix} 0.2 & 0.12 & 0.15 & 0.18 & 0.1 \\ 0.17 & 0.11 & 0 & 0.19 & 0.28 \\ 0.11 & 0.11 & 0.12 & 0.46 & 0.12 \\ 0.1 & 0.14 & 0.18 & 0.17 & 0.19 \\ 0.16 & 0.18 & 0.02 & 0.1 & 0.3 \end{bmatrix};$$

$$(I_5 - A)^{-1} = \begin{bmatrix} 1.71263 & 0.591172 & 0.48902 & 0.874429 & 0.802307 \\ 0.674374 & 1.57739 & 0.299797 & 0.793251 & 0.994 \\ 0.720319 & 0.692703 & 1.54595 & 1.29152 & 0.995559 \\ 0.630957 & 0.636651 & 0.514937 & 1.88665 & 0.945164 \\ 0.675586 & 0.651481 & 0.306599 & 0.710271 & 2.03102 \end{bmatrix};$$

$$X=(I_5-A)^{-1}D=\begin{bmatrix}495.858\\463.861\\679.199\\512.930\\468.156\end{bmatrix}$$

## Application 2:

**1.** $A=\begin{bmatrix}2&3\\1&2\end{bmatrix}$, $A^{-1}=\begin{bmatrix}2&-3\\-1&2\end{bmatrix}$

(a) $\begin{bmatrix}2&-3\\-1&2\end{bmatrix}\begin{bmatrix}51\\30\end{bmatrix}=\begin{bmatrix}12\\9\end{bmatrix}$, $\begin{bmatrix}2&-3\\-1&2\end{bmatrix}\begin{bmatrix}27\\16\end{bmatrix}=\begin{bmatrix}6\\5\end{bmatrix}$, $\begin{bmatrix}2&-3\\-1&2\end{bmatrix}\begin{bmatrix}75\\47\end{bmatrix}=\begin{bmatrix}9\\19\end{bmatrix}$,

$\begin{bmatrix}2&-3\\-1&2\end{bmatrix}\begin{bmatrix}19\\10\end{bmatrix}=\begin{bmatrix}8\\1\end{bmatrix}$, $\begin{bmatrix}2&-3\\-1&2\end{bmatrix}\begin{bmatrix}48\\26\end{bmatrix}=\begin{bmatrix}18\\4\end{bmatrix}$. $\begin{matrix}12&9&6&5&9&19&8&1&18&4\\L&I&F&E&I&S&H&A&R&D\end{matrix}$.

*Answer*: "LIFE IS HARD"

(b) $\begin{bmatrix}2&-3\\-1&2\end{bmatrix}\begin{bmatrix}70\\45\end{bmatrix}=\begin{bmatrix}5\\20\end{bmatrix}$, $\begin{bmatrix}2&-3\\-1&2\end{bmatrix}\begin{bmatrix}103\\62\end{bmatrix}=\begin{bmatrix}20\\21\end{bmatrix}$, $\begin{bmatrix}2&-3\\-1&2\end{bmatrix}\begin{bmatrix}58\\38\end{bmatrix}=\begin{bmatrix}2\\18\end{bmatrix}$,

$\begin{bmatrix}2&-3\\-1&2\end{bmatrix}\begin{bmatrix}102\\61\end{bmatrix}=\begin{bmatrix}21\\20\end{bmatrix}$, $\begin{bmatrix}2&-3\\-1&2\end{bmatrix}\begin{bmatrix}88\\57\end{bmatrix}=\begin{bmatrix}5\\26\end{bmatrix}$.

$\begin{matrix}5&20&20&21&2&18&21&20&5&26\\E&T&T&U&B&R&U&T&E&Z\end{matrix}$. *Answer*: "ET TU BRUTE"

**3.** $\begin{bmatrix}2&-3\\-1&2\end{bmatrix}\begin{bmatrix}11\\7\end{bmatrix}=\begin{bmatrix}1\\3\end{bmatrix}$, $\begin{bmatrix}2&-3\\-1&2\end{bmatrix}\begin{bmatrix}84\\51\end{bmatrix}=\begin{bmatrix}15\\18\end{bmatrix}$, $\begin{bmatrix}2&-3\\-1&2\end{bmatrix}\begin{bmatrix}51\\28\end{bmatrix}=\begin{bmatrix}18\\5\end{bmatrix}$, $\begin{bmatrix}2&-3\\-1&2\end{bmatrix}\begin{bmatrix}66\\43\end{bmatrix}=\begin{bmatrix}3\\20\end{bmatrix}$,

$\begin{bmatrix}2&-3\\-1&2\end{bmatrix}\begin{bmatrix}44\\29\end{bmatrix}=\begin{bmatrix}1\\14\end{bmatrix}$, $\begin{bmatrix}2&-3\\-1&2\end{bmatrix}\begin{bmatrix}107\\65\end{bmatrix}=\begin{bmatrix}19\\23\end{bmatrix}$, $\begin{bmatrix}2&-3\\-1&2\end{bmatrix}\begin{bmatrix}64\\41\end{bmatrix}=\begin{bmatrix}5\\18\end{bmatrix}$

$\begin{matrix}1&3&15&18&18&5&3&20&1&14&19&23&5&18\\A&C&O&R&R&E&C&T&A&N&S&W&E&R\end{matrix}$.

*Answer*: "A CORRECT ANSWER"

**5.** (a) $\begin{matrix}M&E&E&T&M&E&A&T&T&H&E&C&A&S&B&A&H\\13&5&5&20&13&5&1&20&20&8&5&3&1&19&2&1&8\end{matrix}$

(I) $\begin{bmatrix}2&3\\1&2\end{bmatrix}\begin{bmatrix}13\\5\end{bmatrix}=\begin{bmatrix}41\\23\end{bmatrix}$, $\begin{bmatrix}2&3\\1&2\end{bmatrix}\begin{bmatrix}5\\20\end{bmatrix}=\begin{bmatrix}70\\45\end{bmatrix}$, $\begin{bmatrix}2&3\\1&2\end{bmatrix}\begin{bmatrix}13\\5\end{bmatrix}=\begin{bmatrix}41\\23\end{bmatrix}$, $\begin{bmatrix}2&3\\1&2\end{bmatrix}\begin{bmatrix}1\\20\end{bmatrix}=\begin{bmatrix}62\\41\end{bmatrix}$

$\begin{bmatrix}2&3\\1&2\end{bmatrix}\begin{bmatrix}20\\8\end{bmatrix}=\begin{bmatrix}64\\36\end{bmatrix}$, $\begin{bmatrix}2&3\\1&2\end{bmatrix}\begin{bmatrix}5\\3\end{bmatrix}=\begin{bmatrix}19\\11\end{bmatrix}$, $\begin{bmatrix}2&3\\1&2\end{bmatrix}\begin{bmatrix}1\\19\end{bmatrix}=\begin{bmatrix}59\\39\end{bmatrix}$, $\begin{bmatrix}2&3\\1&2\end{bmatrix}\begin{bmatrix}2\\1\end{bmatrix}=\begin{bmatrix}7\\4\end{bmatrix}$

$$\begin{bmatrix}2&3\\1&2\end{bmatrix}\begin{bmatrix}8\\26\end{bmatrix}=\begin{bmatrix}94\\60\end{bmatrix}.$$

*Answer*: 41 23 70 45 41 23 62 41 64 36 19 11 59 39 7 4 94 60

(II) $\begin{bmatrix}1&0&0\\3&1&5\\-2&0&1\end{bmatrix}\begin{bmatrix}13\\5\\5\end{bmatrix}=\begin{bmatrix}13\\69\\-21\end{bmatrix}$, $\begin{bmatrix}1&0&0\\3&1&5\\-2&0&1\end{bmatrix}\begin{bmatrix}20\\13\\5\end{bmatrix}=\begin{bmatrix}20\\98\\-35\end{bmatrix}$, $\begin{bmatrix}1&0&0\\3&1&5\\-2&0&1\end{bmatrix}\begin{bmatrix}1\\20\\20\end{bmatrix}=\begin{bmatrix}1\\123\\18\end{bmatrix}$,

$\begin{bmatrix}1&0&0\\3&1&5\\-2&0&1\end{bmatrix}\begin{bmatrix}8\\5\\3\end{bmatrix}=\begin{bmatrix}8\\44\\-13\end{bmatrix}$, $\begin{bmatrix}1&0&0\\3&1&5\\-2&0&1\end{bmatrix}\begin{bmatrix}1\\19\\2\end{bmatrix}=\begin{bmatrix}1\\32\\0\end{bmatrix}$, $\begin{bmatrix}1&0&0\\3&1&5\\-2&0&1\end{bmatrix}\begin{bmatrix}1\\8\\26\end{bmatrix}=\begin{bmatrix}1\\141\\24\end{bmatrix}$

*Answer*: 13 69 –21 20 98 –35 1 123 18 8 44 –13 1 32 0 1 141 24

(b)

| T | O | M | O | R | R | O | W | N | E | V | E | R | C | O | M | E | S |
|---|---|---|---|---|---|---|---|---|---|---|---|---|---|---|---|---|---|
| 20 | 15 | 13 | 15 | 18 | 18 | 15 | 23 | 14 | 5 | 22 | 5 | 18 | 3 | 15 | 13 | 5 | 19 |

(I) $\begin{bmatrix}2&3\\1&2\end{bmatrix}\begin{bmatrix}20\\15\end{bmatrix}=\begin{bmatrix}85\\50\end{bmatrix}$, $\begin{bmatrix}2&3\\1&2\end{bmatrix}\begin{bmatrix}13\\15\end{bmatrix}=\begin{bmatrix}71\\43\end{bmatrix}$, $\begin{bmatrix}2&3\\1&2\end{bmatrix}\begin{bmatrix}18\\18\end{bmatrix}=\begin{bmatrix}90\\54\end{bmatrix}$, $\begin{bmatrix}2&3\\1&2\end{bmatrix}\begin{bmatrix}15\\23\end{bmatrix}=\begin{bmatrix}99\\61\end{bmatrix}$

$\begin{bmatrix}2&3\\1&2\end{bmatrix}\begin{bmatrix}14\\5\end{bmatrix}=\begin{bmatrix}43\\24\end{bmatrix}$, $\begin{bmatrix}2&3\\1&2\end{bmatrix}\begin{bmatrix}22\\5\end{bmatrix}=\begin{bmatrix}59\\32\end{bmatrix}$, $\begin{bmatrix}2&3\\1&2\end{bmatrix}\begin{bmatrix}18\\3\end{bmatrix}=\begin{bmatrix}45\\24\end{bmatrix}$, $\begin{bmatrix}2&3\\1&2\end{bmatrix}\begin{bmatrix}15\\13\end{bmatrix}=\begin{bmatrix}69\\41\end{bmatrix}$

$\begin{bmatrix}2&3\\1&2\end{bmatrix}\begin{bmatrix}5\\19\end{bmatrix}=\begin{bmatrix}67\\43\end{bmatrix}$.

*Answer*: 85 50 71 43 90 54 99 61 43 24 59 32 45 24 69 41 67 43

(II) $\begin{bmatrix}1&0&0\\3&1&5\\-2&0&1\end{bmatrix}\begin{bmatrix}20\\15\\13\end{bmatrix}=\begin{bmatrix}20\\140\\-27\end{bmatrix}$, $\begin{bmatrix}1&0&0\\3&1&5\\-2&0&1\end{bmatrix}\begin{bmatrix}15\\18\\18\end{bmatrix}=\begin{bmatrix}15\\153\\-12\end{bmatrix}$,

$\begin{bmatrix}1&0&0\\3&1&5\\-2&0&1\end{bmatrix}\begin{bmatrix}15\\23\\14\end{bmatrix}=\begin{bmatrix}15\\138\\-16\end{bmatrix}$, $\begin{bmatrix}1&0&0\\3&1&5\\-2&0&1\end{bmatrix}\begin{bmatrix}5\\22\\5\end{bmatrix}=\begin{bmatrix}5\\62\\-5\end{bmatrix}$, $\begin{bmatrix}1&0&0\\3&1&5\\-2&0&1\end{bmatrix}\begin{bmatrix}18\\3\\15\end{bmatrix}=\begin{bmatrix}18\\132\\-21\end{bmatrix}$,

$\begin{bmatrix}1&0&0\\3&1&5\\-2&0&1\end{bmatrix}\begin{bmatrix}13\\5\\19\end{bmatrix}=\begin{bmatrix}13\\139\\-7\end{bmatrix}$

*Answer*: 20 140 –27 15 153 –12 15 138 –16 5 62 –5 18 132 –21 13 139 –7

(c)

| T | H | E | M | I | S | S | I | O | N | I | S |
|---|---|---|---|---|---|---|---|---|---|---|---|
| 20 | 8 | 5 | 13 | 9 | 19 | 19 | 9 | 15 | 14 | 9 | 19 |

| I | M | P | O | S | S | I | B | L | E |
|---|---|---|---|---|---|---|---|---|---|
| 9 | 13 | 16 | 15 | 19 | 19 | 9 | 2 | 12 | 5 |

(I) $\begin{bmatrix}2&3\\1&2\end{bmatrix}\begin{bmatrix}20\\8\end{bmatrix}=\begin{bmatrix}64\\36\end{bmatrix}$, $\begin{bmatrix}2&3\\1&2\end{bmatrix}\begin{bmatrix}5\\13\end{bmatrix}=\begin{bmatrix}49\\31\end{bmatrix}$, $\begin{bmatrix}2&3\\1&2\end{bmatrix}\begin{bmatrix}9\\19\end{bmatrix}=\begin{bmatrix}75\\47\end{bmatrix}$, $\begin{bmatrix}2&3\\1&2\end{bmatrix}\begin{bmatrix}19\\9\end{bmatrix}=\begin{bmatrix}65\\37\end{bmatrix}$

$$\begin{bmatrix}2&3\\1&2\end{bmatrix}\begin{bmatrix}15\\14\end{bmatrix}=\begin{bmatrix}72\\43\end{bmatrix}, \begin{bmatrix}2&3\\1&2\end{bmatrix}\begin{bmatrix}9\\19\end{bmatrix}=\begin{bmatrix}75\\47\end{bmatrix}, \begin{bmatrix}2&3\\1&2\end{bmatrix}\begin{bmatrix}9\\13\end{bmatrix}=\begin{bmatrix}57\\35\end{bmatrix}, \begin{bmatrix}2&3\\1&2\end{bmatrix}\begin{bmatrix}16\\15\end{bmatrix}=\begin{bmatrix}77\\46\end{bmatrix}$$

$$\begin{bmatrix}2&3\\1&2\end{bmatrix}\begin{bmatrix}19\\19\end{bmatrix}=\begin{bmatrix}95\\57\end{bmatrix}, \begin{bmatrix}2&3\\1&2\end{bmatrix}\begin{bmatrix}9\\2\end{bmatrix}=\begin{bmatrix}24\\13\end{bmatrix}, \begin{bmatrix}2&3\\1&2\end{bmatrix}\begin{bmatrix}12\\5\end{bmatrix}=\begin{bmatrix}39\\22\end{bmatrix}.$$

*Answer*: 64 36 49 31 75 47 65 37 72 43 75 47 57 35 77 46 95 57 24 13 39 22

(II) $$\begin{bmatrix}1&0&0\\3&1&5\\-2&0&1\end{bmatrix}\begin{bmatrix}20\\8\\5\end{bmatrix}=\begin{bmatrix}20\\93\\-35\end{bmatrix}, \begin{bmatrix}1&0&0\\3&1&5\\-2&0&1\end{bmatrix}\begin{bmatrix}13\\9\\19\end{bmatrix}=\begin{bmatrix}13\\143\\-7\end{bmatrix}, \begin{bmatrix}1&0&0\\3&1&5\\-2&0&1\end{bmatrix}\begin{bmatrix}19\\9\\15\end{bmatrix}=\begin{bmatrix}19\\141\\-23\end{bmatrix},$$

$$\begin{bmatrix}1&0&0\\3&1&5\\-2&0&1\end{bmatrix}\begin{bmatrix}14\\9\\19\end{bmatrix}=\begin{bmatrix}14\\146\\-9\end{bmatrix}, \begin{bmatrix}1&0&0\\3&1&5\\-2&0&1\end{bmatrix}\begin{bmatrix}9\\13\\16\end{bmatrix}=\begin{bmatrix}9\\120\\-2\end{bmatrix}, \begin{bmatrix}1&0&0\\3&1&5\\-2&0&1\end{bmatrix}\begin{bmatrix}15\\19\\19\end{bmatrix}=\begin{bmatrix}15\\159\\-11\end{bmatrix},$$

$$\begin{bmatrix}1&0&0\\3&1&5\\-2&0&1\end{bmatrix}\begin{bmatrix}9\\2\\12\end{bmatrix}=\begin{bmatrix}9\\89\\-6\end{bmatrix}, \begin{bmatrix}1&0&0\\3&1&5\\-2&0&1\end{bmatrix}\begin{bmatrix}5\\26\\26\end{bmatrix}=\begin{bmatrix}5\\171\\16\end{bmatrix}.$$

*Answer*: 20 93 –35 13 143 –7 19 141 –23 14 146 –9 9 120 –2 15 159 –11 9 89 –6 5 171 16

## Application 3:

**1.** Solve $X = D + CX$, where $X=\begin{bmatrix}x_1\\x_2\end{bmatrix}$, $D=\begin{bmatrix}2000\\1000\end{bmatrix}$, and $C=\begin{bmatrix}1/9&3/9\\3/9&1/9\end{bmatrix}$. If $(I_2-C)$ is invertible, the solution is given by $X=(I_2-C)^{-1}D$. $(I_2-C)=\begin{bmatrix}8/9&-1/3\\-1/3&8/9\end{bmatrix}$;

$$\left[\begin{array}{cc|cc}8/9&-1/3&1&0\\-1/3&8/9&0&1\end{array}\right]\xrightarrow[R_1=\frac{9}{8}r_1]{}\left[\begin{array}{cc|cc}1&-3/8&9/8&0\\-1/3&8/9&0&1\end{array}\right]$$

$$\xrightarrow[R_2=r_2+\frac{1}{3}r_1]{}\left[\begin{array}{cc|cc}1&-3/8&9/8&0\\0&55/72&3/8&1\end{array}\right]\xrightarrow[R_2=\frac{72}{55}r_2]{}\left[\begin{array}{cc|cc}1&-3/8&9/8&0\\0&1&27/55&72/55\end{array}\right]$$

$$\xrightarrow[R_1=r_1+\frac{3}{8}r_2]{}\left[\begin{array}{cc|cc}1&0&72/55&27/55\\0&1&27/55&72/55\end{array}\right]$$

$$(I_2-C)^{-1}=\begin{bmatrix}72/55&27/55\\27/55&72/55\end{bmatrix}, \text{ so } \begin{bmatrix}x_1\\x_2\end{bmatrix}=\begin{bmatrix}72/55&27/55\\27/55&72/55\end{bmatrix}\begin{bmatrix}2000\\1000\end{bmatrix}=\begin{bmatrix}34200/11\\25200/11\end{bmatrix}\approx\begin{bmatrix}3109.09\\2290.91\end{bmatrix}.$$

$x_1=3109.09, x_2=2290.91$. The table of direct costs and indirect costs can now be computed:

| Dept. | Total Costs | Direct Costs | Indirect Costs $S_1$ | Indirect Costs $S_2$ |
|---|---|---|---|---|
| $S_1$ | \$3109.09 | \$2000 | \$345.45 | \$763.64 |
| $S_2$ | \$2290.91 | \$1000 | \$1036.36 | \$254.55 |
| $P_1$ | \$3354.54 | \$2500 | \$345.45 | \$509.09 |
| $P_2$ | \$2790.91 | \$1500 | \$1036.36 | \$254.55 |
| $P_3$ | \$3854.54 | \$3000 | \$345.45 | \$509.09 |
| Totals: | | \$10000 | \$3109.07 | \$2290.92 |

Service charges allocated to $P_1$, $P_2$, $P_3$:

$\$345.45+\$509.09+\$1036.36+\$254.55+\$345.45+\$509.09=\$2999.99$

Direct costs of $S_1$ & $S_2$: $\$2000+\$1000=\$3000$

*Note*: $\$2999.99 \approx \$3000$.

## Application 4:

**1.** $A^T = \begin{bmatrix} 4 & 3 \\ 1 & 1 \\ 2 & 0 \end{bmatrix}$ **3.** $A^T = \begin{bmatrix} 1 & 0 & 1 \\ 11 & 12 & 4 \end{bmatrix}$ **5.** $A^T = [8 \quad 6 \quad 3]$

**7.** $A^T A = \begin{bmatrix} 3 & 5 & 6 & 7 \\ 1 & 1 & 1 & 1 \end{bmatrix} \begin{bmatrix} 3 & 1 \\ 5 & 1 \\ 6 & 1 \\ 7 & 1 \end{bmatrix} = \begin{bmatrix} 119 & 21 \\ 21 & 4 \end{bmatrix}$; $A^T Y = \begin{bmatrix} 3 & 5 & 6 & 7 \\ 1 & 1 & 1 & 1 \end{bmatrix} \begin{bmatrix} 10 \\ 13 \\ 15 \\ 16 \end{bmatrix} = \begin{bmatrix} 297 \\ 54 \end{bmatrix}$.

Solve $\begin{bmatrix} 119 & 21 \\ 21 & 4 \end{bmatrix} \begin{bmatrix} m \\ b \end{bmatrix} = \begin{bmatrix} 297 \\ 54 \end{bmatrix}$: $\left[\begin{array}{cc|c} 119 & 21 & 297 \\ 21 & 4 & 54 \end{array}\right] \xrightarrow[R_1=\frac{1}{119}r_1]{} \left[\begin{array}{cc|c} 1 & 3/17 & 297/119 \\ 21 & 4 & 54 \end{array}\right]$

$\xrightarrow[R_2=r_2-21r_1]{} \left[\begin{array}{cc|c} 1 & 3/17 & 297/119 \\ 0 & 5/17 & 27/17 \end{array}\right] \xrightarrow[R_2=\frac{17}{5}r_2]{} \left[\begin{array}{cc|c} 1 & 3/17 & 297/119 \\ 0 & 1 & 27/5 \end{array}\right]$

$\xrightarrow[R_1=r_1-\frac{3}{17}r_2]{} \left[\begin{array}{cc|c} 1 & 0 & 54/35 \\ 0 & 1 & 27/5 \end{array}\right]$: $m=\dfrac{54}{35}, b=\dfrac{27}{5}$.

(a) $y=\dfrac{54}{35}x+\dfrac{27}{5}$

(b) $y=\dfrac{54}{35}(8)+\dfrac{27}{5}=\dfrac{621}{35}\approx 17.743$ thousand units.

**9.** $$A^T A = \begin{bmatrix} 10 & 17 & 11 & 18 & 21 \\ 1 & 1 & 1 & 1 & 1 \end{bmatrix} \begin{bmatrix} 10 & 1 \\ 17 & 1 \\ 11 & 1 \\ 18 & 1 \\ 21 & 1 \end{bmatrix} = \begin{bmatrix} 1275 & 77 \\ 77 & 5 \end{bmatrix}$$

$$A^T Y = \begin{bmatrix} 10 & 17 & 11 & 18 & 21 \\ 1 & 1 & 1 & 1 & 1 \end{bmatrix} \begin{bmatrix} 50 \\ 61 \\ 55 \\ 60 \\ 70 \end{bmatrix} = \begin{bmatrix} 4692 \\ 296 \end{bmatrix}. \text{ Solve } \begin{bmatrix} 1275 & 77 \\ 77 & 5 \end{bmatrix} \begin{bmatrix} m \\ b \end{bmatrix} = \begin{bmatrix} 4692 \\ 296 \end{bmatrix}:$$

$$\left[\begin{array}{cc|c} 1275 & 77 & 4692 \\ 77 & 5 & 296 \end{array}\right] \xrightarrow[R_1 = \frac{1}{1275} r_1]{} \left[\begin{array}{cc|c} 1 & 77/1275 & 92/25 \\ 77 & 5 & 296 \end{array}\right]$$

$$\xrightarrow[R_2 = r_2 - 77r_1]{} \left[\begin{array}{cc|c} 1 & 77/1275 & 92/25 \\ 0 & 446/1275 & 316/25 \end{array}\right] \xrightarrow[R_2 = \frac{1275}{446} r_2]{} \left[\begin{array}{cc|c} 1 & 77/1275 & 92/25 \\ 0 & 1 & 8058/223 \end{array}\right]$$

$$\xrightarrow[R_1 = r_1 - \frac{77}{1275} r_2]{} \left[\begin{array}{cc|c} 1 & 0 & 334/223 \\ 0 & 1 & 8058/223 \end{array}\right]: m = \frac{334}{223}, b = \frac{8058}{223}.$$

$$y = \frac{334}{223}x + \frac{8058}{223}$$

**11.** (a) $$A^T = \begin{bmatrix} 1 & 1 & 3 \\ 1 & 0 & 2 \\ 2 & 1 & 3 \end{bmatrix} \neq \begin{bmatrix} 1 & 1 & 2 \\ 1 & 0 & 1 \\ 3 & 2 & 3 \end{bmatrix} = A; \text{ not symmetric}$$

(b) $$A^T = \begin{bmatrix} 0 & 1 & 3 \\ 1 & 4 & 7 \\ 3 & 7 & 5 \end{bmatrix} = \begin{bmatrix} 0 & 1 & 3 \\ 1 & 4 & 7 \\ 3 & 7 & 5 \end{bmatrix} = A; \text{ symmetric}$$

(c) $$A^T = \begin{bmatrix} 1 & 2 & 3 \\ 2 & 4 & 5 \\ 3 & 5 & 1 \\ 0 & 0 & 0 \end{bmatrix} \neq \begin{bmatrix} 1 & 2 & 3 & 0 \\ 2 & 4 & 5 & 0 \\ 3 & 5 & 1 & 0 \end{bmatrix} = A; \text{ not symmetric}$$

A symmetric matrix must be square: If $A$ is $m \times n$ then $A^T$ is $n \times m$. If $A$ is symmetric, then $A^T = A$, so $A^T$ and $A$ must have the same dimensions. Thus $m = n$ and $A$ is square.

## Chapter 2 Review

### True or False

**1.** True **3.** False **5.** False **7.** False

### Fill in the Blank

**1.** $3\times 2$ **3.** rows, columns **5.** $3\times 3$

### Review Exercises

**1.** $\begin{bmatrix} -2+1 & 0+3 & 7+9 \\ 1+2 & 8+7 & 3+5 \\ 2+3 & 4+6 & 21+8 \end{bmatrix} = \begin{bmatrix} -1 & 3 & 16 \\ 3 & 15 & 8 \\ 5 & 10 & 29 \end{bmatrix}$

**3.** $3\cdot\begin{bmatrix} -1 & 3 & 16 \\ 3 & 15 & 8 \\ 5 & 10 & 29 \end{bmatrix} = \begin{bmatrix} -3 & 9 & 48 \\ 9 & 45 & 24 \\ 15 & 30 & 87 \end{bmatrix}$

**5.** $\begin{bmatrix} -6+5 & 0+15 & 21+45 \\ 3+10 & 24+35 & 9+25 \\ 6+15 & 12+30 & 63+40 \end{bmatrix} = \begin{bmatrix} -1 & 15 & 66 \\ 13 & 59 & 34 \\ 21 & 42 & 103 \end{bmatrix}$

**7.** $2(5A) = 10A = \begin{bmatrix} -20 & 0 & 70 \\ 10 & 80 & 30 \\ 20 & 40 & 210 \end{bmatrix}$

**9.** $\begin{bmatrix} -4+1/2-0 & 0+3/2-3 & 14+9/2-6 \\ 2+1-0 & 16+7/2-15 & 6+5/2-3 \\ 4+3/2-24 & 8+3-21 & 42+4-27 \end{bmatrix} = \begin{bmatrix} -7/2 & -3/2 & 25/2 \\ 3 & 9/2 & 11/2 \\ -37/2 & -10 & 19 \end{bmatrix}$

**11.** $\begin{bmatrix} -2+0+21 & -6+0+42 & -18+0+56 \\ 1+16+9 & 3+56+18 & 9+40+24 \\ 2+8+63 & 6+28+126 & 18+20+168 \end{bmatrix} = \begin{bmatrix} 19 & 36 & 38 \\ 26 & 77 & 73 \\ 73 & 160 & 206 \end{bmatrix}$

**13.** $\begin{bmatrix} 3 & 3 & 2 \\ 1 & -1 & 2 \\ 1 & 2 & -13 \end{bmatrix}\begin{bmatrix} 0 & 1 & 2 \\ 0 & 5 & 1 \\ 8 & 7 & 9 \end{bmatrix} = \begin{bmatrix} 0+0+16 & 3+15+14 & 6+3+18 \\ 0+0+16 & 1-5+14 & 2-1+18 \\ 0+0-104 & 1+10-91 & 2+2-117 \end{bmatrix} = \begin{bmatrix} 16 & 32 & 27 \\ 16 & 10 & 19 \\ -104 & -80 & -113 \end{bmatrix}$

**15.** $BC = \begin{bmatrix} 0+0+72 & 1+15+63 & 2+3+81 \\ 0+0+40 & 2+35+35 & 4+7+45 \\ 0+0+64 & 3+30+56 & 6+6+72 \end{bmatrix}$; $A(BC) = \begin{bmatrix} -2 & 0 & 7 \\ 1 & 8 & 3 \\ 2 & 4 & 21 \end{bmatrix}\begin{bmatrix} 72 & 79 & 86 \\ 40 & 72 & 56 \\ 64 & 89 & 84 \end{bmatrix}$

$= \begin{bmatrix} -144+0+448 & -158+0+623 & -172+0+588 \\ 72+320+192 & 79+576+267 & 86+448+252 \\ 144+160+1344 & 158+288+1869 & 172+224+1764 \end{bmatrix} = \begin{bmatrix} 304 & 465 & 416 \\ 584 & 922 & 786 \\ 1648 & 2315 & 2160 \end{bmatrix}$

**17.** $\left[\begin{array}{cc|cc}3&0&1&0\\-2&1&0&1\end{array}\right]\xrightarrow[R_1=r_1+r_2]{}\left[\begin{array}{cc|cc}1&1&1&1\\-2&1&0&1\end{array}\right]\xrightarrow[R_2=r_2+2r_1]{}\left[\begin{array}{cc|cc}1&1&1&1\\0&3&2&3\end{array}\right]$

$\xrightarrow[R_2=\frac{1}{3}r_2]{}\left[\begin{array}{cc|cc}1&1&1&1\\0&1&2/3&1\end{array}\right]\xrightarrow[R_1=r_1-r_2]{}\left[\begin{array}{cc|cc}1&0&1/3&0\\0&1&2/3&1\end{array}\right];\ \begin{bmatrix}3&0\\-2&1\end{bmatrix}^{-1}=\begin{bmatrix}1/3&0\\2/3&1\end{bmatrix}$

**19.** $\left[\begin{array}{ccc|ccc}1&2&3&1&0&0\\2&4&5&0&1&0\\3&5&6&0&0&1\end{array}\right]\xrightarrow[\substack{R_2=r_2-2r_1\\R_3=r_3-3r_1}]{}\left[\begin{array}{ccc|ccc}1&2&3&1&0&0\\0&0&-1&-2&1&0\\0&-1&-3&-3&0&1\end{array}\right]$

$\xrightarrow[\substack{R_2=-r_3\\R_3=-r_2}]{}\left[\begin{array}{ccc|ccc}1&2&3&1&0&0\\0&1&3&3&0&-1\\0&0&1&2&-1&0\end{array}\right]\xrightarrow[R_1=r_1-2r_2]{}\left[\begin{array}{ccc|ccc}1&0&-3&-5&0&2\\0&1&3&3&0&-1\\0&0&1&2&-1&0\end{array}\right]$

$\xrightarrow[\substack{R_1=r_1+3r_3\\R_2=r_2-3r_3}]{}\left[\begin{array}{ccc|ccc}1&0&0&1&-3&2\\0&1&0&-3&3&-1\\0&0&1&2&-1&0\end{array}\right];\ \begin{bmatrix}1&2&3\\2&4&5\\3&5&6\end{bmatrix}^{-1}=\begin{bmatrix}1&-3&2\\-3&3&-1\\2&-1&0\end{bmatrix}$

**21.** $\left[\begin{array}{ccc|ccc}4&3&-1&1&0&0\\0&2&2&0&1&0\\3&-1&0&0&0&1\end{array}\right]\xrightarrow[R_1=r_1-r_3]{}\left[\begin{array}{ccc|ccc}1&4&-1&1&0&-1\\0&2&2&0&1&0\\3&-1&0&0&0&1\end{array}\right]$

$\xrightarrow[\substack{R_3=r_3-3r_1\\R_2=\frac{1}{2}r_2}]{}\left[\begin{array}{ccc|ccc}1&4&-1&1&0&-1\\0&1&1&0&1/2&0\\0&-13&3&-3&0&4\end{array}\right]\xrightarrow[\substack{R_1=r_1-4r_2\\R_3=r_3+13r_2}]{}\left[\begin{array}{ccc|ccc}1&0&-5&1&-2&-1\\0&1&1&0&1/2&0\\0&0&16&-3&13/2&4\end{array}\right]$

$\xrightarrow[R_3=\frac{1}{16}r_3]{}\left[\begin{array}{ccc|ccc}1&0&-5&1&-2&-1\\0&1&1&0&1/2&0\\0&0&1&-3/16&13/32&1/4\end{array}\right]\xrightarrow[\substack{R_1=r_1+5r_3\\R_2=r_2-r_3}]{}\left[\begin{array}{ccc|ccc}1&0&0&1/16&1/32&1/4\\0&1&0&3/16&3/32&-1/4\\0&0&1&-3/16&13/32&1/4\end{array}\right]$

;

$\begin{bmatrix}4&3&-1\\0&2&2\\3&-1&0\end{bmatrix}^{-1}=\begin{bmatrix}1/16&1/32&1/4\\3/16&3/32&-1/4\\-3/16&13/32&1/4\end{bmatrix}$

**23.** $\left[\begin{array}{ccc|ccc}1&2&-3&1&0&0\\4&6&2&0&1&0\\-3&-6&9&0&0&1\end{array}\right]\xrightarrow[\substack{R_2=r_2-4r_1\\R_3=r_3+3r_1}]{}\left[\begin{array}{ccc|ccc}1&2&-3&1&0&0\\0&-2&14&-4&1&0\\\mathbf{0}&\mathbf{0}&\mathbf{0}&3&0&1\end{array}\right]$; no inverse

**25.** $$\left[\begin{array}{cc|c}-5 & 2 & -2\\ -3 & 3 & 4\end{array}\right] \xrightarrow[R_1=-\frac{1}{5}r_1]{} \left[\begin{array}{cc|c}1 & -2/5 & 2/5\\ -3 & 3 & 4\end{array}\right] \xrightarrow[R_2=r_2+3r_1]{} \left[\begin{array}{cc|c}1 & -2/5 & 2/5\\ 0 & 9/5 & 26/5\end{array}\right]$$

$$\xrightarrow[R_2=\frac{5}{9}r_2]{} \left[\begin{array}{cc|c}1 & -2/5 & 2/5\\ 0 & 1 & 26/9\end{array}\right] \xrightarrow[R_1=r_1+\frac{2}{5}r_2]{} \left[\begin{array}{cc|c}1 & 0 & 14/9\\ 0 & 1 & 26/9\end{array}\right];\ x=\frac{14}{9}, y=\frac{26}{9}$$

**27.** $$\left[\begin{array}{ccc|c}1 & 2 & 5 & 6\\ 3 & 7 & 12 & 23\\ 1 & 4 & 0 & 25\end{array}\right] \xrightarrow[\substack{R_2=r_2-3r_1\\ R_3=r_3-r_1}]{} \left[\begin{array}{ccc|c}1 & 2 & 5 & 6\\ 0 & 1 & -3 & 5\\ 0 & 2 & -5 & 19\end{array}\right] \xrightarrow[\substack{R_1=r_1-2r_2\\ R_3=r_3-2r_2}]{} \left[\begin{array}{ccc|c}1 & 0 & 11 & -4\\ 0 & 1 & -3 & 5\\ 0 & 0 & 1 & 9\end{array}\right]$$

$$\xrightarrow[\substack{R_1=r_1-11r_3\\ R_2=r_2+3r_3}]{} \left[\begin{array}{ccc|c}1 & 0 & 0 & -103\\ 0 & 1 & 0 & 32\\ 0 & 0 & 1 & 9\end{array}\right];\ x=-103, y=32, z=9$$

**29.** $$\left[\begin{array}{ccc|c}1 & 2 & 7 & 2\\ 3 & 7 & 18 & -1\\ 1 & 4 & 2 & -13\end{array}\right] \xrightarrow[\substack{R_2=r_2-3r_1\\ R_3=r_3-r_1}]{} \left[\begin{array}{ccc|c}1 & 2 & 7 & 2\\ 0 & 1 & -3 & -7\\ 0 & 2 & -5 & -15\end{array}\right] \xrightarrow[\substack{R_1=r_1-2r_2\\ R_3=r_3-2r_2}]{} \left[\begin{array}{ccc|c}1 & 0 & 13 & 16\\ 0 & 1 & -3 & -7\\ 0 & 0 & 1 & -1\end{array}\right]$$

$$\xrightarrow[\substack{R_1=r_1-13r_3\\ R_2=r_2+3r_3}]{} \left[\begin{array}{ccc|c}1 & 0 & 0 & 29\\ 0 & 1 & 0 & -10\\ 0 & 0 & 1 & -1\end{array}\right];\ x=29, y=-10, z=-1$$

**31.** $$\left[\begin{array}{ccc|c}2 & -1 & 1 & 1\\ 1 & 1 & -1 & 2\\ 3 & -1 & 1 & 0\end{array}\right] \xrightarrow[\substack{R_1=r_2\\ R_2=r_1}]{} \left[\begin{array}{ccc|c}1 & 1 & -1 & 2\\ 2 & -1 & 1 & 1\\ 3 & -1 & 1 & 0\end{array}\right] \xrightarrow[\substack{R_2=r_2-2r_1\\ R_3=r_3-3r_1}]{} \left[\begin{array}{ccc|c}1 & 1 & -1 & 2\\ 0 & -3 & 3 & -3\\ 0 & -4 & 4 & -6\end{array}\right]$$

$$\xrightarrow[R_2=-\frac{1}{3}r_2]{} \left[\begin{array}{ccc|c}1 & 1 & -1 & 2\\ 0 & 1 & -1 & 1\\ 0 & -4 & 4 & -6\end{array}\right] \xrightarrow[R_3=r_3+4r_2]{} \left[\begin{array}{ccc|c}1 & 1 & -1 & 2\\ 0 & 1 & -1 & 1\\ \mathbf{0} & \mathbf{0} & \mathbf{0} & \mathbf{-2}\end{array}\right];\ \text{no solution.}$$

**33.** $$\left[\begin{array}{ccc|c}0 & 1 & -2 & 6\\ 3 & 2 & -1 & 2\\ 4 & 0 & 3 & -1\end{array}\right] \xrightarrow[\substack{R_1=r_3\\ R_2=r_1\\ R_3=r_2}]{} \left[\begin{array}{ccc|c}4 & 0 & 3 & -1\\ 0 & 1 & -2 & 6\\ 3 & 2 & -1 & 2\end{array}\right] \xrightarrow[R_1=r_1-r_3]{} \left[\begin{array}{ccc|c}1 & -2 & 4 & -3\\ 0 & 1 & -2 & 6\\ 3 & 2 & -1 & 2\end{array}\right]$$

$$\xrightarrow[R_3=r_3-3r_1]{} \left[\begin{array}{ccc|c}1 & -2 & 4 & -3\\ 0 & 1 & -2 & 6\\ 0 & 8 & -13 & 11\end{array}\right] \xrightarrow[\substack{R_1=r_1+2r_2\\ R_3=r_3-8r_2}]{} \left[\begin{array}{ccc|c}1 & 0 & 0 & 9\\ 0 & 1 & -2 & 6\\ 0 & 0 & 3 & -37\end{array}\right]$$

$$\xrightarrow[R_3=\frac{1}{3}r_3]{}\left[\begin{array}{ccc|c}1&0&0&9\\0&1&-2&6\\0&0&1&-37/3\end{array}\right]\xrightarrow[R_2=r_2+2r_3]{}\left[\begin{array}{ccc|c}1&0&0&9\\0&1&0&-56/3\\0&0&1&-37/3\end{array}\right];$$

$x=9, y=-\dfrac{56}{3}, z=-\dfrac{37}{3}$.

**35.** $$\left[\begin{array}{ccc|c}1&-3&0&5\\0&3&1&0\\2&-1&2&2\end{array}\right]\xrightarrow[R_3=r_3-2r_1]{}\left[\begin{array}{ccc|c}1&-3&0&5\\0&3&1&0\\0&5&2&-8\end{array}\right]\xrightarrow[\substack{R_2=r_3\\R_3=r_2}]{}\left[\begin{array}{ccc|c}1&-3&0&5\\0&5&2&-8\\0&3&1&0\end{array}\right]$$

$$\xrightarrow[R_2=r_2-2r_3]{}\left[\begin{array}{ccc|c}1&-3&0&5\\0&-1&0&-8\\0&3&1&0\end{array}\right]\xrightarrow[\substack{R_1=r_1-3r_2\\R_3=r_3+3r_2}]{}\left[\begin{array}{ccc|c}1&0&0&29\\0&-1&0&-8\\0&0&1&-24\end{array}\right]\xrightarrow[R_2=-r_2]{}\left[\begin{array}{ccc|c}1&0&0&29\\0&1&0&8\\0&0&1&-24\end{array}\right];$$

$x=29, y=8, z=-24$

**37.** $$\left[\begin{array}{ccc|c}3&1&-2&3\\1&-2&1&4\end{array}\right]\xrightarrow[R_1=r_1-3r_2]{}\left[\begin{array}{ccc|c}0&7&-5&-9\\1&-2&1&4\end{array}\right]\xrightarrow[\substack{R_1=r_2\\R_2=r_1}]{}\left[\begin{array}{ccc|c}1&-2&1&4\\0&7&-5&-9\end{array}\right]$$

$$\xrightarrow[R_2=\frac{1}{7}r_2]{}\left[\begin{array}{ccc|c}1&-2&1&4\\0&1&-5/7&-9/7\end{array}\right]\xrightarrow[R_1=r_1+2r_2]{}\left[\begin{array}{ccc|c}1&0&-3/7&10/7\\0&1&-5/7&-9/7\end{array}\right]:$$

$$\begin{cases}x-\dfrac{3}{7}z=\dfrac{10}{7}, & x=\dfrac{10}{7}+\dfrac{3}{7}z\\[2mm] y-\dfrac{5}{7}z=-\dfrac{9}{7}, & y=-\dfrac{9}{7}+\dfrac{5}{7}z\end{cases}$$

Three sample solutions: $x=1, y=-2, z=-1$; $x=4, y=3, z=6$; $x=7, y=8, z=13$

**39.** $$\left[\begin{array}{ccc|c}1&2&-1&5\\2&-1&2&0\end{array}\right]\xrightarrow[R_2=r_2-2r_1]{}\left[\begin{array}{ccc|c}1&2&-1&5\\0&-5&4&-10\end{array}\right]\xrightarrow[R_2=-\frac{1}{5}r_2]{}\left[\begin{array}{ccc|c}1&2&-1&5\\0&1&-4/5&2\end{array}\right]$$

$$\xrightarrow[R_1=r_1-2r_2]{}\left[\begin{array}{ccc|c}1&0&3/5&1\\0&1&-4/5&2\end{array}\right]:\begin{cases}x+\dfrac{3}{5}z=1, & x=1-\dfrac{3}{5}z\\[2mm] y-\dfrac{4}{5}z=2, & y=2+\dfrac{4}{5}z\end{cases}.$$

Three sample solutions: $x=4, y=-2, z=-5$; $x=1, y=2, z=0$; $x=-2, y=6, z=5$

**41.** $$\left[\begin{array}{cc|c}2&-1&6\\1&-2&0\\3&-1&6\end{array}\right]\xrightarrow[\substack{R_1=r_2\\R_2=r_1}]{}\left[\begin{array}{cc|c}1&-2&0\\2&-1&6\\3&-1&6\end{array}\right]\xrightarrow[\substack{R_2=r_2-2r_1\\R_3=r_3-3r_1}]{}\left[\begin{array}{cc|c}1&-2&0\\0&3&6\\0&5&6\end{array}\right]$$

$$\xrightarrow[R_3=r_3-\frac{5}{3}r_2]{}\left[\begin{array}{cc|c}1 & -2 & 0\\ 0 & 3 & 6\\ \mathbf{0} & \mathbf{0} & \mathbf{-4}\end{array}\right]$$; no solution.

**43.** $AB=\begin{bmatrix}x-y & x+y\\ z-w & z+x\end{bmatrix}$; $BA=\begin{bmatrix}x+z & y+w\\ -x+z & -y+w\end{bmatrix}$. From $AB = BA$, we obtain

$$\begin{cases}x-y=x+z, & x+y=y+w\\ z-w=-x+z, & z+x=-y+w\end{cases}, \text{ or } \begin{cases}(1) & -y= & z\\ (2) & x = & w\\ (3) & -w= & -x\\ (4) & z+x=-y+w\end{cases}$$

From (2) and (3) $x = w$. Substituting in (4), $z+x=-y+x \Rightarrow z=-y$, which agrees with (1). Thus, $x$, $y$, $z$, $w$ must satisfy $x = w$ and $y=-z$ if $AB = BA$.

**45.** Let $x$ and $y$ represent the number of creams and caramels, respectively, in a box of candy. Then $x+y=50$. The cost of producing a box of candy is $0.10x+0.05y$ dollars. In order to break even, $0.10x+0.05y=4$. Solve $\begin{cases}x+ \quad y=50\\ 01.x+0.05y=\ 4\end{cases}$:

$$\left[\begin{array}{cc|c}1 & 1 & 50\\ 0.1 & 0.05 & 4\end{array}\right]\xrightarrow[R_2=r_2-0.1r_1]{}\left[\begin{array}{cc|c}1 & 1 & 50\\ 0 & -0.05 & -1\end{array}\right]\xrightarrow[R_2=-20r_2]{}\left[\begin{array}{cc|c}1 & 1 & 50\\ 0 & 1 & 20\end{array}\right]\xrightarrow[R_1=r_1-r_2]{}\left[\begin{array}{cc|c}1 & 0 & 30\\ 0 & 1 & 20\end{array}\right];$$

$x=30, y=20$. A box of candy should contain 30 creams and 20 caramels. To obtain a profit, increase the number of caramels (the less expensive of the two candies to produce).

**47.** Let $x$, $y$ and $z$ represent the number of pounds of peanuts, almonds and cashews, respectively, in the 100 pound mixture. Then $x+y+z=100$. If sold separately, the nuts would sell for $2x+6y+5z$ dollars, while the mixture will sell for a total of \$400. Solve $\begin{cases}x+\ y+\ z=100\\ 2x+6y+5z=400\end{cases}$:

$$\left[\begin{array}{ccc|c}1 & 1 & 1 & 100\\ 2 & 6 & 5 & 400\end{array}\right]\xrightarrow[R_2=r_2-2r_1]{}\left[\begin{array}{ccc|c}1 & 1 & 1 & 100\\ 0 & 4 & 3 & 200\end{array}\right]$$

$$\xrightarrow[R_2=\frac{1}{4}r_2]{}\left[\begin{array}{ccc|c}1 & 1 & 1 & 100\\ 0 & 1 & 3/4 & 50\end{array}\right]\xrightarrow[R_1=r_1-r_2]{}\left[\begin{array}{ccc|c}1 & 0 & 1/4 & 50\\ 0 & 1 & 3/4 & 50\end{array}\right];$$

$x=50-\frac{1}{4}z, y=50-\frac{3}{4}z$, $x\geq 0, y\geq 0, z\geq 0$. Possible mixtures include:

| | | | | | | |
|---|---|---|---|---|---|---|
| $50-\frac{1}{4}z$: peanuts | 45 | 43 | 41 | 39 | 37 | 35 |
| $50-\frac{3}{4}z$: almonds | 35 | 29 | 23 | 17 | 11 | 5 |
| $z$: cashews | 20 | 28 | 36 | 44 | 52 | 60 |

**49.** Let $x$, $y$ and $z$ represent the amounts the couple invests in Treasury bills, corporate bonds and junk bonds, respectively. Then $x+y+z=40,000$ dollars and the annual yield of the investment is $0.06x+0.08y+0.10z$ dollars.

(a) Solve $\begin{cases} x+\quad y+\quad z=40000 \\ 0.06x+0.08y+0.10z=\ 2500 \end{cases}$:

$$\left[\begin{array}{ccc|c} 1 & 1 & 1 & 40000 \\ 0.06 & 0.08 & 0.10 & 2500 \end{array}\right] \xrightarrow[R_2=r_2-0.06r_1]{} \left[\begin{array}{ccc|c} 1 & 1 & 1 & 40000 \\ 0 & 0.02 & 0.04 & 100 \end{array}\right]$$

$$\xrightarrow[R_2=50r_2]{} \left[\begin{array}{ccc|c} 1 & 1 & 1 & 40000 \\ 0 & 1 & 2 & 5000 \end{array}\right] \xrightarrow[R_1=r_1-r_2]{} \left[\begin{array}{ccc|c} 1 & 0 & -1 & 35000 \\ 0 & 1 & 2 & 5000 \end{array}\right];$$

$x=35000+z, y=5000-2z, x\geq 0, y\geq 0, z\geq 0$.

| | | | | |
|---|---|---|---|---|
| $35000+z$: treasury bills | \$36000 | \$36500 | \$37000 | \$37500 |
| $5000-2z$: corporate bonds | \$3000 | \$2000 | \$1000 | \$0 |
| $z$: junk bonds | \$1000 | \$1500 | \$2000 | \$2500 |

(b) Solve $\begin{cases} x+\quad y+\quad z=40000 \\ 0.06x+0.08y+0.10z=\ 3000 \end{cases}$:

$$\left[\begin{array}{ccc|c} 1 & 1 & 1 & 40000 \\ 0.06 & 0.08 & 0.10 & 3000 \end{array}\right] \xrightarrow[R_2=r_2-0.06r_1]{} \left[\begin{array}{ccc|c} 1 & 1 & 1 & 40000 \\ 0 & 0.02 & 0.04 & 600 \end{array}\right]$$

$$\xrightarrow[R_2=50r_2]{} \left[\begin{array}{ccc|c} 1 & 1 & 1 & 40000 \\ 0 & 1 & 2 & 30000 \end{array}\right] \xrightarrow[R_1=r_1-r_2]{} \left[\begin{array}{ccc|c} 1 & 0 & -1 & 10000 \\ 0 & 1 & 2 & 30000 \end{array}\right];$$

$x=10000+z, y=30000-2z, x\geq 0, y\geq 0, z\geq 0$.

| | | | | |
|---|---|---|---|---|
| $10000+z$: treasury bills | \$10000 | \$15000 | \$20000 | \$25000 |
| $30000-2z$: corporate bonds | \$30000 | \$20000 | \$10000 | \$0 |
| $z$: junk bonds | \$0 | \$5000 | \$10000 | \$15000 |

(c) Solve $\begin{cases} x+\quad y+\quad z=40000 \\ 0.06x+0.08y+0.10z=\ 3500 \end{cases}$:

$$\left[\begin{array}{ccc|c} 1 & 1 & 1 & 40000 \\ 0.06 & 0.08 & 0.10 & 3500 \end{array}\right] \xrightarrow[R_2=r_2-0.06r_1]{} \left[\begin{array}{ccc|c} 1 & 1 & 1 & 40000 \\ 0 & 0.02 & 0.04 & 1100 \end{array}\right]$$

$$\xrightarrow[R_2=50r_2]{} \left[\begin{array}{ccc|c} 1 & 1 & 1 & 40000 \\ 0 & 1 & 2 & 55000 \end{array}\right] \xrightarrow[R_1=r_1-r_2]{} \left[\begin{array}{ccc|c} 1 & 0 & -1 & -15000 \\ 0 & 1 & 2 & 55000 \end{array}\right];$$

$x=-15000+z, y=55000-2z, x\geq 0, y\geq 0, z\geq 0$.

| | | | | |
|---|---|---|---|---|
| $-15000+z$: treasury bills | \$0 | \$5000 | \$10000 | \$12500 |
| $55000-2z$: corporate bonds | \$25000 | \$15000 | \$5000 | \$0 |
| $z$: junk bonds | \$15000 | \$20000 | \$25000 | \$27500 |

## Professional Exam Questions

**1.** b

**3.** d

# Chapter 3

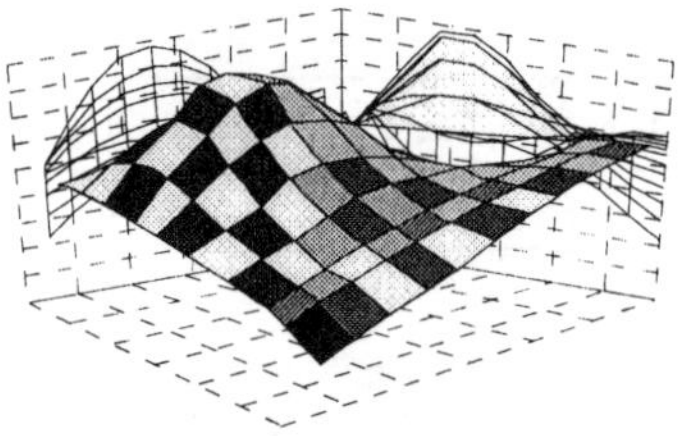

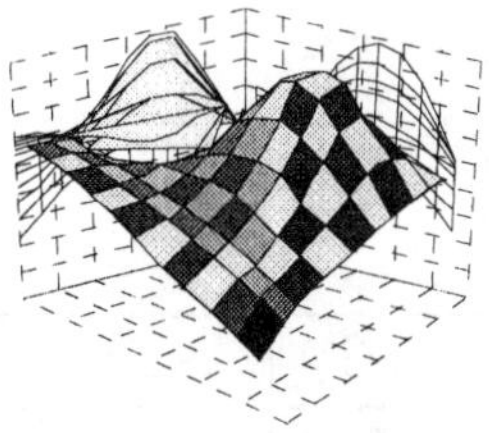

# Linear Programming: Geometric Approach

## 3.1 Linear Inequalities

**1.** $x \geq 0$

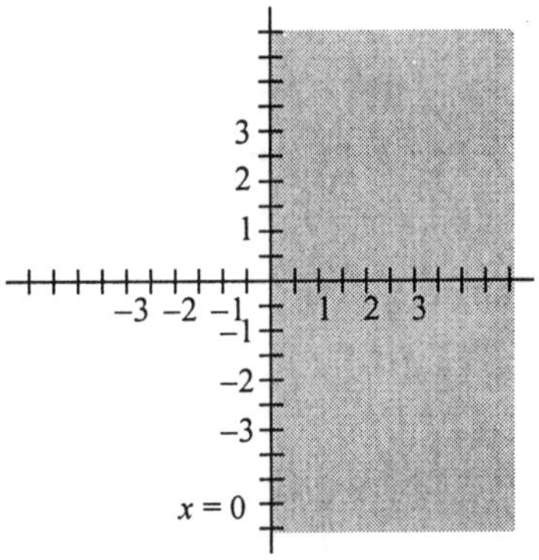


**3.** $x \leq 4$

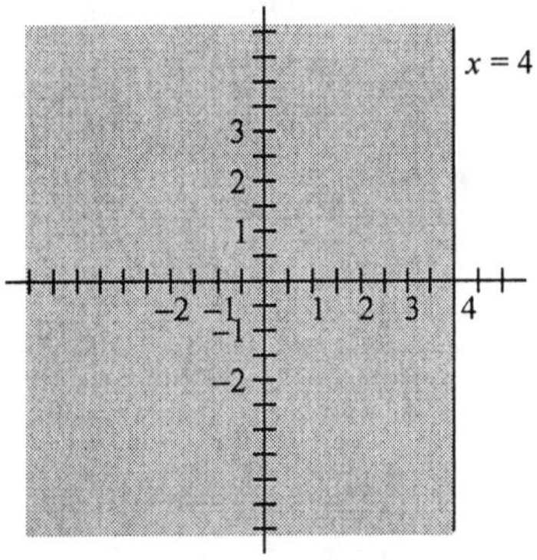


**5.** $y \geq 1$

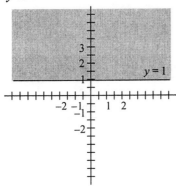

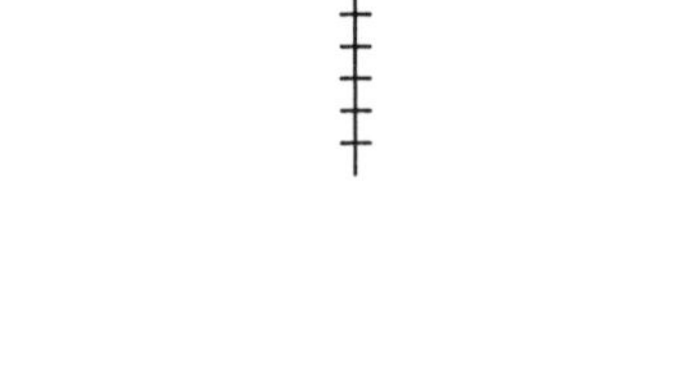


**7.** $2x + 3y \leq 6$

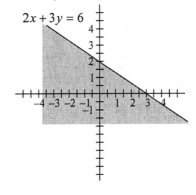

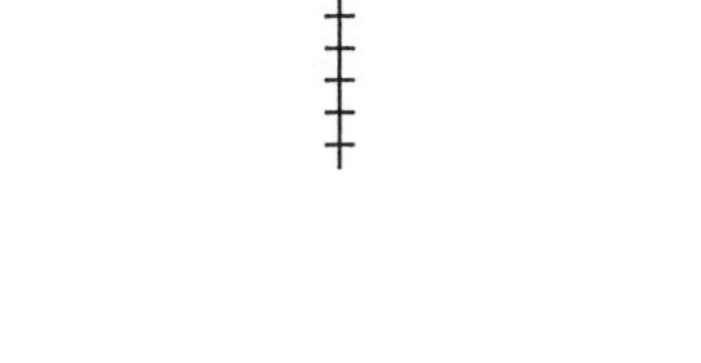

**9.** $5x + y \leq 10$

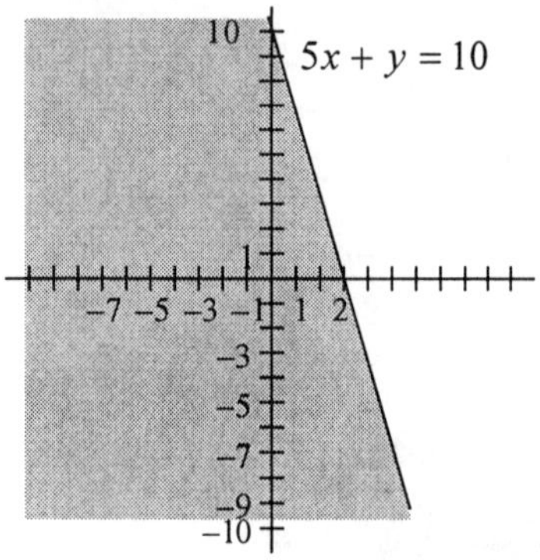


**11.** $x + 5y \leq 5$

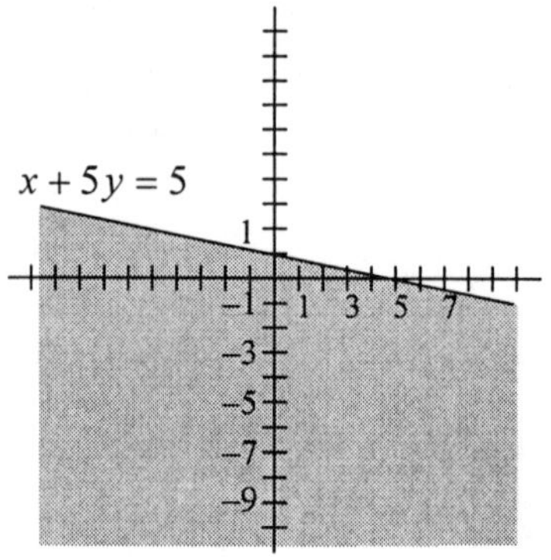


**13.**

| point | $-10x+3y$ | $\leq 0?$ | $-3x+2y$ | $\geq 0?$ | $2x+y$ | $\leq 15?$ | part of graph? |
|---|---|---|---|---|---|---|---|
| (3, 8) | –6 | yes | 7 | yes | 14 | yes | yes |
| (12, 9) | –93 | yes | –18 | no | 33 | no | no |
| (5, 1) | –47 | yes | –13 | no | 11 | yes | no |

**15.**

| point | $10x+3y$ | $\geq 0?$ | $3x+2y$ | $\geq 0?$ | $x+y$ | $\leq 15?$ | part of graph? |
|---|---|---|---|---|---|---|---|
| (5, –8) | 26 | yes | –1 | no | –3 | yes | no |
| (10, 10) | 130 | yes | 50 | yes | 20 | no | no |
| (5, 1) | 53 | yes | 17 | yes | 6 | yes | yes |

**17.**

| point | $2y-10x$ | $\geq 0?$ | $2y-x$ | $\geq 0?$ | $y+6x$ | $\geq 15?$ | part of graph? |
|---|---|---|---|---|---|---|---|
| (5, 3) | –44 | no | 1 | yes | 33 | yes | no |
| (6, 12) | –36 | no | 18 | yes | 48 | yes | no |
| (6, 1) | -58 | no | –4 | no | 37 | yes | no |

**19.**

| region | test point | $5x-4y$ | $\leq 8?$ | $2x+5y$ | $\leq 23?$ |
|---|---|---|---|---|---|
| (a) | (0, –10) | 40 | no | –50 | yes |
| (b) | (0, 0) | 0 | yes | 0 | yes |
| (c) | (0, 10) | –40 | yes | 50 | no |
| (d) | (100, 0) | 500 | no | ·200 | no |

*Answer*: b

**21.**

| region | test point | $2x-3y$ | $\leq -3?$ | $4x+6y$ | $\leq 30?$ |
|---|---|---|---|---|---|
| (a) | (0, 10) | –30 | yes | 60 | no |
| (b) | (–10, 0) | –20 | yes | –40 | yes |
| (c) | (0, 0) | 0 | no | 0 | yes |
| (d) | (10, 0) | 20 | no | 40 | no |

*Answer*: b

**23.**

| region | test point | $5x-3y$ | $\le 3$? | $2x+6y$ | $\le 30$? |
|---|---|---|---|---|---|
| (a) | (0, 10) | –30 | yes | 60 | no |
| (b) | (0, 0) | 0 | yes | 0 | yes |
| (c) | (0, –10) | 30 | no | –60 | yes |
| (d) | (100, 0) | 500 | no | 200 | no |

*Answer*: b

**25.**

| region | test point | $5x-4y$ | $\le 0$? | $2x+4y$ | $\le 28$? |
|---|---|---|---|---|---|
| (a) | (10, 5) | 30 | no | 40 | no |
| (b) | (10, 0) | 50 | no | 20 | yes |
| (c) | (–10, 0) | –50 | yes | –20 | yes |
| (d) | (0, 10) | –40 | yes | 40 | no |

*Answer*: c

**27.** Intercepts:
$x+y=2$: (0, 2) and (2, 0)
Bounded. Corner points: (0, 0), (0, 2), (2, 0)

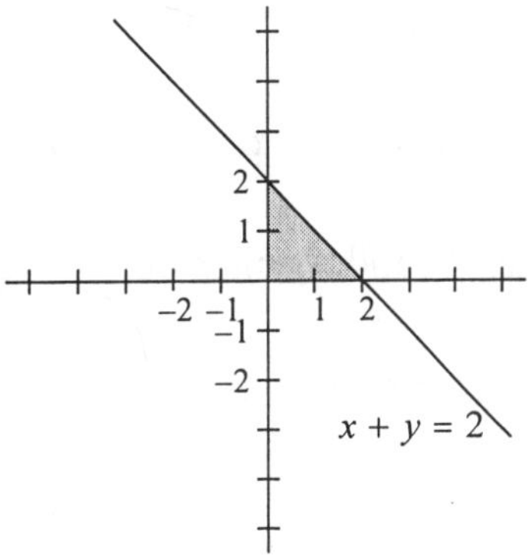


**29.** Intercepts:
$L_1$ : $x+y=2$ : (0, 2) and (2, 0)
$L_2$ : $2x+3y=6$ : (0, 2) and (3, 0)
Bounded. Corner points: (0, 2), (3, 0), (2, 0)

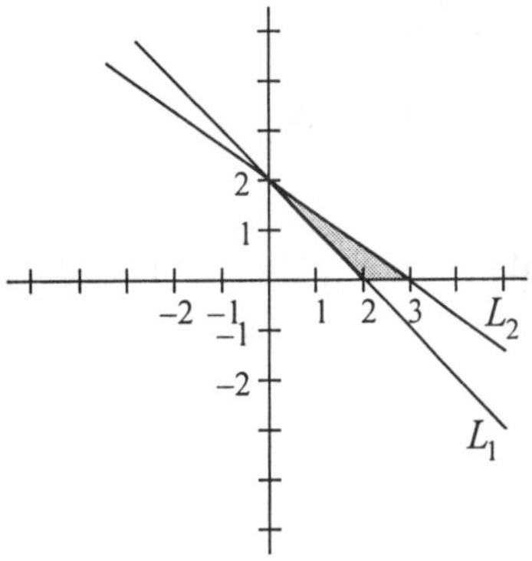


**31.** Intercepts :
$L_1$: $x+y=2$: (0, 2) & (2, 0)
$L_2$: $x+y=8$: (0, 8) & (8, 0)
$L_3$: $2x+y=10$: (0, 10) & (5, 0)
Intersections:

$$\begin{aligned} L_2 &: \begin{cases} x+y=8 \\ 2x+y=10 \end{cases} \\ L_3 &: \end{aligned}$$

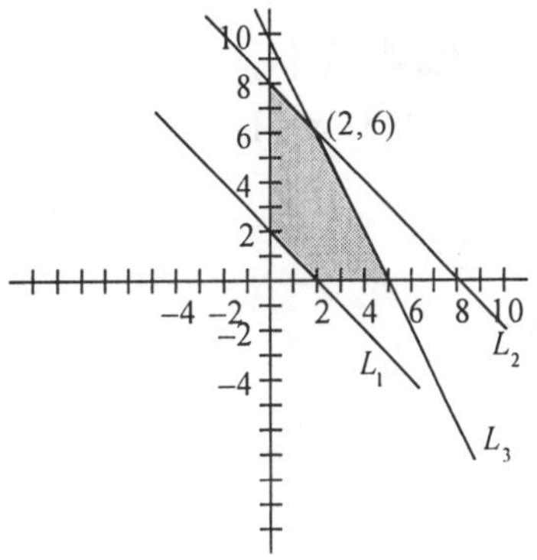

$$\begin{array}{r} -2L_2: \begin{cases} -2x-2y=-16 \\ \end{cases} \\ L_3: \begin{cases} \;\;\,2x+\;\;y=\;\;10 \end{cases} \\ \hline \end{array}$$

$-y=-6; y=6$. Using $L_2$: $x+6=8$; $x=2$. Lines $L_2$ and $L_3$ intersect at (2, 6)

Bounded. Corner points: (0, 2), (0, 8), (2, 6), (5, 0), (2, 0)

**33.** Intercepts:

$L_1$: $x+y=2$: (0, 2) & (2, 0)

$L_2$: $2x+3y=12$: (0, 4) & (6, 0)

$L_3$: $3x+y=12$: (0, 12) & (4, 0)

Intersection:

$$\begin{array}{l} L_2: \\ L_3: \end{array}\begin{cases} 2x+3y=12 \\ 3x+\;\;y=12 \end{cases}$$

$$\begin{array}{r} L_2: \\ -3L_3: \end{array}\begin{cases} \;\;\,2x+3y=\;\;12 \\ -9x-3y=-36 \end{cases}$$

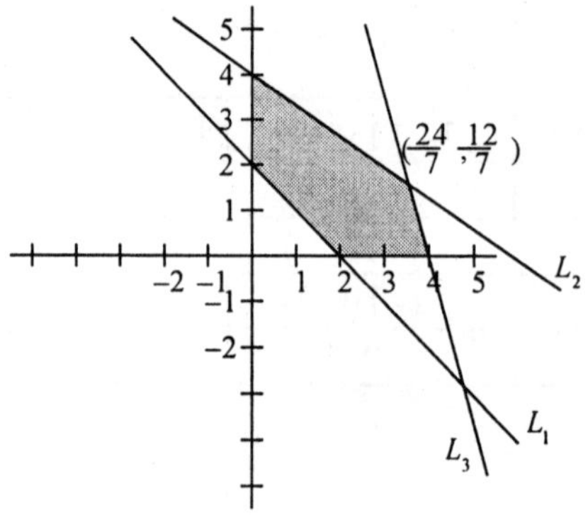


$-7x=-24$; $x=\dfrac{24}{7}$. Using $L_2$: $2\left(\dfrac{24}{7}\right)+3y=12$; $\dfrac{48}{7}+3y=12$; $3y=\dfrac{36}{7}$;

$y=\dfrac{12}{7}$. Lines $L_2$ and $L_3$ intersect at $\left(\dfrac{24}{7}, \dfrac{12}{7}\right)$.

Bounded. Corner points: (0, 2), (0, 4), $\left(\dfrac{24}{7}, \dfrac{12}{7}\right)$, (4, 0), (2, 0)

**35.** Intercepts:

$L_1$: $x+2y=1$: $\left(0, \dfrac{1}{2}\right)$ & (1, 0)

$L_2$: $x+2y=10$: (0, 5) & (10, 0)

Bounded.

Corner points: $\left(0, \dfrac{1}{2}\right)$, (0, 5), (10, 0), (1, 0)

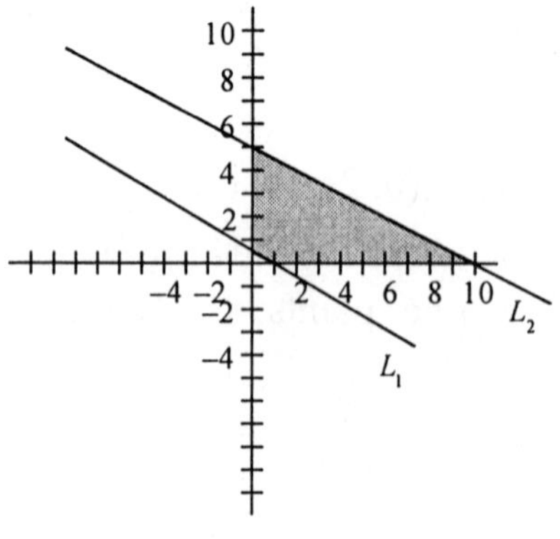


**37** Intercepts:

$L_1$: $x+2y=2$: (0, 1) & (2, 0)

$L_2$: $x+y=4$: (0, 4) & (4, 0)

$L_3$: $3x+y=3$: (0, 3) & (1, 0)

Intersection:

$$\begin{array}{l} L_1: \\ L_3: \end{array}\begin{cases} \;\;x+2y=2 \\ 3x+\;\;y=3 \end{cases}$$

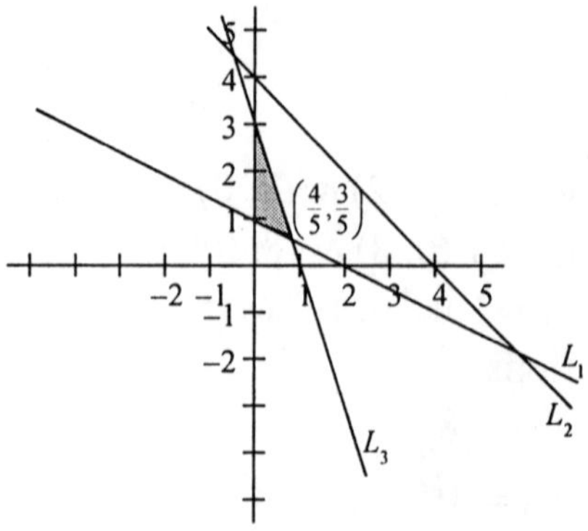

$$\begin{array}{r} L_1: \\ -2L_3: \end{array}\left\{\begin{array}{l} x+2y=\ \ 2 \\ -6x-2y=-6 \end{array}\right.$$

$-5x=-4$; $x=\frac{4}{5}$. Using $L_1$: $\frac{4}{5}+2y=2$; $2y=\frac{6}{5}$; $y=\frac{3}{5}$.

Lines $L_1$ and $L_3$ intersect at $\left(\frac{4}{5},\frac{3}{5}\right)$.

Bounded. Corner points: (0, 1), (0, 3), $\left(\frac{4}{5},\frac{3}{5}\right)$

**39.** (a) Let $x$ and $y$ represent the number of packages of the low-grade and high-grade mixture, respectively.

$$\begin{cases} 4x+8y \le 60\cdot 16 = 960 \\ 12x+8y \le 90\cdot 16 = 1440 \\ x\ge 0\ \&\ y\ge 0 \end{cases}$$

or $\begin{cases} x+2y\le 240 \\ 3x+2y\le 360 \\ x\ge 0\ \&\ y\ge 0 \end{cases}$.

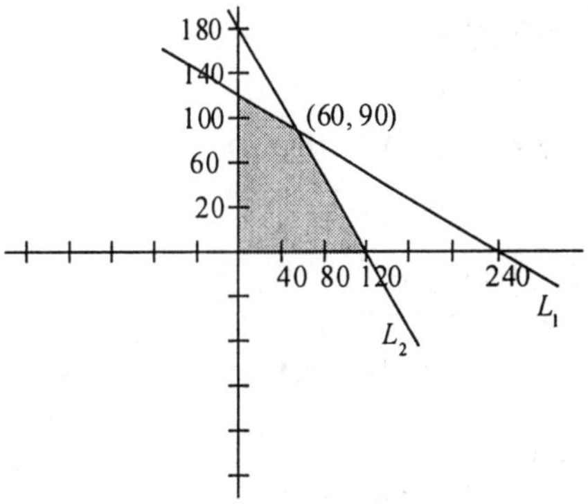


(b) Intercepts:

$L_1: x+2y=240$: (0, 120) & (240, 0)

$L_2: 3x+2y=360$: (0, 180) & (120, 0)

Intersection:

$\begin{array}{r} L_1: \\ L_2: \end{array}\left\{\begin{array}{l} x+2y=240 \\ 3x+2y=360 \end{array}\right.$ $L_2-L_1: 2x=120$; $x=60$; Using $L_1$: $60+2y=240$; $2y=180$; $y=90$. Lines $L_1$ and $L_2$ intersect at (60, 90).

The corner points are: (0, 0), (0, 120), (60, 90), (120, 0)

**41.** (a) $\begin{cases} 3x+2y\le\ 80 \text{ (maximum grinding hours} = 2\cdot 40 = 80) \\ 4x+3y\le 120 \text{ (maximum finishing hours} = 3\cdot 40 = 120) \\ x\ge 0\ \&\ y\ge 0 \end{cases}$

(b) Intercepts:

$L_1$: $3x+2y=80$: (0, 40) & $\left(\frac{80}{3},0\right)$

$L_2$: $4x+3y=120$: (0, 40) & (30, 0)

The three corner points are:

(0, 0), (0, 40), $\left(\frac{80}{3},0\right)$.

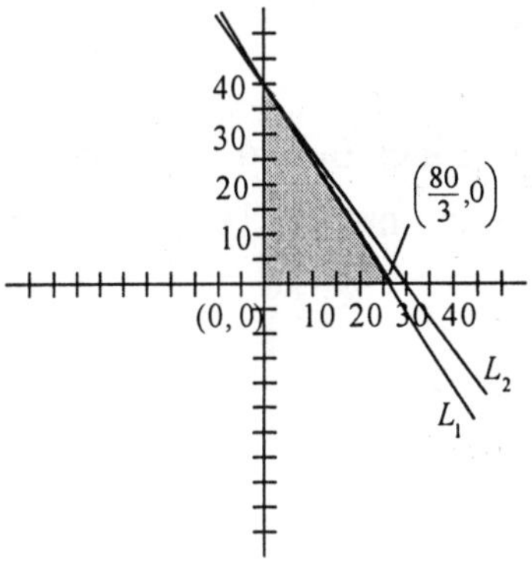

**43.** (a) Measuring $x$ and $y$ in thousands of dollars, clearly $x \geq 0$ and $y \geq 0$. Also, $x \geq 15$, $y \leq 10$ (from your recommendations), and $x + y \leq 25$ (since they have up to \$25 thousand dollars to invest). Thus, $x$ and $y$ should satisfy the system $\begin{cases} x \geq 15 \\ y \leq 10 \\ x + y \leq 25 \\ y \geq 0 \end{cases}$.

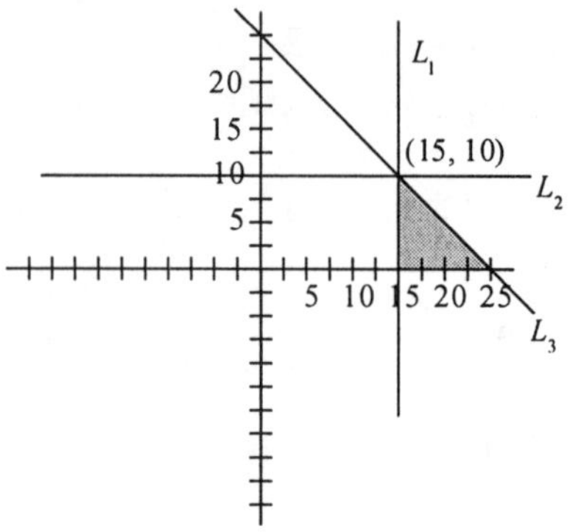


(b) Intercepts:
$L_1$: $x = 15$ : (15, 0)
$L_2$: $y = 10$ : (0, 10)
$L_3$: $x + y = 25$ : (25, 0), (0, 25)
The corner points are (15, 0), (15, 10), (25, 0).

(c) (15, 0): invest \$15,000 in Treasury bills and do not invest the remaining \$10,000.
(15, 10): invest \$15,000 in Treasury bills and \$10,000 in corporate bonds.
(25, 0): invest all \$25,000 in Treasury bills.

**45.** (a) Let $x$ and $y$ represent the number of units of the first and second grains, resepectively.

Then $\begin{cases} x + 2y \geq 5 \\ 5x + y \geq 16 \\ x \geq 0, y \geq 0 \end{cases}$

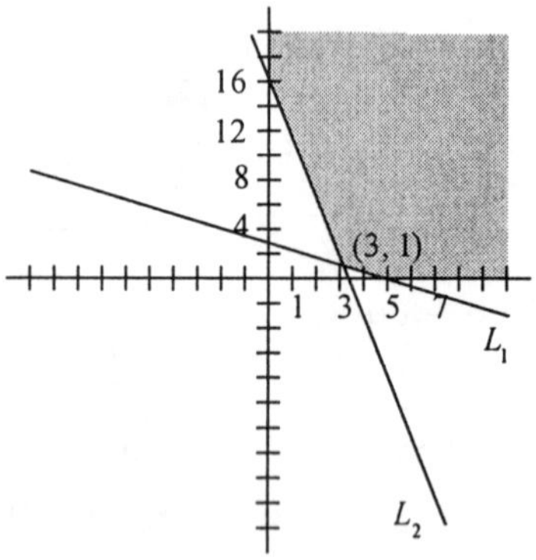


(b) Intercepts:

$L_1$: $x + 2y = 5$ : $\left(0, \dfrac{5}{2}\right)$, (5, 0)

$L_2$: $5x + y = 16$ : (0, 16), $\left(\dfrac{16}{5}, 0\right)$

Intersection:

$$\begin{matrix} L_1: \begin{cases} x + 2y = 5 \\ 5x + y = 16 \end{cases} & L_1: \begin{cases} x + 2y = 5 \\ -10x - 2y = -32 \end{cases} \\ L_2: & -2L_2: \end{matrix}$$

$-9x = -27$; $x = 3$. Using $L_1$ : $3 + 2y = 5$; $2y = 2$; $y = 1$. Lines $L_1$ and $L_2$ intersect at (3, 1).

Corner points: (0, 16), (3, 1), (5, 0)

**47.** (a) Let $x$ and $y$ represent the number of ounces of food A and food B, respectively.

Then: $\begin{cases} x \geq 0, y \geq 0 \\ 5x + 4y \geq 85 \text{ (carbohydrates)} \\ 3x + 3y \geq 70 \text{ (fat)} \\ 2x + 3y \geq 50 \text{ (protein)} \end{cases}$

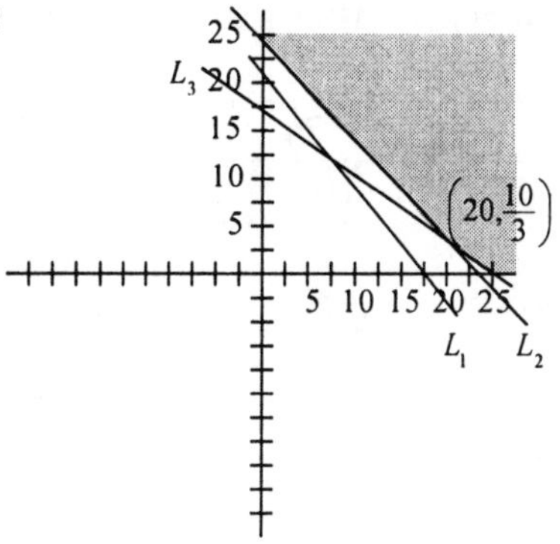


(b) Intercepts:

$L_1$: $5x + 4y = 85$ : $\left(0, \frac{85}{4}\right)$, (17, 0)

$L_2$: $3x + 3y = 70$ : $\left(0, \frac{70}{3}\right)$, $\left(\frac{70}{3}, 0\right)$

$L_3$: $2x + 3y = 50$ : $\left(0, \frac{50}{3}\right)$, (25, 0).

Intersection:

$L_2 - L_3$: $x = 20$. Using $L_3$ : $40 + 3y = 50$; $3y = 10$; $y = \frac{10}{3}$.

Lines $L_2$ and $L_3$ intersect at $\left(20, \frac{10}{3}\right)$.

Corner points: $\left(0, \frac{70}{3}\right)$, $\left(20, \frac{10}{3}\right)$, (25, 0).

## Technology Exercises

**1.** Corner points: (–1, 2), (2, –1), (2, 2)

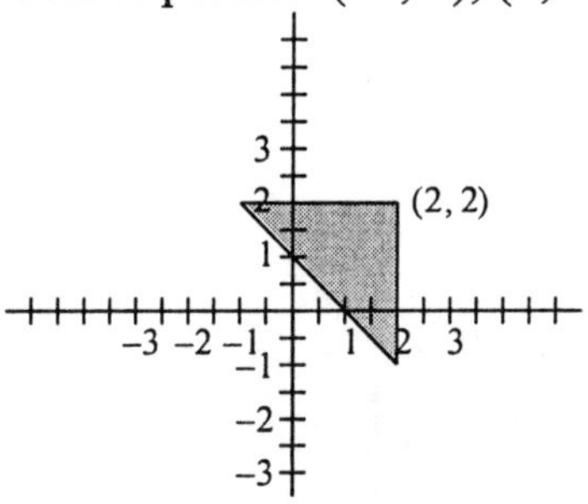


**3.** Corner points: (1, 0), (1, 1.33), (3, 0)

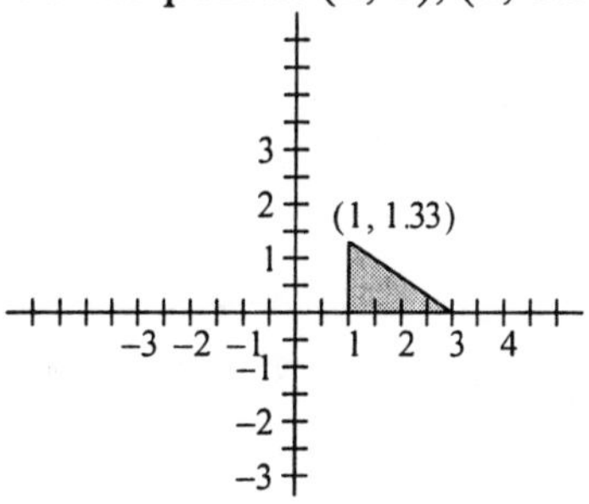


**5.** Corner points: (0, –1), (0, 5), (3, 2)

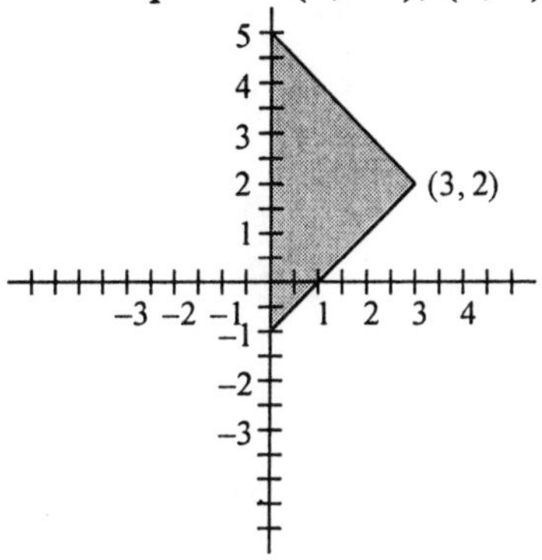

## 3.2 A Geometric Approach to Linear Programming Problems

**1.**

| corner point: | (2, 2) | (2, 7) | (7, 8) | (8, 1) |
|---|---|---|---|---|
| $z = 2x + 3y$ | 10 | 25 | 38 | 19 |

maximum $z$-value: 38
minimum $z$-value: 10

**3.**

| corner point: | (2, 2) | (2, 7) | (7, 8) | (8, 1) |
|---|---|---|---|---|
| $z = x + y$ | 4 | 9 | 15 | 9 |

maximum $z$-value: 15
minimum $z$-value: 4

**5.**

| corner point: | (2, 2) | (2, 7) | (7, 8) | (8, 1) |
|---|---|---|---|---|
| $z = x + 6y$ | 14 | 44 | 55 | 14 |

maximum $z$-value: 55
minimum $z$-value: 14

minimum occurs at any point on the line segment joining (2, 2) and (8,1).

**7.**

| corner point: | (2, 2) | (2, 7) | (7, 8) | (8, 1) |
|---|---|---|---|---|
| $z = 3x + 4y$ | 14 | 34 | 53 | 28 |

maximum $z$-value: 53
minimum $z$-value: 14

**9.**

| corner point: | (2, 2) | (2, 7) | (7, 8) | (8, 1) |
|---|---|---|---|---|
| $z = 10x + y$ | 22 | 27 | 78 | 81 |

maximum $z$-value: 81
minimum $z$-value: 22

**11.** $4x + 3y = 12$ has intercepts (0, 4) and (3, 0). The region described by these constraints (pictured at the right) has corner points (0, 4), (3, 0) and (13, 0). (Note that the region is unbounded above.)

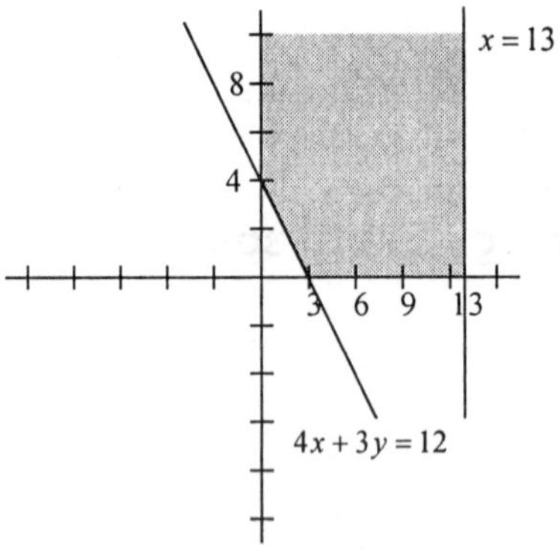


**13.** $x + y = 15$ has intercepts (0, 15) and (15, 0). The region described by these constraints (pictured at the right) has corner points (0, 0), (0, 10), (5, 10) and (15, 0).

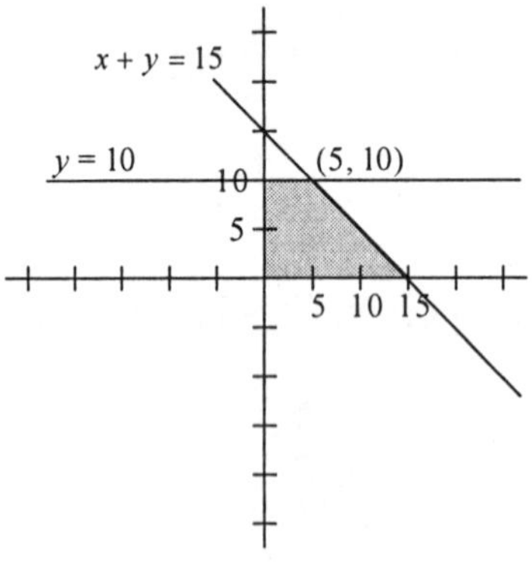

**15.** $4x+3y=12$ has intercepts (0, 4) and (3, 0). The region described by these constraints (pictured at the right) has corner points (0, 4), (0, 8), (10, 8), (10, 0) and (3, 0).

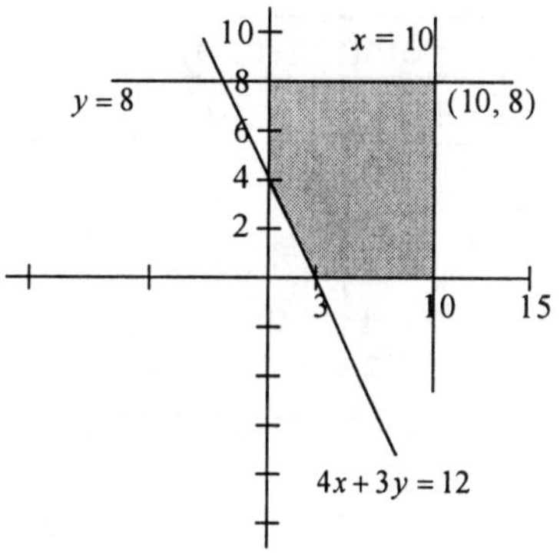


**17.**

| corner point: | (0, 1) | (0, 2) | (1, 1) |
|---|---|---|---|
| $z=5x+7y$ | 7 | 14 | 12 |

maximum $z$-value: 14

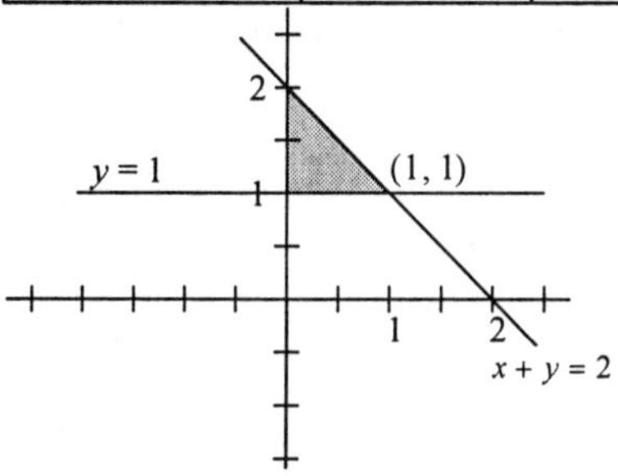


**19.**

| corner point: | (0, 2) | (2, 0) | (3, 0) |
|---|---|---|---|
| $z=5x+7y$ | 14 | 10 | 15 |

maximum $z$-value: 15

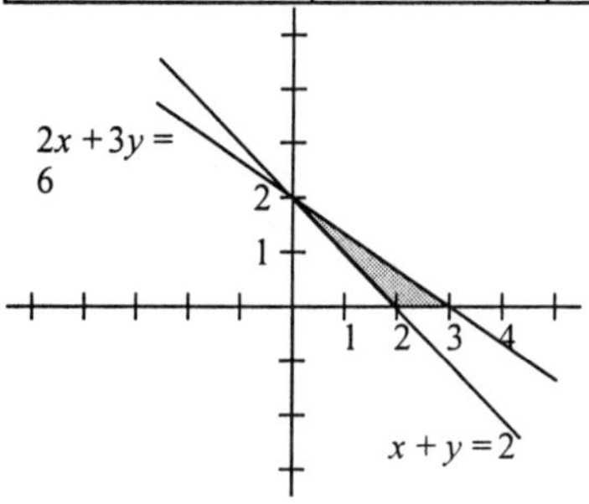


**21.**

| corner point: | (0, 2) | (0, 8) | (2, 6) | (5, 0) | (2, 0) |
|---|---|---|---|---|---|
| $z=5x+7y$ | 14 | 56 | 52 | 25 | 10 |

maximum $z$-value: 56

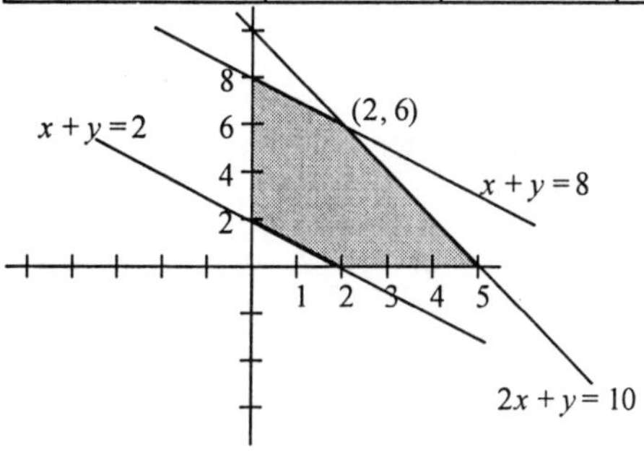

**23.**

| corner point: | (6, 0) | (6, 4) | (10, 0) |
|---|---|---|---|
| $z = 5x + 7y$ | 30 | 58 | 50 |

maximum $z$ value: 58

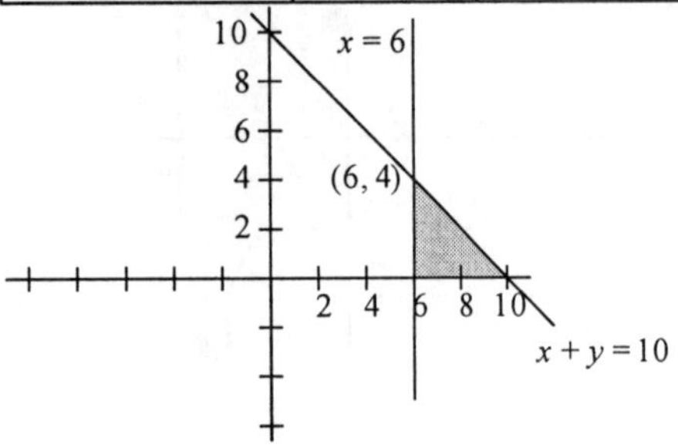


**25.**

| corner point: | (0, 0) | (1, 1) | (2, 0) |
|---|---|---|---|
| $z = 2x + 3y$ | 0 | 5 | 4 |

minimum $z$-value: 0

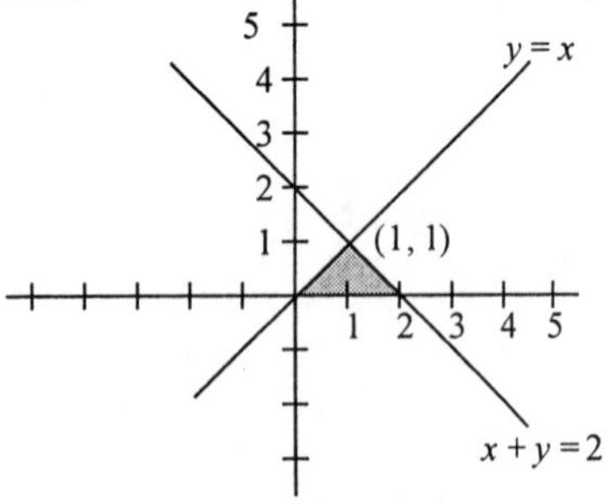


**27.**

| corner point: | (0, 2) | (0, 4) | (3, 3) | (4, 0) | (2, 0) |
|---|---|---|---|---|---|
| $z = 2x + 3y$ | 6 | 12 | 15 | 8 | 4 |

minimum $z$-value: 4

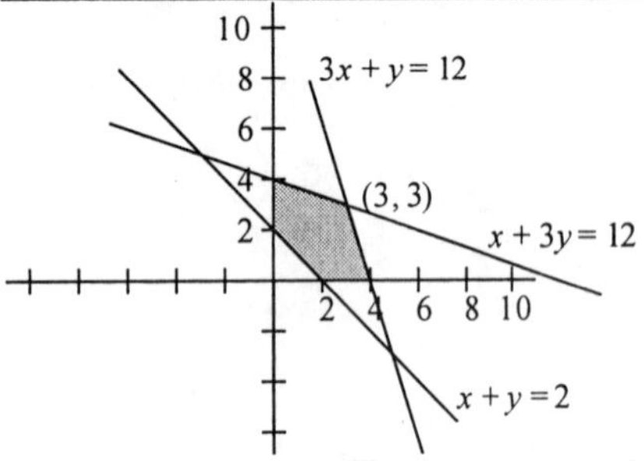


**29.**

| corner point: | (0, 2) | (3, 0) | (2, 0) |
|---|---|---|---|
| $z = 2x + 3y$ | 6 | 6 | 4 |

minimum $z$-value: 4

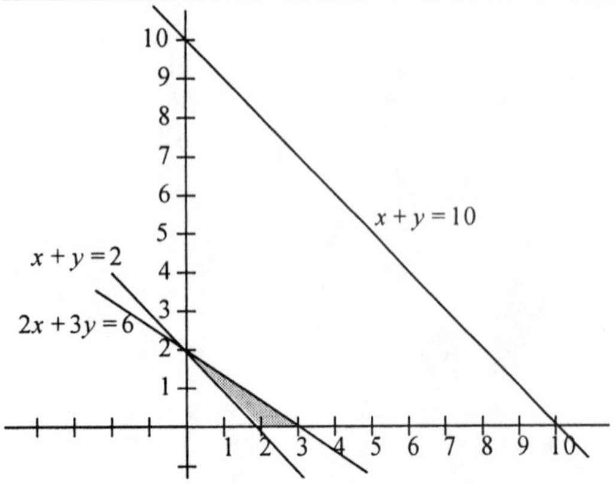

**31.**

| corner point: | (0, 0.5) | (0, 5) | (2, 4) | (0.2, 0.4) |
|---|---|---|---|---|
| $z = 2x + 3y$ | 1.5 | 15 | 16 | 1.6 |

minimum $z$-value: 1.5

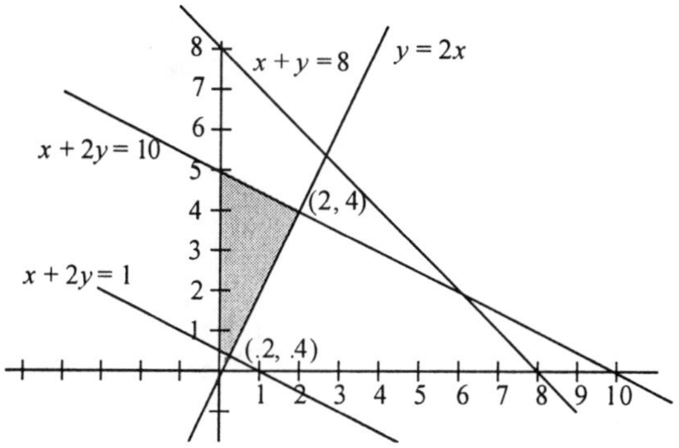


For Problems 33–40, the constraints define the region (pictured at the right) whose corner points are (0, 10), (10, 0) and $\left(\frac{10}{3}, \frac{10}{3}\right)$.

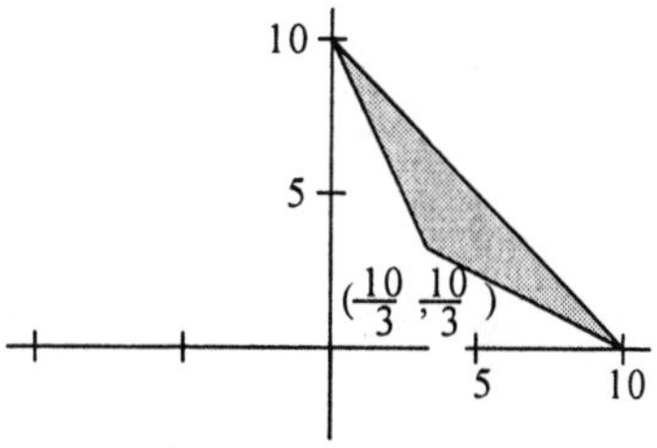


**33.**

| corner point: | (0, 10) | (10, 0) | $\left(\frac{10}{3}, \frac{10}{3}\right)$ |
|---|---|---|---|
| $z = x + y$ | 10 | 10 | $\frac{20}{3}$ |

maximum $z$-value: 10 (at any point on the line segment joining (0,10) and (10,0)).

minimum $z$-value: $\frac{20}{3}$

**35.**

| corner point: | (0, 10) | (10, 0) | $\left(\frac{10}{3}, \frac{10}{3}\right)$ |
|---|---|---|---|
| $z = 5x + 2y$ | 20 | 50 | $\frac{70}{3}$ |

maximum $z$-value: 50

minimum $z$-value: 20

**37.**

| corner point: | (0, 10) | (10, 0) | $\left(\frac{10}{3}, \frac{10}{3}\right)$ |
|---|---|---|---|
| $z = 3x + 4y$ | 40 | 30 | $\frac{70}{3}$ |

maximum $z$-value: 40

minimum $z$-value: $\frac{70}{3}$

**39.**

| corner point: | (0, 10) | (10, 0) | $\left(\frac{10}{3}, \frac{10}{3}\right)$ |
|---|---|---|---|
| $z = 10x + y$ | 10 | 100 | $\frac{110}{3}$ |

maximum $z$-value: 100

minimum $z$-value: 10

**41.**

| corner point: | (0, 3) | (4, 4) | (3, 0) |
|---|---|---|---|
| $z = 18x + 30y$ | 90 | 192 | 54 |

maximum $z$-value: 192
minimum $z$-value: 54

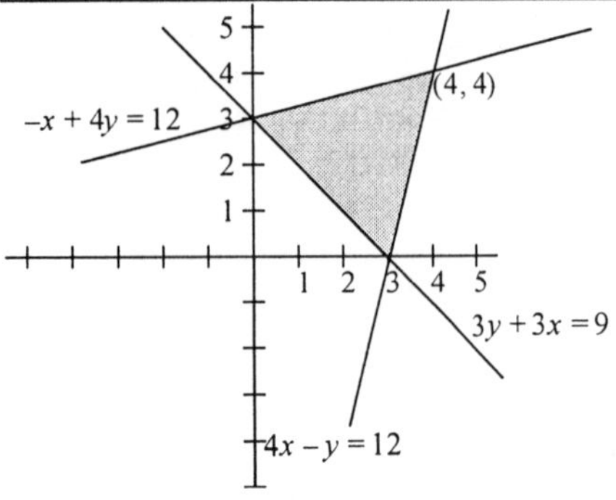


**43.**

| corner point: | (0, 2) | (4, 5) | (3, 0) |
|---|---|---|---|
| $z = 7x + 6y$ | 12 | 58 | 21 |

maximum $z$-value: 58
minimum $z$-value: 12

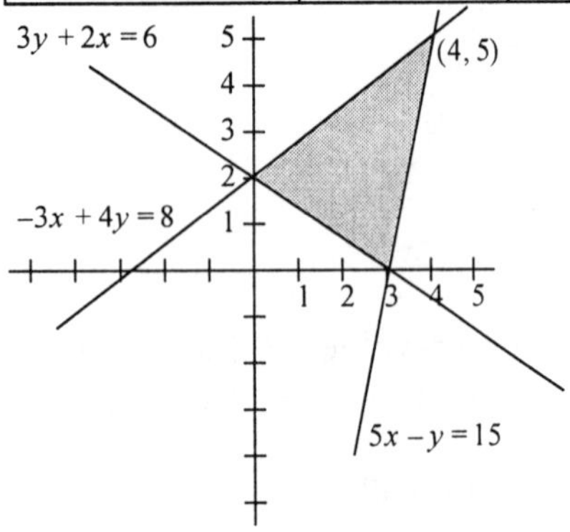


**45.**

| corner point: | (0, 5) | (0, 7) | (3, 10) | (15, 10) | (15, 0) | (3, 0) |
|---|---|---|---|---|---|---|
| $z = -20x + 30y$ | 150 | 210 | 240 | 0 | −300 | −60 |

maximum $z$-value: 240

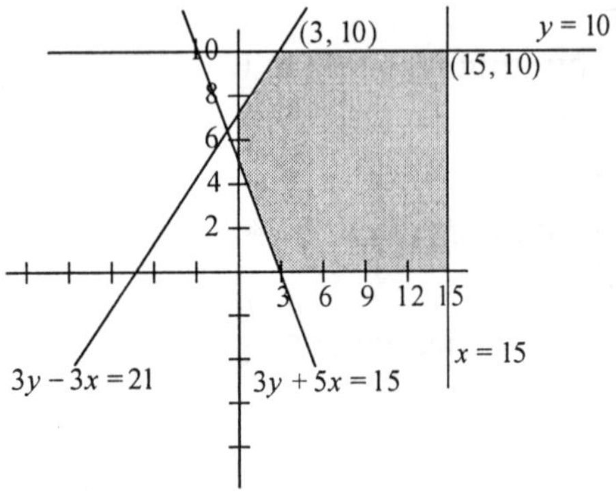

**47.**

| corner point: | (0, 3) | (0, 7) | (2, 10) | (15, 10) | (15, 0) | (3, 0) |
|---|---|---|---|---|---|---|
| $z = -12x + 24y$ | 72 | 168 | 216 | 60 | −180 | −36 |

maximum $z$-value: 216

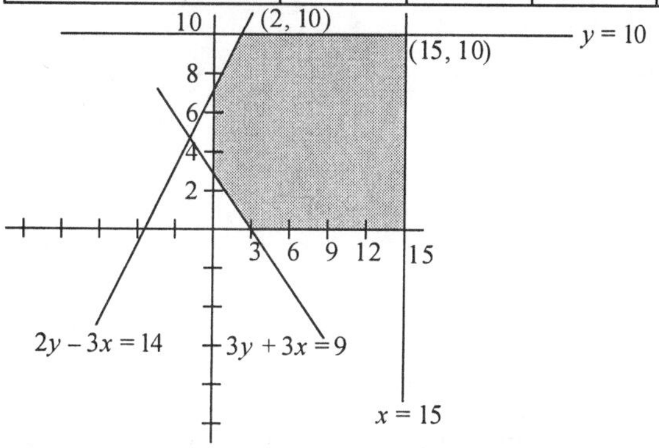


**49.** With $x$ and $y$ defined as in Example 5, the new objective function is $z = 0.30x + 0.40y$. The constraints are unchanged from Example 5.

| corner point: | (0, 0) | (0, 150) | (160, 0) | (90, 105) |
|---|---|---|---|---|
| $P = 0.30x + 0.40y$ | 0 | 60 | 48 | 69 |

A maximum profit of \$69 is obtained if 90 packages of the low-grade mixture and 105 packages of the high-grade mixture are made.

## 3.3 Applications

**1.** Let $x$ and $y$ represent the acres of soybeans and corn, respectively. The objective is to maximize the profit $P = 300x + 150y$. The constraints are imposed by the amount of land available, cultivation costs, and workdays required:

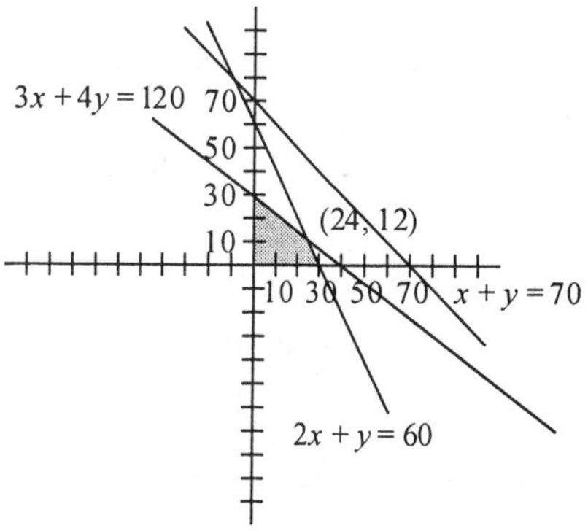


$$\begin{cases} \text{available land: } x + y \le 70 \\ \text{cultivation cost: } 60x + 30y \le 1800 \text{ or } 2x + y \le 60. \\ \text{workdays required: } 3x + 4y \le 120 \end{cases}, \quad x \ge 0, y \ge 0$$

The region described by these constraints is bounded with corner points (0, 0), (0, 30), (24, 12) and (30, 0).

| corner point: | (0, 0) | (0, 30) | (24, 12) | (30, 0) |
|---|---|---|---|---|
| $P = 300x + 150y$ | 0 | 4500 | 9000 | 9000 |

The maximum profit is \$9000, obtained when the number of acres of soybeans ($x$) and the number of acres of corn ($y$) satisfy $2x + y = 60$ and no more than 12 acres of corn are planted.

**3.** Let $x$ represent the number of units of the first product manufactured and $y$ the number of units of the second product manufactured. The objective is to maximize the profit $P = 40x + 60y$ dollars. The constraints are imposed by the number of hours each of the three machines are available:

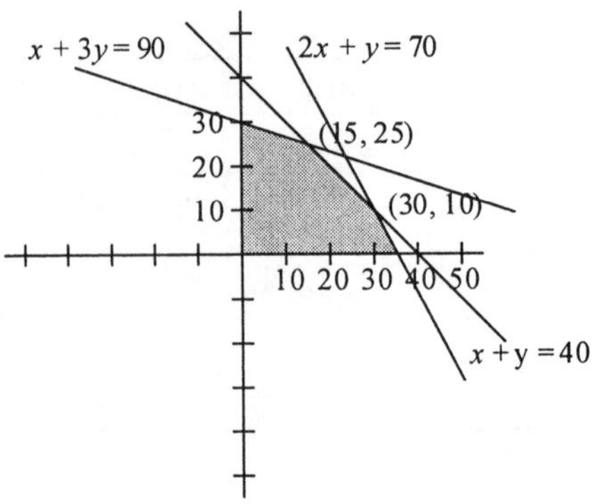


$$\begin{cases} \text{first machine}: 2x + y \le 70 \\ \text{second machine}: x + 2y \le 40 \\ \text{third machine}: x + 3y \le 90, x \ge 0, y \ge 0. \end{cases}$$

The region described by these constraints is bounded with corner points (0, 0), (0, 30), (15, 25), (30, 10) and (35, 0).

| corner point: | (0, 0) | (0, 30) | (15, 25) | (30, 10) | (35, 0) |
|---|---|---|---|---|---|
| $P = 40x + 60y$ | 0 | 1800 | 2100 | 1800 | 1400 |

A maximum profit of \$2100 is obtained if 15 units of the first product and 25 units of the second product are manufactured.

**5.** Let $x$ represent the amount invested in the first security and $y$ the amount invested in the second security. The objective is to maximize income $I = 0.1x + 0.08y$ dollars. The constraints are imposed by the consultant's beliefs:

$$\begin{cases} \text{invest up to \$30,000}: \quad x + y \le 30{,}000 \\ \text{amt. invested in 1}^{\text{st}} \text{ security} \\ \text{is at most } \frac{1}{3} \text{ that invested} \\ \text{in 2}^{\text{nd}} \text{ security}: \quad x \le \frac{1}{3}y \\ x \ge 0, y \ge 0 \end{cases}$$

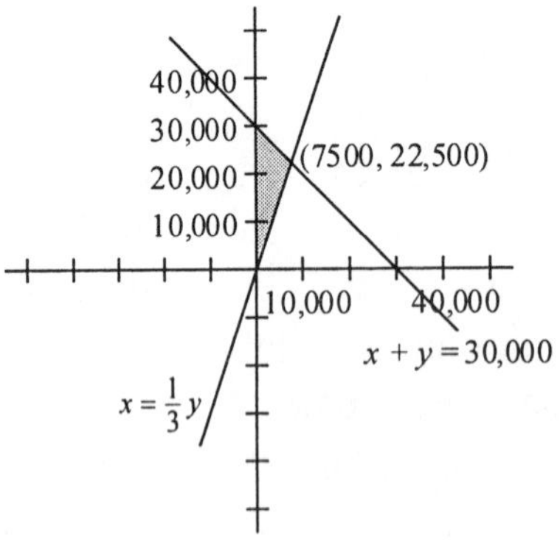


The region described by these constraints is bounded with corner points (0, 0), (0, 30,000) and (7,500, 22,500).

| corner points | (0, 0) | (0, 30,000) | (7,500, 22,500) |
|---|---|---|---|
| $I = 0.1x + 0.08y$ | 0 | 2,400 | 2,550 |

The maximum income is \$2,550, obtained when \$7,500 is invested in the first security and \$22,500 is invested in the second security.

**7.** Let $x$ represent the number of weeks the first repairperson is employed, $y$ the number of weeks the second repairperson is employed.
The objective is to minimize the labor cost $C = 250x + 220y$. The constraints are imposed by the weeks required of each repairperson on the various types of appliances:

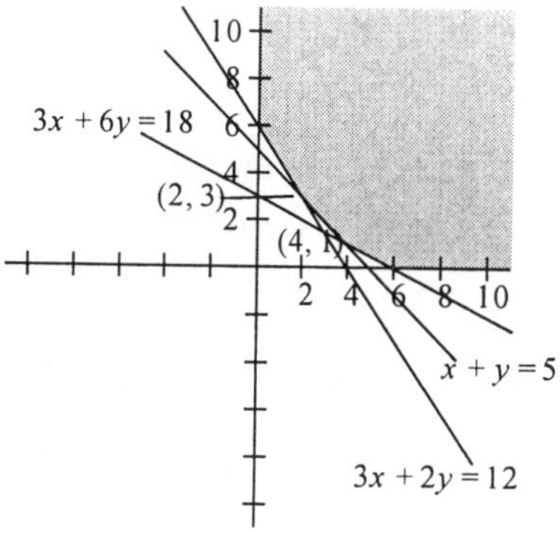


$$\begin{cases} \text{vacuum cleaners: } x + y \geq 5 \\ \text{TV sets: } 3x + 2y \geq 12 \\ \text{VCRs: } 3x + 6y \geq 18 \\ x \geq 0, y \geq 0 \end{cases}$$

The region described by these constraints is unbounded with corner points (0, 6), (2, 3), (4, 1) and (6, 0).

| corner point: | (0, 6) | (2, 3) | (4, 1) | (6, 0) |
|---|---|---|---|---|
| $C = 250x + 220y$ | 1320 | 1160 | 1220 | 1500 |

Note that the cost increases in the direction of unboundedness. The minimum labor cost is \$1160, obtained when the first repairperson is employed for 2 weeks and the second repairperson is employed for 3 weeks.

**9.** Let $x$ represent the number of units of item A and $y$ the number of units of item B. The objective is to maximize profit $P = 1.5x + 1y$. The constraints are imposed by government pollution standards.

$$\begin{cases} 2x + 4y \leq 3000 & \text{(maximum cu. ft. of carbon monoxide)} \\ 6x + 3y \leq 5400 & \text{(maximum cu. ft. of sulfur dioxide)} \\ x \geq 0, y \geq 0 \end{cases} \quad \text{or} \quad \begin{cases} x + 2y \leq 1500 \\ 2x + y \leq 1800 \\ x \geq 0, y \geq 0 \end{cases}$$

The region described by these constraints is bounded with corner points (0, 0), (0, 750), (700, 400), (900, 0).

| corner point: | (0, 0) | (0, 750) | (700, 400) | (900, 0) |
|---|---|---|---|---|
| $P = 1.5x + 1y$ | 0 | 750 | 1450 | 1350 |

The maximum profit is \$1,450; obtained when 700 units of item A, and 400 units of item B are produced.

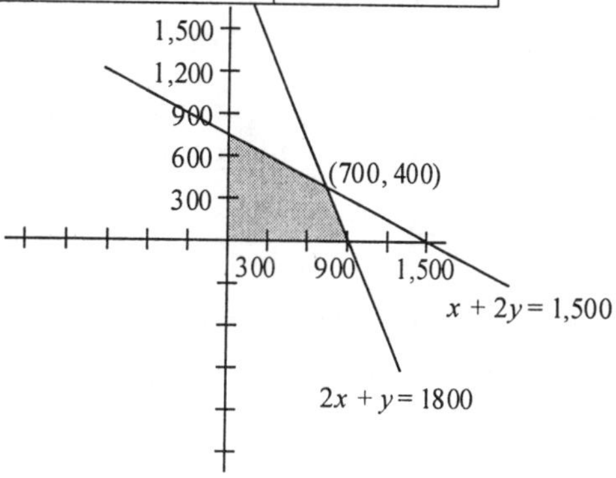

**11.** Let $x$ represent ounces of Supplement I and $y$ ounces of Supplement II.
The problem is to minimize cost $C = (\$0.03)x + (\$0.04)y$, subject to $x \geq 0$, $y \geq 0$,

$$\begin{cases} 5x+25y \geq 50 & \text{or} \quad x+5y \geq 10 \quad \text{(vitamin 1)} \\ 25x+10y \geq 100 & \text{or} \quad 5x+2y \geq 20 \quad \text{(vitamin 2)} \\ 10x+10y \geq 60 & \text{or} \quad x+y \geq 6 \quad \text{(vitamin 3)} \\ 35x+20y \geq 180 & \text{or} \quad 7x+4y \geq 36 \quad \text{(vitamin 4)} \end{cases}$$

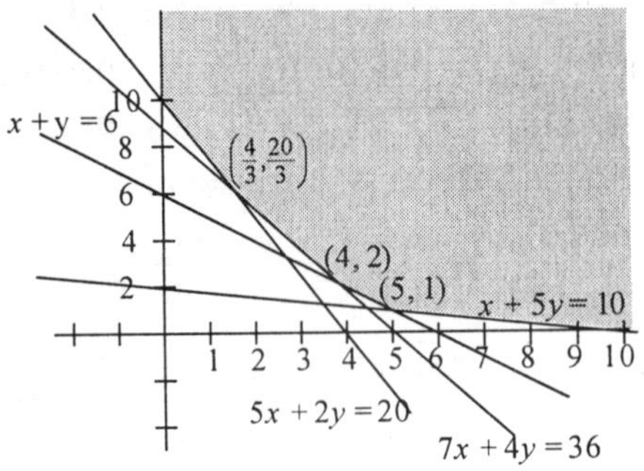


The region described by these constraints is unbounded with corner points (0, 10), (10, 0), (5, 1), (4, 2), $\left(\frac{4}{3}, \frac{20}{3}\right)$.

| corner points | (0, 10) | (10, 0) | (5, 1) | (4, 2) | $\left(\frac{4}{3}, \frac{20}{3}\right)$ |
|---|---|---|---|---|---|
| $C = (\$0.03)x + (\$0.04)y$ | .40 | .30 | .19 | .20 | .31 |

The minimum cost is \$0.19 when he adds 5 ounces of Supplement I and 1 ounce of Supplement II to each 100 ounces of feed.

**13.** Let $x$ represent the number of rolls of high-grade carpet made, $y$ the number of rolls of low-grade carpet made *and* $H$ selling price of high-grade carpet in dollars per roll. The objective is to maximize income $I = \text{(revenue)} - \text{(cost)}$.

Revenue $= Hx + 300y$; Cost = material cost + labor cost;

$$\begin{aligned} \text{Material cost} &= (\text{wool cost}) + (\text{nylon cost}) \\ &= \left(5\$/\text{yd}^2 \cdot 20\,\text{yd}^2/\text{roll} \cdot x \text{ rolls}\right) \\ &\quad + \left(2\$/\text{yd}^2 \cdot 40\,\text{yd}^2/\text{roll} \cdot x \text{ rolls} + 2\$/\text{yd}^2 \cdot 40\,\text{yd}^2/\text{roll} \cdot y \text{ rolls}\right) \\ &= 100x + 80x + 80y = 180x + 80y \end{aligned}$$

$$\begin{aligned} \text{Labor cost} &= 6\$/\text{hr} \cdot 40\,\text{hrs/roll} \cdot x \text{ rolls} + 6\$/\text{hr} \cdot 20\,\text{hrs/roll} \cdot y \text{ rolls} \\ &= 240x + 120y \end{aligned}$$

Total cost $= (180x + 80y) + (240x + 120y) = 420x + 200y$.

Income $I = (Hx + 300y) - (420x + 200y) = (H - 420)x + 100y$.

The constraints are imposed by the availability of raw materials and of labor:

$$\begin{cases} \text{wool}: 20x \leq 1200 \\ \text{nylon}: 40x + 40y \leq 1000 \\ \text{workhours}: 40x + 20y \leq 800 \\ x \geq 0, y \geq 0 \end{cases} \quad \text{or} \quad \begin{cases} x \leq 60 \\ x + y \leq 25 \\ 2x + y \leq 40 \\ x \geq 0, y \geq 0 \end{cases}.$$

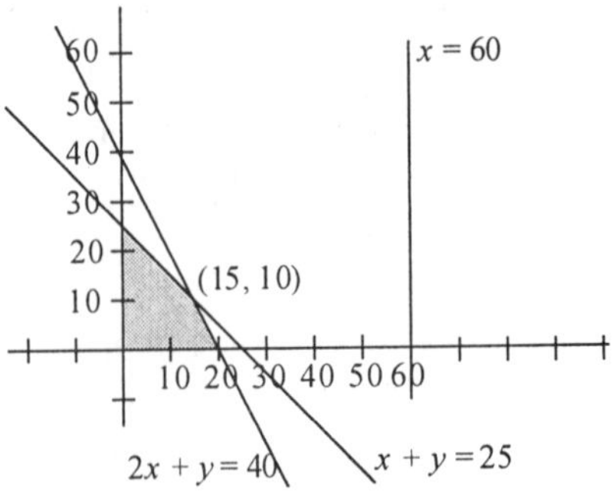


The region described by these constraints is bounded with corner points (0, 0), (0, 25), (15, 10) and (20, 0).

| corner point: | (0, 0) | (0, 25) | (15, 10) | (20, 0) |
|---|---|---|---|---|
| $Z = (H - 420)x + 100y$ | 0 | 2500 | $15H - 5300$ | $20H - 8400$ |

If income is maximized at a point where some rolls of each type are made, it must occur when $(x, y) = (15, 10)$, so $2500 \leq 15H - 5300$ and $20H - 8400 \leq 15H - 5300$. Thus, $\$520 \leq H \leq \$620$. (Note that this can also be found by comparing the slopes of constant lines for the objective function versus the slope of the edges between (0, 25) & (15, 10) and between (15, 10) & (20, 0).)

## Technology Exercises

**1.**

| corner point: | (1, 2) | (4, 5) | (6, 3) | (3, 0) |
|---|---|---|---|---|
| $z = 3.5x + 1.25y$ | 6 | 20.25 | 24.75 | 10.5 |

maximum $z$-value: 24.75
minimum $z$-value: 6

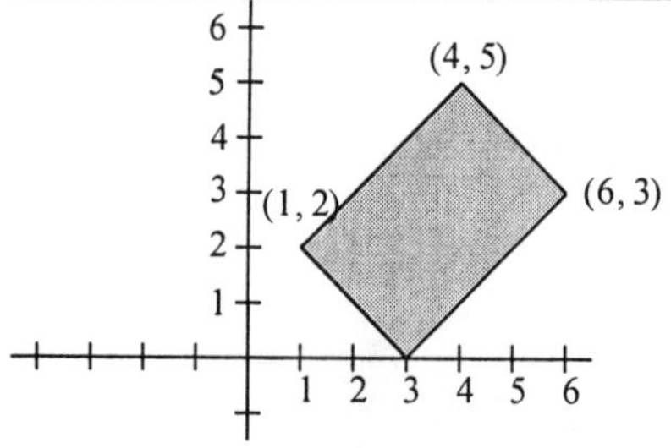


**3.**

| corner point: | (0.55, 5.27) | (4, 7) | (5.33, 4.33) | (1.5, 0.5) |
|---|---|---|---|---|
| $z = 3.5x + 1.25y$ | 8.51 | 22.75 | 24.07 | 5.875 |

maximum $z$-value: 24.07
minimum $z$-value: 5.875

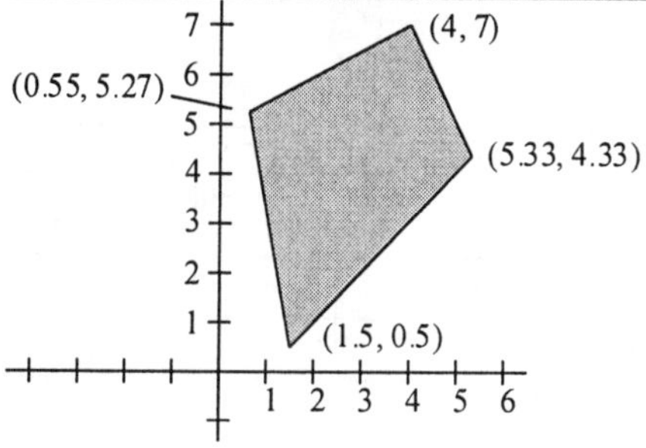


**5.**

| corner point: | (0.55, 5.27) | (4, 7) | (6.67, 1.67) | (5, 0) | (1.6, 0) |
|---|---|---|---|---|---|
| $z = 3.5x + 1.25y$ | 8.51 | 22.75 | 25.43 | 17.5 | 5.6 |

maximum $z$-value: 25.43
minimum $z$-value: 5.6

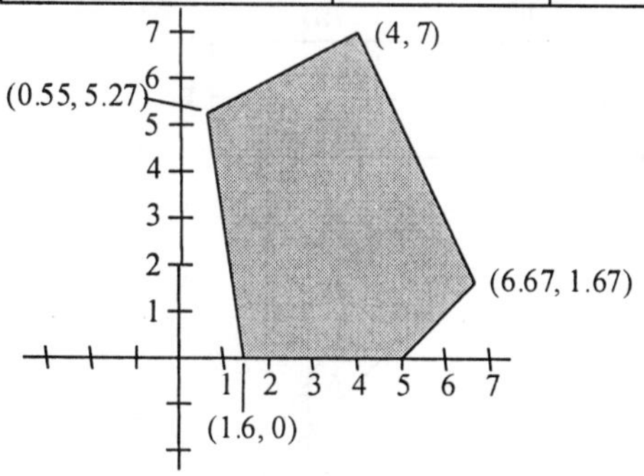

## Chapter 3 Review

### True or False

**1.** True

**3.** True

**5.** True

### Fill in the Blank

**1.** half plane

**3.** feasible

**5.** corner point

### Review Exercises

**1.** $x+3y \le 0$

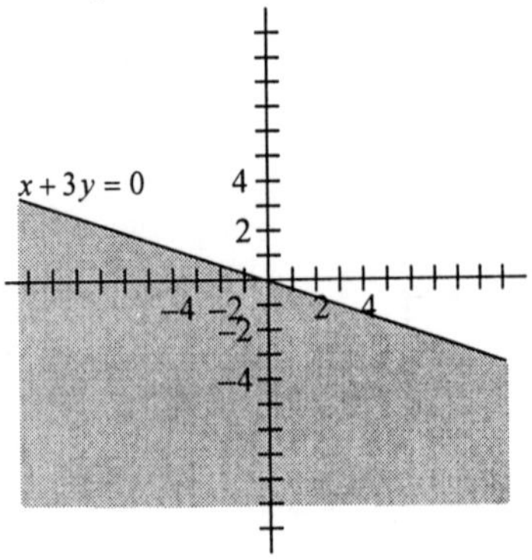


**3.** $5x+y \ge 10$

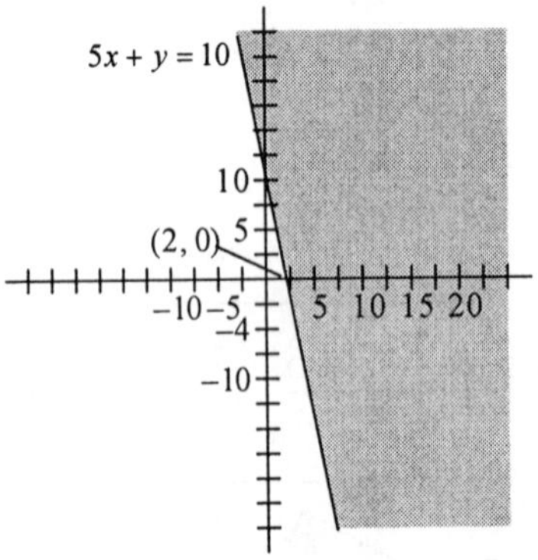


**5.**

| point | $7y+10x$ | $\le 0$? | $2y+9x$ | $\ge 0$? | $y+3x$ | $\le 15$? | part of graph? |
|---|---|---|---|---|---|---|---|
| $P_1=(4,-3)$ | 19 | no | 30 | yes | 9 | yes | no |
| $P_2=(2,-6)$ | –22 | yes | 6 | yes | 0 | yes | yes |
| $P_3=(8,-3)$ | 59 | no | 66 | yes | 21 | no | no |

$P_2$ is part of the graph.

**7.**

| region | test point | $6x-4y$ | $\le 12$? | $3x+2y$ | $\le 18$? |
|---|---|---|---|---|---|
| (a) | (–100, 0) | –600 | yes | –300 | yes |
| (b) | (0, 100) | –400 | yes | 200 | no |
| (c) | (100, 0) | 600 | no | 300 | no |
| (d) | (0, –100) | 400 | no | –200 | yes |

*Answer*: a

**9.** Intercepts:

$L_1: 3x+2y=12$ : (0, 6), (4,0)

$L_2: x+y=4$ : (0, 4), (4,0)

Bounded. Corner points: (0, 4), (0, 6), (4, 0)

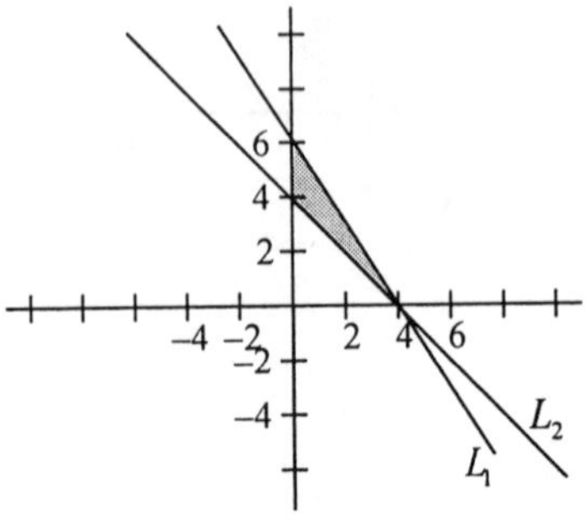

**11.** Intercepts:

$L_1: x+2y=4 : (0, 2), (4, 0)$

$L_2: 3x+y=6 : (0, 6), (2, 0)$

Intersection:

$$\begin{aligned} L_1: &\quad x+2y=4 \\ -2L_2: &\; -6x-2y=-12 \\ \hline &\quad -5x=-8;\ x=\frac{8}{5}. \end{aligned}$$

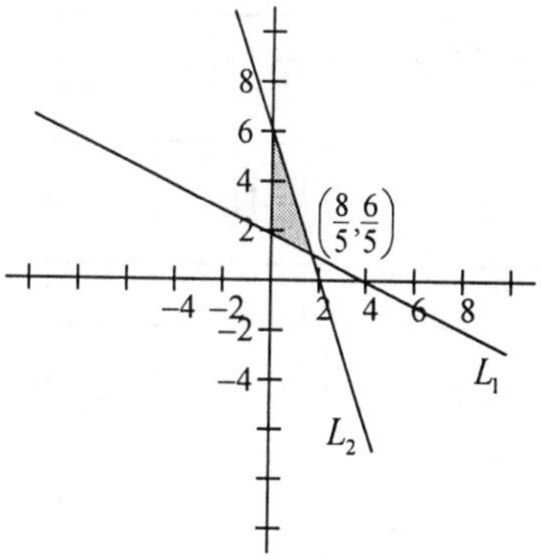


Using $L_1$: $\frac{8}{5}+2y=4;\ 2y=\frac{12}{5};\ y=\frac{6}{5}$. Lines $L_1$ and $L_2$ intersect at $\left(\frac{8}{5},\frac{6}{5}\right)$.

Bounded. Corner points: (0, 2), (0, 6), $\left(\frac{8}{5},\frac{6}{5}\right)$.

**13.** Intercepts:

$L_1: 3x+2y=6 : (0, 3), (2, 0)$

$L_2: 3x+2y=12 : (0, 6), (4, 0)$

$L_3: x+2y=8 : (0, 4), (8, 0)$

Intersection:

$$\begin{aligned} L_2: &\; 3x+2y=12 \\ -L_3: &\; -x-2y=-8 \\ \hline &\quad 2x=4;\ x=2. \end{aligned}$$

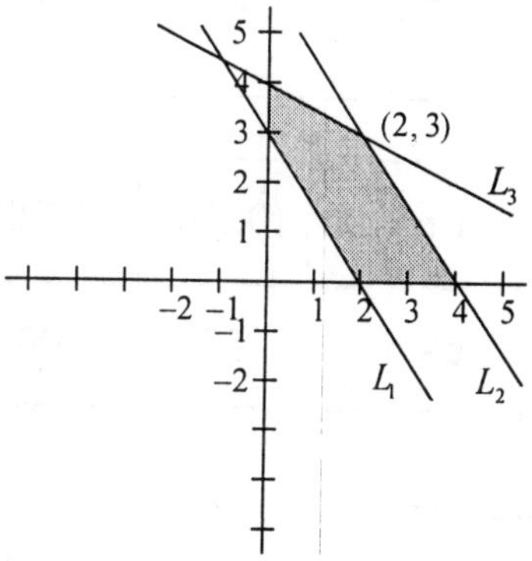


Using $L_3$: $2+2y=8;\ 2y=6;\ y=3$. Lines $L_2$ and $L_3$ intersect at (2, 3).

Bounded. Corner points: (0, 3), (0, 4), (2, 3), (4, 0), (2, 0)

For Problems 15–22, the feasible region is bounded with corner points (0, 10), (0, 20), $\left(\frac{40}{3},\frac{40}{3}\right)$, (20, 0), (10, 0).

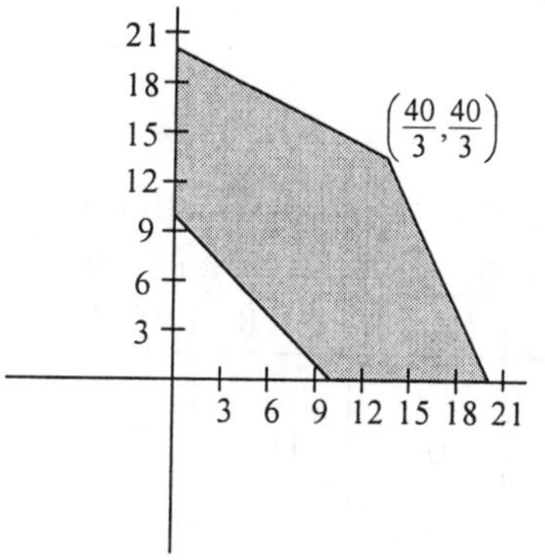


Intercepts:

$L_1: x+2y=40 : (0, 20), (40, 0)$

$L_2: 2x+y=40 : (0, 40), (20, 0)$

$L_3: x+y=10 : (0, 10), (10, 0)$

Intersection:

$$\begin{aligned} L_1: &\quad x+2y=40 \\ -2L_2: &\; -4x-2y=-80 \\ \hline \end{aligned}$$

$$-3x=-40;\ x=\frac{40}{3}.\ \text{Using } L_1: \frac{40}{3}+2y=40;\ 2y=\frac{80}{3};\ y=\frac{40}{3}.$$

Lines $L_1$ and $L_2$ intersect at $\left(\frac{40}{3},\frac{40}{3}\right)$.

**15.**

| corner point: | (0, 10) | (0, 20) | $\left(\frac{40}{3}, \frac{40}{3}\right)$ | (20, 0) | (10, 0) |
|---|---|---|---|---|---|
| $z = x + y$ | 10 | 20 | $\frac{80}{3} = 26\frac{2}{3}$ | 20 | 10 |

maximum $z$-value: $\frac{80}{3}$

**17.**

| corner point: | (0, 10) | (0, 20) | $\left(\frac{40}{3}, \frac{40}{3}\right)$ | (20, 0) | (10, 0) |
|---|---|---|---|---|---|
| $z = 5x + 2y$ | 20 | 40 | $\frac{280}{3} = 93\frac{1}{3}$ | 100 | 50 |

minimum $z$-value: 20

**19.**

| corner point: | (0, 10) | (0, 20) | $\left(\frac{40}{3}, \frac{40}{3}\right)$ | (20, 0) | (10, 0) |
|---|---|---|---|---|---|
| $z = 2x + y$ | 10 | 20 | 40 | 40 | 20 |

maximum $z$-value: 40 (at any point on the line segment joining $\left(\frac{40}{3}, \frac{40}{3}\right)$ and (20, 0).)

**21.**

| corner point: | (0, 10) | (0, 20) | $\left(\frac{40}{3}, \frac{40}{3}\right)$ | (20, 0) | (10, 0) |
|---|---|---|---|---|---|
| $z = 2x + 5y$ | 50 | 100 | $\frac{280}{3} = 93\frac{1}{3}$ | 40 | 20 |

minimum $z$-value: 20

**23.** Intercepts:
$L_1 : x = 5$: (5, 0)
$L_2 : y = 8$: (0, 5)
$L_3 : 3x + 4y = 12$: (0, 3), (4, 0)
Intersection:
$\begin{matrix} x & = 5 \\ y & = 8 \end{matrix}$ : (5, 8)

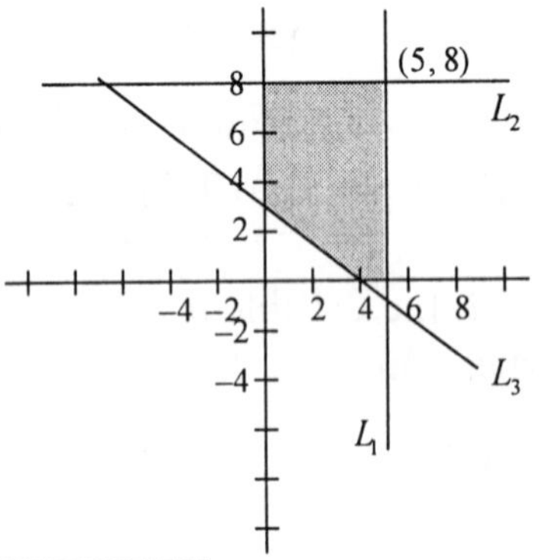


The feasible region is bounded with corner points (0, 3), (0, 8), (5, 8), (5, 0), (4, 0).

| corner point: | (0, 3) | (0, 8) | (5, 8) | (5, 0) | (4, 0) |
|---|---|---|---|---|---|
| $z = 15x + 20y$ | 60 | 160 | 235 | 75 | 60 |

maximum $z$-value: 235
minimum $z$-value: 60
(at any point on the line segment joining (0, 3) and (4, 0).)

**25.** Intercepts:

$L_1 : x = 5$: (5, 0)

$L_2 : y = 6$: (0, 6)

$L_3 : 2x + 3y = 22$: $\left(0, 7\frac{1}{3}\right)$, (11, 0)

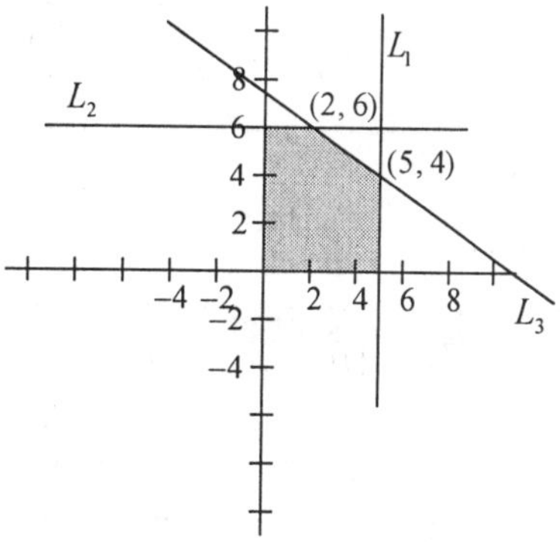


Intersections:

$$\begin{cases} x = 5 \\ 2x + 3y = 22 \end{cases} : (5, 4)$$

$$\begin{cases} 2x + 3y = 22 \\ y = 6 \end{cases} : (2, 6)$$

The feasible region is bounded with corner points (0, 0), (0, 6), (2, 6), (5, 4) and (5, 0).

| corner point: | (0, 0) | (0, 6) | (2, 6) | (5, 4) | (5, 0) |
|---|---|---|---|---|---|
| $z = 15x + 20y$ | 0 | 120 | 150 | 155 | 75 |

maximum $z$-value: 155
minimum $z$-value: 0

**27.** Intercepts:

$L_1 : x = 9 : (9, 0)$

$L_2 : y = 8 : (0, 8)$

$L_3 : x + y = 3 : (0, 3), (3, 0)$

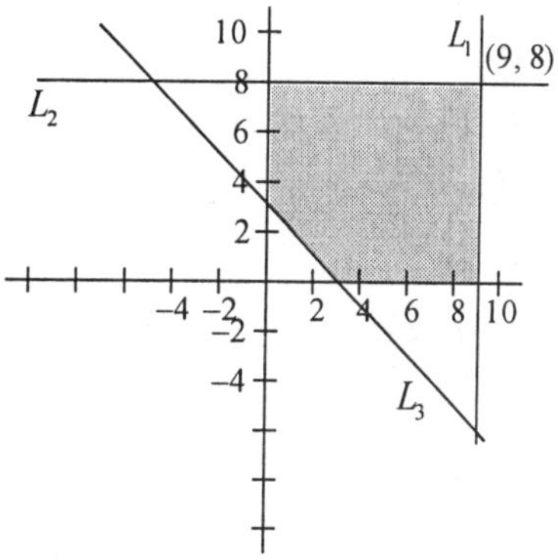


Intersection:

$$\begin{matrix} L_1 : \\ L_2 : \end{matrix} \begin{cases} x = 9 \\ y = 8 \end{cases} : (9,8)$$

The feasible region is bounded with corner points (0, 3), (0, 8), (9, 8), (9, 0) and (3, 0).

| corner point: | (0, 3) | (0, 8) | (9, 8) | (9, 0) | (3, 0) |
|---|---|---|---|---|---|
| $z = 2x + 3y$ | 9 | 24 | 42 | 18 | 6 |

maximum $z$-value: 42

**29.** Intercepts:

$L_1 : x = 8 : (8, 0)$

$L_2 : y = 8 : (0, 8)$

$L_3 : x + y = 1 : (0, 1), (1, 0)$

$L_4 : y = 2x : (0, 0)$

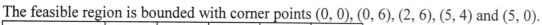

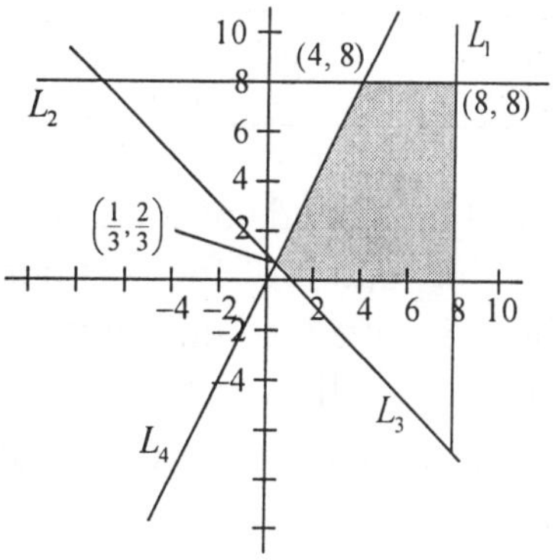


Intersection:

$$\begin{matrix} L_1 : \\ L_2 : \end{matrix} \begin{cases} x = 8 \\ y = 8 \end{cases} : (8, 8) \qquad \begin{matrix} L_2 : \\ L_4 : \end{matrix} \begin{cases} y = 8 \\ y = 2x \end{cases} : (4, 8)$$

$$\begin{aligned} L_3 &: \begin{cases} x+y=1 \\ \underline{\;\;-y=-2x} \end{cases} \\ -L_4 &: \end{aligned}$$

$$x = 1-2x;\; 3x=1;\; x=\frac{1}{3}.\text{ Using } L_3: \frac{1}{3}+y=1;\; y=\frac{2}{3}.$$

Lines $L_3$ and $L_4$ intersect at $\left(\frac{1}{3},\frac{2}{3}\right)$.

The feasible region is bounded with corner points $\left(\frac{1}{3},\frac{2}{3}\right)$,(4, 8), (8, 8), (8, 0) and (1, 0).

| corner point: | $\left(\frac{1}{3},\frac{2}{3}\right)$ | (4, 8) | (8, 8) | (8, 0) | (1, 0) |
|---|---|---|---|---|---|
| $z = x+2y$ | $\frac{5}{3}$ | 20 | 24 | 8 | 1 |

maximum $z$-value: 24

**31.** Intercepts:
$L_1$: $x = 10$: (10, 0)
$L_2$: $y = 8$: (0, 8)
$L_3$: $x + 2y = 8$: (0, 4), (8, 0)
$L_4$: $y = x$ : (0, 0)

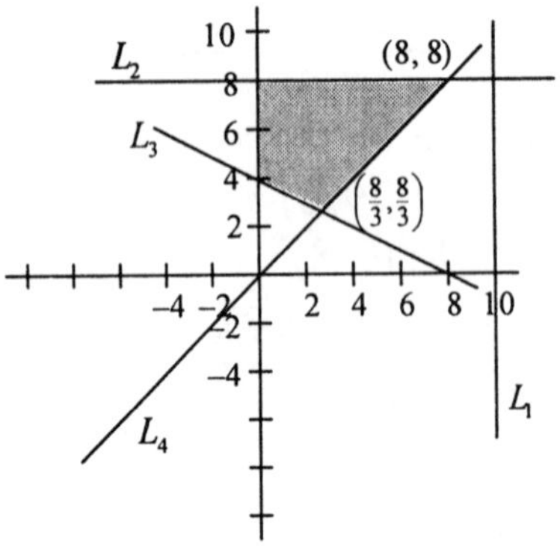


Intersections:

$$\begin{aligned} L_2 &: \begin{cases} y=8 \\ y=x \end{cases} : (8, 8) \\ L_4 &: \end{aligned}$$

$$L_3: \begin{cases} x+2y=8 \\ \quad\;\; y=x \end{cases} : x+2x=8;\; 3x=8;\; x=\frac{8}{3};\; y=\frac{8}{3}.$$
$L_4$: Lines $L_3$ and $L_4$ intersect at $\left(\frac{8}{3},\frac{8}{3}\right)$.

The feasible region is bounded with corner points (0, 4), (0, 8), (8, 8) and $\left(\frac{8}{3},\frac{8}{3}\right)$.

| corner point: | (0, 4) | (0, 8) | (8, 8) | $\left(\frac{8}{3},\frac{8}{3}\right)$ |
|---|---|---|---|---|
| $z = 3x+2y$ | 8 | 16 | 40 | $\frac{40}{3} = 13\frac{1}{3}$ |

Minimum $z$-value: 8

**33.** Let $x$ represent the pounds of food A purchased per month, $y$ the pounds of foods B purchased per month. The objective is to minimize the cost $C = 1.30x + 0.80y$ dollars. The constraints are imposed by the amounts of carbohydrates, protein and fat Katy requires each month:

$$\begin{cases} \text{carbohydrates}: 5x+2y \geq 60 \\ \qquad\text{protein}: 3x+2y \geq 45 \\ \qquad\qquad\text{fat}: 4x+\;\; y \geq 30 \\ x \geq 0,\; y \geq 0 \end{cases}$$

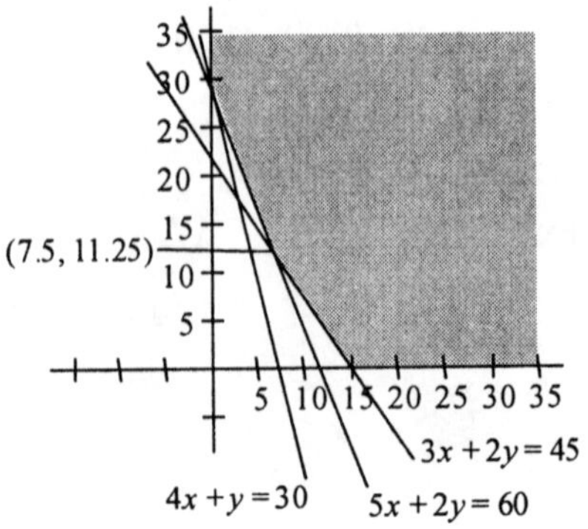

The region described by these constraints is unbounded with corner points (0, 30), $\left(7\frac{1}{2}, 11\frac{1}{4}\right)$ and (15, 0).

| corner point: | (0, 30) | $\left(7\frac{1}{2}, 11\frac{1}{4}\right)$ | (15, 0) |
|---|---|---|---|
| $z = 1.3x + 0.8y$ | 24 | 18.75 | 19.5 |

(Note that the cost increases in the direction of the unboundedness, i.e., for large $x$- and/or large $y$-values.) The minimum cost is \$18.75 per month, when Katy buys $7\frac{1}{2}$ pounds of food A and $11\frac{1}{4}$ pounds of food B.

**35.** Let $x$ and $y$ represent the number of downhill skis and cross-country skis, respectively. The objective is to maximize profit: $P = 70x + 50y$ dollars. The constraints are imposed by the available time for manufacturing and finishing.

$$\begin{cases} \text{manufacturing}: 2x + y \le 40 \\ \text{finishing}: \quad x + y \le 32 \\ x \ge 0, y \ge 0 \end{cases}$$

The region described by these constraints is bounded with corner points (0, 0), (0, 32), (8, 24) and (20, 0).

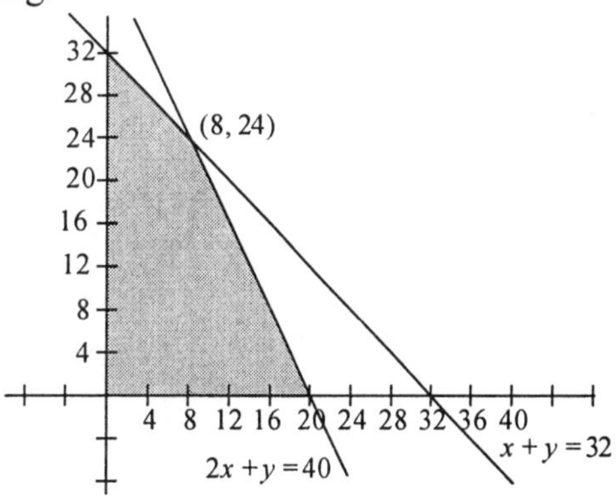


| corner point: | (0, 0) | (0, 32) | (8, 24) | (20, 0) |
|---|---|---|---|---|
| $P = 70x + 50y$ | \$0 | \$1600 | \$1760 | \$1400 |

The maximum profit is \$1,760 when the manufacturer makes 8 downhill skis and 24 cross-country skis.

**37.** Let $x$ and $y$ represent the number of pounds of type I explosive and type II explosive, respectively. The objective is to maximize profit: $P = 60x + 40y$ dollars. The constraints are imposed by the availability of storage and mixing and packaging work-hours.

$$\begin{cases} \text{type I storage}: \quad x \le 100 \\ \text{type II storage}: \quad y \le 150 \\ \text{mixing}: 60x + 40y \le 7200 \text{ or } 3x + 2y \le 360 \\ \text{packaging}: 70x + 40y \le 7800 \text{ or } 7x + 4y \le 780 \end{cases}$$

The region described by these constraints is bounded with corner points (0, 0), (0, 150), (20, 150), (60, 90), (100, 20) and (100, 0).

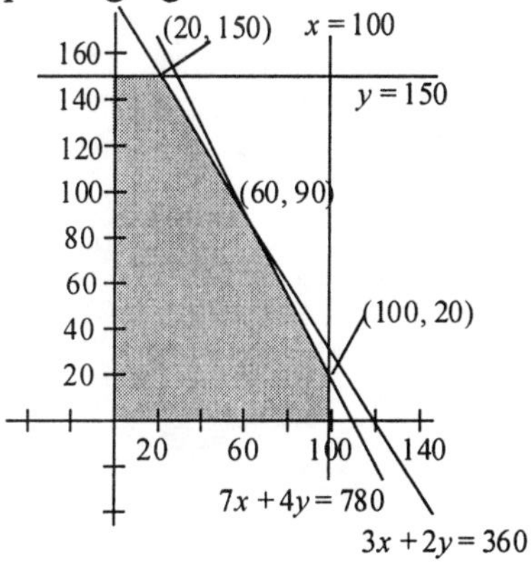

| corner point: | (0, 0) | (0, 150) | (20, 150) | (60, 90) | (100, 20) | (100, 0) |
|---|---|---|---|---|---|---|
| $P = 60x + 40y$ | \$0 | \$6000 | \$7200 | \$7200 | \$6800 | \$6000 |

The maximum profit is \$7,200. It occurs when 20 to 60 pounds of Type I explosives are produced, while 150 to 90 pounds of Type II explosives are produced, satisfying $3x + 2y = 360$.

## Professional Exam Questions

**1.** b

**3.** c

**5.** d

**7.** c

**9.** b

**11.** b; net income $= (\text{income}) - (\text{overhead}) = (27.5A + 75B) - (19A + 51B) = 8.5A + 24B$.

**13.** c

# Chapter 4

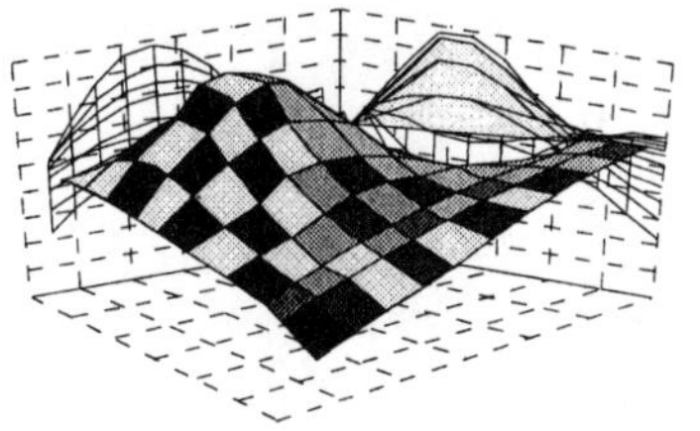

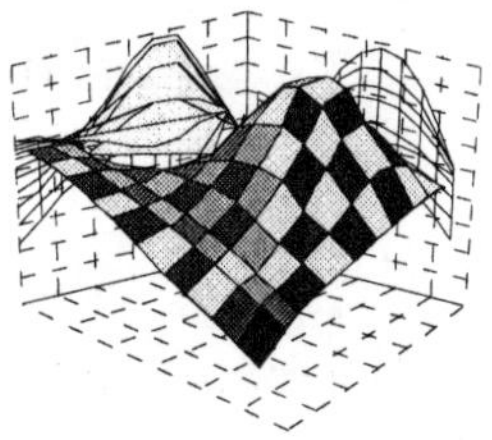

# Linear Programming: Simplex Method

## 4.1 The Simplex Tableau; Pivoting

**1.** In standard form.

**3.** Not in standard form: Condition 1 is violated for $x_2$ and $x_3$.

**5.** Not in standard form: Condition 2 is violated by the second constraint; condition 1 is violated for $x_3$.

**7.** Not in standard form: Condition 2 is violated by the first constraint.

**9.** In standard form.

**11.** The first constraint cannot be modified in a way that puts the problem in standard form.

**13.** The second constraint cannot be modified in a way that puts the problem in standard form.

**15.** Modify the problem to read: Maximize $P = 2x_1 + x_2 + 3x_3$ subject to the constraints

$$\begin{cases} x_1 - x_2 - x_3 \le 6 \\ -2x_1 + 3x_2 \le 12 \\ x_3 \le 2 \\ x_1 \ge 0, x_2 \ge 0, x_3 \ge 0 \end{cases}.$$

**17.** System with slack variables:
$$\begin{cases} P - 2x_1 - x_2 - 3x_3 = 0 \\ 5x_1 + 2x_2 + x_3 + s_1 = 20 \\ 6x_1 + x_2 + 4x_3 + s_2 = 24 \\ x_1 + x_2 + 4x_3 + s_3 = 16 \end{cases}$$

Initial simplex tableau:

| BV | $P$ | $x_1$ | $x_2$ | $x_3$ | $s_1$ | $s_2$ | $s_3$ | RHS |
|---|---|---|---|---|---|---|---|---|
| $s_1$ | 0 | 5 | 2 | 1 | 1 | 0 | 0 | 20 |
| $s_2$ | 0 | 6 | 1 | 4 | 0 | 1 | 0 | 24 |
| $s_3$ | 0 | 1 | 1 | 4 | 0 | 0 | 1 | 16 |
| $P$ | 1 | −2 | −1 | −3 | 0 | 0 | 0 | 0 |

**19.** System with slack variables:
$$\begin{cases} P - 3x_1 - 5x_2 = 0 \\ 2.2x_1 - 1.8x_2 + s_1 = 5 \\ 0.8x_1 + 1.2x_2 + s_2 = 2.5 \\ x_1 + x_2 + s_3 = 0.1 \end{cases}$$

Initial simplex tableau:

| BV | $P$ | $x_1$ | $x_2$ | $s_1$ | $s_2$ | $s_3$ | RHS |
|---|---|---|---|---|---|---|---|
| $s_1$ | 0 | 2.2 | −1.8 | 1 | 0 | 0 | 5 |
| $s_2$ | 0 | 0.8 | 1.2 | 0 | 1 | 0 | 2.5 |
| $s_3$ | 0 | 1 | 1 | 0 | 0 | 1 | 0.1 |
| $P$ | 1 | −3 | −5 | 0 | 0 | 0 | 0 |

**21.** System with slack variables:
$$\begin{cases} P - 2x_1 - 3x_2 - x_3 = 0 \\ x_1 + x_2 + x_3 + s_1 = 50 \\ 3x_1 + 2x_2 + x_3 + s_2 = 10 \end{cases}$$

Initial simplex tableau:

| BV | $P$ | $x_1$ | $x_2$ | $x_3$ | $s_1$ | $s_2$ | RHS |
|---|---|---|---|---|---|---|---|
| $s_1$ | 0 | 1 | 1 | 1 | 1 | 0 | 50 |
| $s_2$ | 0 | 3 | 2 | 1 | 0 | 1 | 10 |
| $P$ | 1 | −2 | −3 | −1 | 0 | 0 | 0 |

**23.** System with slack variables:
$$\begin{cases} P - 3x_1 - 4x_2 - 2x_3 = 0 \\ 3x_1 + 2x_2 + 4x_3 + s_1 = 5 \\ x_1 + x_2 + s_2 = 5 \\ 2x_1 - x_2 + x_3 + s_3 = 6 \end{cases}$$

Initial simplex tableau:

$$\begin{array}{c} BV \\ s_1 \\ s_2 \\ s_3 \\ P \end{array} \begin{array}{c} \begin{array}{ccccccc|c} P & x_1 & x_2 & x_3 & s_1 & s_2 & s_3 & RHS \end{array} \\ \left[\begin{array}{ccccccc|c} 0 & 3 & 1 & 4 & 1 & 0 & 0 & 5 \\ 0 & 1 & 1 & 0 & 0 & 1 & 0 & 5 \\ 0 & 2 & -1 & 1 & 0 & 0 & 1 & 6 \\ \hline 1 & -3 & -4 & -2 & 0 & 0 & 0 & 0 \end{array}\right] \end{array}$$

**25.** Modified version:

Maximize $P = x_1 + 2x_2 + 5x_3$

subject to the constraints:

$x_1 - 2x_2 - 3x_3 \le 10$

$-3x_1 - x_2 + x_3 \le 12$

$x_1 \ge 0, x_2 \ge 0, x_3 \ge 0$

System with slack variables:

$$\begin{aligned} P - x_1 - 2x_2 - 5x_3 \qquad &= 0 \\ x_1 - 2x_2 - 3x_3 + s_1 \qquad &= 10 \\ -3x_1 - x_2 + x_3 \qquad + s_2 &= 12 \end{aligned}$$

Initial Simplex tableau:

$$\begin{array}{c} BV \\ s_1 \\ s_2 \\ P \end{array} \begin{array}{c} \begin{array}{cccccc|c} P & x_1 & x_2 & x_3 & s_1 & s_2 & RHS \end{array} \\ \left[\begin{array}{cccccc|c} 0 & 1 & -2 & -3 & 1 & 0 & 10 \\ 0 & -3 & -1 & 1 & 0 & 1 & 12 \\ \hline 1 & -1 & -2 & -5 & 0 & 0 & 0 \end{array}\right] \end{array}$$

**27.** Modified version:

Maximize $P = 2x_1 + 3x_2 + x_3 + 6x_4$

subject to the constraints:

$-x_1 + x_2 + 2x_3 + x_4 \le 10$

$-x_1 + x_2 - x_3 + x_4 \le 8$

$x_1 + x_2 + x_3 + x_4 \le 9$

$x_1 \ge 0, x_2 \ge 0, x_3 \ge 0, x_4 \ge 0$

System with slack variables:

$$\begin{aligned} P - 2x_1 - 3x_2 - x_3 - 6x_4 \qquad &= 0 \\ -x_1 + x_2 + 2x_3 + x_4 + s_1 \qquad &= 10 \\ -x_1 + x_2 - x_3 + x_4 \qquad + s_2 \qquad &= 8 \\ x_1 + x_2 + x_3 + x_4 \qquad + s_3 &= 9 \end{aligned}$$

Initial Simplex tableau:

$$\begin{array}{c} BV \\ s_1 \\ s_2 \\ s_3 \\ P \end{array} \begin{array}{c} \begin{array}{cccccccc|c} P & x_1 & x_2 & x_3 & x_4 & s_1 & s_2 & s_3 & RHS \end{array} \\ \left[\begin{array}{cccccccc|c} 0 & -1 & 1 & 2 & 1 & 1 & 0 & 0 & 10 \\ 0 & -1 & 1 & -1 & 1 & 0 & 1 & 0 & 8 \\ 0 & 1 & 1 & 1 & 1 & 0 & 0 & 1 & 9 \\ \hline 1 & -2 & -3 & -1 & -6 & 0 & 0 & 0 & 0 \end{array}\right] \end{array}$$

**29.** New tableau:

$$\begin{array}{c} BV \\ x_2 \\ s_2 \\ P \end{array} \begin{array}{c} \begin{array}{ccccc|c} P & x_1 & x_2 & s_1 & s_2 & RHS \end{array} \\ \left[\begin{array}{ccccc|c} 0 & 1/2 & 1 & 1/2 & 0 & 150 \\ 0 & 2 & 0 & -1 & 1 & 180 \\ \hline 1 & 0 & 0 & 1 & 0 & 300 \end{array}\right] \end{array}$$

New system: $\begin{cases} x_2 = -\frac{1}{2}x_1 - \frac{1}{2}s_1 + 150 \\ s_2 = -2x_1 + s_1 + 180 \\ P = -s_1 + 300 \end{cases}$. Current values: $P = 300$, $x_2 = 150$, $s_2 = 180$.

**31.** New tableau:

| BV | P | $x_1$ | $x_2$ | $x_3$ | $s_1$ | $s_2$ | $s_3$ | RHS |
|---|---|---|---|---|---|---|---|---|
| $s_1$ | 0 | −2 | 0 | 0 | 1 | 0 | −1 | 6 |
| $s_2$ | 0 | 5/4 | −3/2 | 0 | 0 | 1 | −1/4 | 55/2 |
| $x_3$ | 0 | 3/4 | 1/2 | 1 | 0 | 0 | 1/4 | 9/2 |
| P | 1 | 5/4 | −1/2 | 0 | 0 | 0 | 3/4 | 27/2 |

New system: $\begin{cases} s_1 = 2x_1 + s_3 + 6 \\ s_2 = -\frac{5}{4}x_1 + \frac{3}{2}x_2 + \frac{1}{4}s_3 + \frac{55}{2} \\ x_3 = -\frac{3}{4}x_1 - \frac{1}{2}x_2 - \frac{1}{4}s_3 + \frac{9}{2} \\ P = -\frac{5}{4}x_1 + \frac{1}{2}x_2 - \frac{3}{4}s_3 + \frac{27}{2} \end{cases}$.

Current values: $P = \frac{27}{2}$, $s_1 = 6$, $s_2 = \frac{55}{2}$, $x_3 = \frac{9}{2}$.

**33.** New tableau:

| BV | P | $x_1$ | $x_2$ | $x_3$ | $x_4$ | $s_1$ | $s_2$ | $s_3$ | $s_4$ | RHS |
|---|---|---|---|---|---|---|---|---|---|---|
| $s_1$ | 0 | −3 | 0 | 1 | 0 | 1 | 0 | 0 | 0 | 20 |
| $x_4$ | 0 | 2 | 0 | 0 | 1 | 0 | 1 | 0 | 0 | 24 |
| $s_3$ | 0 | 0 | −3 | 1 | 0 | 0 | 0 | 1 | 0 | 28 |
| $s_4$ | 0 | −2 | −3 | 0 | 0 | 0 | −1 | 0 | 1 | 0 |
| P | 1 | 7 | −2 | −3 | 0 | 0 | 4 | 0 | 0 | 96 |

New system: $\begin{cases} s_1 = 3x_1 - x_3 + 20 \\ x_4 = -2x_1 - s_2 + 24 \\ s_3 = 3x_2 - x_3 + 28 \\ s_4 = 2x_1 + 3x_2 + s_2 \\ P = -7x_1 + 2x_2 + 3x_3 - 4s_2 + 96 \end{cases}$.

Current values: $P = 96$, $s_1 = 20$, $x_4 = 24$, $s_3 = 28$, $s_4 = 0$.

## Technology Exercises

**1.**
$$\begin{array}{c} BV \\ s_1 \\ x_2 \\ P \end{array}\begin{array}{c} \begin{array}{cccccc} P & x_1 & x_2 & s_1 & s_2 & RHS \end{array} \\ \left[\begin{array}{ccccc|c} 0 & \frac{7}{3} & 0 & 1 & -\frac{2}{3} & 100 \\ 0 & \frac{1}{3} & 1 & 0 & \frac{1}{3} & 50 \\ \hline 1 & -1 & 0 & 0 & 1 & 150 \end{array}\right]\end{array}$$

**3.**
$$\begin{array}{c} BV \\ x_2 \\ s_2 \\ P \end{array}\begin{array}{c} \begin{array}{cccccc} P & x_1 & x_2 & s_1 & s_2 & RHS \end{array} \\ \left[\begin{array}{ccccc|c} 0 & \frac{1}{4} & 1 & \frac{1}{4} & 0 & 25 \\ 0 & \frac{3}{4} & 0 & -\frac{5}{4} & 1 & 10 \\ \hline 1 & -\frac{5}{4} & 0 & \frac{3}{4} & 0 & 75 \end{array}\right]\end{array}$$

## 4.2 The Simplex Method: Solving Maximum Problems in Standard Form

**1.** b; It requires further pivoting. The pivot element is the entry 1 in the row corresponding to $s_1$, and the column corresponding to $x_1$.

**3.** a; It is a final tableau. The solution is $P = \frac{256}{7}, x_1 = \frac{32}{7}, x_2 = 0$.

**5.** c; The problem is unbounded and there is no solution.
(See the column corresponding to $x_2$.)

**7.** b; It requires further pivoting. The pivot element is the entry 1 in the row corresponding to $x_1$, and the column corresponding to $s_1$.

**9.**
$$\begin{array}{c} BV \\ s_1 \\ s_2 \\ P \end{array}\begin{array}{c} \begin{array}{cccccc} P & x_1 & x_2 & s_1 & s_2 & RHS \end{array} \\ \left[\begin{array}{ccccc|c} 0 & 2 & \textcircled{3} & 1 & 0 & 12 \\ 0 & 3 & 1 & 0 & 1 & 12 \\ \hline 1 & -5 & -7 & 0 & 0 & 0 \end{array}\right]\end{array} \rightarrow \begin{array}{c} BV \\ x_2 \\ s_2 \\ P \end{array}\begin{array}{c} \begin{array}{cccccc} P & x_1 & x_2 & s_1 & s_2 & RHS \end{array} \\ \left[\begin{array}{ccccc|c} 0 & 2/3 & 1 & 1/3 & 0 & 4 \\ 0 & \textcircled{7/3} & 0 & -1/3 & 1 & 8 \\ \hline 1 & -1/3 & 0 & 7/3 & 0 & 28 \end{array}\right]\end{array}$$

$$\rightarrow \begin{array}{c} BV \\ x_2 \\ x_1 \\ P \end{array}\begin{array}{c} \begin{array}{cccccc} P & x_1 & x_2 & s_1 & s_2 & RHS \end{array} \\ \left[\begin{array}{ccccc|c} 0 & 0 & 1 & 3/7 & -2/7 & 12/7 \\ 0 & 1 & 0 & -1/7 & 3/7 & 24/7 \\ \hline 1 & 0 & 0 & 16/7 & 1/7 & 204/7 \end{array}\right]\end{array}.$$

The maximum value for $P$ is $\frac{204}{7} = 29\frac{1}{7}$, obtained when $x_1 = \frac{24}{7}$ and $x_2 = \frac{12}{7}$.

**11.**
$$\begin{array}{c} BV \\ s_1 \\ s_2 \\ P \end{array}\begin{array}{c} \begin{array}{cccccc} P & x_1 & x_2 & s_1 & s_2 & RHS \end{array} \\ \left[\begin{array}{ccccc|c} 0 & 1 & \textcircled{2} & 1 & 0 & 2 \\ 0 & 2 & 1 & 0 & 1 & 2 \\ \hline 1 & -5 & -7 & 0 & 0 & 0 \end{array}\right]\end{array} \rightarrow \begin{array}{c} BV \\ x_2 \\ s_2 \\ P \end{array}\begin{array}{c} \begin{array}{cccccc} P & x_1 & x_2 & s_1 & s_2 & RHS \end{array} \\ \left[\begin{array}{ccccc|c} 0 & 1/2 & 1 & 1/2 & 0 & 1 \\ 0 & \textcircled{3/2} & 0 & -1/2 & 1 & 1 \\ \hline 1 & -3/2 & 0 & 7/2 & 0 & 7 \end{array}\right]\end{array} \rightarrow$$

$$\begin{array}{c} BV \\ x_2 \\ x_1 \\ P \end{array}\begin{array}{c} \begin{array}{cccccc} P & x_1 & x_2 & s_1 & s_2 & RHS \end{array} \\ \left[\begin{array}{ccccc|c} 0 & 0 & 1 & 2/3 & -1/3 & 2/3 \\ 0 & 1 & 0 & -1/3 & 2/3 & 2/3 \\ \hline 1 & 0 & 0 & 3 & 1 & 8 \end{array}\right]. \end{array}$$

The maximum value for $P$ is 8, obtained when $x_1 = \frac{2}{3}$ and $x_2 = \frac{2}{3}$.

**13.**

$$\begin{array}{c} BV \\ s_1 \\ s_2 \\ s_3 \\ P \end{array}\begin{array}{c} \begin{array}{cccccccc} P & x_1 & x_2 & s_1 & s_2 & s_3 & RHS \end{array} \\ \left[\begin{array}{cccccc|c} 0 & \textcircled{1} & 1 & 1 & 0 & 0 & 2 \\ 0 & 2 & 3 & 0 & 1 & 0 & 12 \\ 0 & 3 & 1 & 0 & 0 & 1 & 12 \\ \hline 1 & -3 & -1 & 0 & 0 & 0 & 0 \end{array}\right] \end{array} \to \begin{array}{c} BV \\ x_1 \\ s_2 \\ s_3 \\ P \end{array}\begin{array}{c} \begin{array}{cccccccc} P & x_1 & x_2 & s_1 & s_2 & s_3 & RHS \end{array} \\ \left[\begin{array}{cccccc|c} 0 & 1 & 1 & 1 & 0 & 0 & 2 \\ 0 & 0 & 1 & -2 & 1 & 0 & 8 \\ 0 & 0 & -2 & -3 & 0 & 1 & 6 \\ \hline 1 & 0 & 2 & 3 & 0 & 0 & 6 \end{array}\right]. \end{array}$$

The maximum value for $P$ is 6, obtained when $x_1 = 2$ and $x_2 = 0$.

**15.**

$$\begin{array}{c} BV \\ s_1 \\ s_2 \\ P \end{array}\begin{array}{c} \begin{array}{cccccccc} P & x_1 & x_2 & x_3 & s_1 & s_2 & RHS \end{array} \\ \left[\begin{array}{cccccc|c} 0 & -2 & 1 & -2 & 1 & 0 & 4 \\ 0 & \textcircled{1} & -2 & 1 & 0 & 1 & 2 \\ \hline 1 & -2 & -1 & -1 & 0 & 0 & 0 \end{array}\right] \end{array} \to \begin{array}{c} BV \\ s_1 \\ x_1 \\ P \end{array}\begin{array}{c} \begin{array}{cccccccc} P & x_1 & x_2 & x_3 & s_1 & s_2 & RHS \end{array} \\ \left[\begin{array}{cccccc|c} 0 & 0 & -3 & 0 & 1 & 2 & 8 \\ 0 & 1 & -2 & 1 & 0 & 1 & 2 \\ \hline 1 & 0 & -5 & 1 & 0 & 2 & 4 \end{array}\right]. \end{array}$$

The problem is unbounded and has no solution.

**17.**

$$\begin{array}{c} BV \\ s_1 \\ s_2 \\ P \end{array}\begin{array}{c} \begin{array}{cccccccc} P & x_1 & x_2 & x_3 & s_1 & s_2 & RHS \end{array} \\ \left[\begin{array}{cccccc|c} 0 & 1 & 2 & 1 & 1 & 0 & 25 \\ 0 & 3 & 2 & \textcircled{3} & 0 & 1 & 30 \\ \hline 1 & -2 & -1 & -3 & 0 & 0 & 0 \end{array}\right] \end{array} \to \begin{array}{c} BV \\ s_1 \\ x_3 \\ P \end{array}\begin{array}{c} \begin{array}{cccccccc} P & x_1 & x_2 & x_3 & s_1 & s_2 & RHS \end{array} \\ \left[\begin{array}{cccccc|c} 0 & 0 & 4/3 & 0 & 1 & -1/3 & 15 \\ 0 & 1 & 2/3 & 1 & 0 & 1/3 & 10 \\ \hline 1 & 1 & 1 & 0 & 0 & 1 & 30 \end{array}\right]. \end{array}$$

The maximum value for $P$ is 30, obtained when $x_1 = 0$, $x_2 = 0$ and $x_3 = 10$.

**19.**

$$\begin{array}{c} BV \\ s_1 \\ s_2 \\ s_3 \\ P \end{array}\begin{array}{c} \begin{array}{cccccccccc} P & x_1 & x_2 & x_3 & x_4 & s_1 & s_2 & s_3 & RHS \end{array} \\ \left[\begin{array}{cccccccc|c} 0 & 2 & 1 & 2 & 3 & 1 & 0 & 0 & 12 \\ 0 & 0 & \textcircled{2} & 1 & 2 & 0 & 1 & 0 & 20 \\ 0 & 2 & 1 & 4 & 0 & 0 & 0 & 1 & 16 \\ \hline 1 & -2 & -4 & -1 & -1 & 0 & 0 & 0 & 0 \end{array}\right] \end{array} \to$$

$$\begin{array}{c} BV \\ s_1 \\ x_2 \\ s_3 \\ P \end{array}\begin{array}{c} \begin{array}{cccccccccc} P & x_1 & x_2 & x_3 & x_4 & s_1 & s_2 & s_3 & RHS \end{array} \\ \left[\begin{array}{cccccccc|c} 0 & \textcircled{2} & 0 & 3/2 & 2 & 1 & -1/2 & 0 & 2 \\ 0 & 0 & 1 & 1/2 & 1 & 0 & 1/2 & 0 & 10 \\ 0 & 2 & 0 & 7/2 & -1 & 0 & -1/2 & 1 & 6 \\ \hline 1 & -2 & 0 & 1 & 3 & 0 & 2 & 0 & 40 \end{array}\right] \end{array} \to$$

$$\begin{array}{c} BV \\ x_1 \\ x_2 \\ s_3 \\ P \end{array} \begin{array}{c} \begin{array}{ccccccccc} P & x_1 & x_2 & x_3 & x_4 & s_1 & s_2 & s_3 & RHS \end{array} \\ \left[\begin{array}{cccccccc|c} 0 & 1 & 0 & 3/4 & 1 & 1/2 & -1/4 & 0 & 1 \\ 0 & 0 & 1 & 1/2 & 1 & 0 & 1/2 & 0 & 10 \\ 0 & 0 & 0 & 2 & -3 & -1 & 0 & 1 & 4 \\ \hline 1 & 0 & 0 & 5/2 & 5 & 1 & 3/2 & 0 & 42 \end{array}\right] \end{array}.$$

The maximum value for $P$ is 42, obtained when $x_1 = 1$, $x_2 = 10$, $x_3 = 0$ and $x_4 = 0$.

**21.**
$$\begin{array}{c} BV \\ s_1 \\ s_2 \\ s_3 \\ P \end{array} \begin{array}{c} \begin{array}{cccccccc} P & x_1 & x_2 & x_3 & s_1 & s_2 & s_3 & RHS \end{array} \\ \left[\begin{array}{ccccccc|c} 0 & \textcircled{1} & 2 & 4 & 1 & 0 & 0 & 20 \\ 0 & 2 & 4 & 4 & 0 & 1 & 0 & 60 \\ 0 & 3 & 4 & 1 & 0 & 0 & 1 & 90 \\ \hline 1 & -2 & -1 & -1 & 0 & 0 & 0 & 0 \end{array}\right] \end{array} \rightarrow \begin{array}{c} BV \\ x_1 \\ s_2 \\ s_3 \\ P \end{array} \begin{array}{c} \begin{array}{cccccccc} P & x_1 & x_2 & x_3 & s_1 & s_2 & s_3 & RHS \end{array} \\ \left[\begin{array}{ccccccc|c} 0 & 1 & 2 & 4 & 1 & 0 & 0 & 20 \\ 0 & 0 & 0 & -4 & -2 & 1 & 0 & 20 \\ 0 & 0 & -2 & -11 & -3 & 0 & 1 & 30 \\ \hline 1 & 0 & 3 & 7 & 2 & 0 & 0 & 40 \end{array}\right] \end{array}.$$

The maximum value for $P$ is 40, obtained when $x_1 = 20$, $x_2 = 0$ and $x_3 = 0$.

**23.**
$$\begin{array}{c} BV \\ s_1 \\ s_2 \\ P \end{array} \begin{array}{c} \begin{array}{cccccccc} P & x_1 & x_2 & x_3 & x_4 & s_1 & s_2 & RHS \end{array} \\ \left[\begin{array}{ccccccc|c} 0 & 5 & 0 & \textcircled{4} & 6 & 1 & 0 & 20 \\ 0 & 4 & 2 & 2 & 8 & 0 & 1 & 40 \\ \hline 1 & -1 & -2 & -4 & 1 & 0 & 0 & 0 \end{array}\right] \end{array} \rightarrow \begin{array}{c} BV \\ x_3 \\ s_2 \\ P \end{array} \begin{array}{c} \begin{array}{cccccccc} P & x_1 & x_2 & x_3 & x_4 & s_1 & s_2 & RHS \end{array} \\ \left[\begin{array}{ccccccc|c} 0 & 5/4 & 0 & 1 & 3/2 & 1/4 & 0 & 5 \\ 0 & 3/2 & \textcircled{2} & 0 & 5 & -1/2 & 1 & 30 \\ \hline 1 & 4 & -2 & 0 & 7 & 1 & 0 & 20 \end{array}\right] \end{array}$$

$$\rightarrow \begin{array}{c} BV \\ x_3 \\ x_2 \\ P \end{array} \begin{array}{c} \begin{array}{cccccccc} P & x_1 & x_2 & x_3 & x_4 & s_1 & s_2 & RHS \end{array} \\ \left[\begin{array}{ccccccc|c} 0 & 5/4 & 0 & 1 & 3/2 & 1/4 & 0 & 5 \\ 0 & 3/4 & 1 & 0 & 5/2 & -1/4 & 1/2 & 15 \\ \hline 1 & 11/2 & 0 & 0 & 12 & 1/2 & 1 & 50 \end{array}\right] \end{array}.$$

The maximum value for $P$ is 50, obtained when $x_1 = 0$, $x_2 = 15$, $x_3 = 5$ and $x_4 = 0$.

**25.** Let $x_1$, $x_2$, $x_3$ represent the number of jeans produced of types I, II, III, respectively. The objective is to maximize the profit $P = 3x_1 + 4.5x_2 + 6x_3$ dollars. The constraints are imposed by the time required for the three manufacturing phases:

$$\begin{cases} \text{cutting}: \ 8x_1 + 12x_2 + 18x_3 \le 5200 \\ \text{sewing}: 12x_1 + 18x_2 + 24x_3 \le 6000 \\ \text{finishing}: \ 4x_1 + \ 8x_2 + 12x_3 \le 2200 \\ \qquad x_1 \ge 0, x_2 \ge 0, x_3 \ge 0 \end{cases}.$$

$$\begin{array}{c} BV \\ s_1 \\ s_2 \\ s_3 \\ P \end{array} \begin{array}{c} \\ \\ \\ \\ \\ \end{array} \left[\begin{array}{ccccccc|c} P & x_1 & x_2 & x_3 & s_1 & s_2 & s_3 & RHS \\ 0 & 8 & 12 & 18 & 1 & 0 & 0 & 5200 \\ 0 & 12 & 18 & 24 & 0 & 1 & 0 & 6000 \\ 0 & 4 & 8 & \textcircled{12} & 0 & 0 & 1 & 2200 \\ \hline 1 & -3 & -4.5 & -6 & 0 & 0 & 0 & 0 \end{array}\right] \rightarrow$$

$$\begin{array}{c} BV \\ s_1 \\ s_2 \\ x_3 \\ P \end{array} \left[\begin{array}{ccccccc|c} P & x_1 & x_2 & x_3 & s_1 & s_2 & s_3 & RHS \\ 0 & 2 & 0 & 0 & 1 & 0 & -\frac{3}{2} & 1900 \\ 0 & \textcircled{4} & 2 & 0 & 0 & 1 & -2 & 1600 \\ 0 & \frac{1}{3} & \frac{2}{3} & 1 & 0 & 0 & \frac{1}{12} & \frac{550}{3} \\ \hline 1 & -1 & -\frac{1}{2} & 0 & 0 & 0 & \frac{1}{2} & 1100 \end{array}\right] \rightarrow \begin{array}{c} BV \\ s_1 \\ x_1 \\ x_3 \\ P \end{array} \left[\begin{array}{ccccccc|c} P & x_1 & x_2 & x_3 & s_1 & s_2 & s_3 & RHS \\ 0 & 0 & -1 & 0 & 1 & -\frac{1}{2} & -\frac{1}{2} & 1100 \\ 0 & 1 & \frac{1}{2} & 0 & 0 & \frac{1}{4} & -\frac{1}{2} & 400 \\ 0 & 0 & \frac{1}{2} & 1 & 0 & -\frac{1}{12} & \frac{1}{4} & 50 \\ \hline 1 & 0 & 0 & 0 & 0 & \frac{1}{4} & 0 & 1500 \end{array}\right].$$

$P = 1500, x_1 = 400, x_2 = 0, x_3 = 50$.

A maximum profit of \$1500 is realized when 400 of Jean I, none of Jean II and 50 of Jean III are manufactured.

**27.** Let $x_1$, $x_2$ and $x_3$, represent the number of units of products A, B, and C, respectively. The objective is to maximize profit $P = x_1 + x_2 + 2x_3$ dollars. The constraints are imposed by the salesperson's cost, selling time, and delivery expense.

$$\begin{cases} \text{cost:} \quad 3x_1 + 5x_2 + 4x_3 \le 500 \text{ dollars} \\ \text{selling time:} \; 10x_1 + 15x_2 + 12x_3 \le 1800 \text{ minutes} \\ \text{delivery expenses:} \quad .5x_1 + x_3 \le 75 \text{ dollars} \\ x_1 \ge 0, x_2 \ge 0, x_3 \ge 0 \end{cases}$$

$$\begin{array}{c} BV \\ s_1 \\ s_2 \\ s_3 \\ P \end{array} \left[\begin{array}{ccccccc|c} P & x_1 & x_2 & x_3 & s_1 & s_2 & s_3 & RHS \\ 0 & 3 & 5 & 4 & 1 & 0 & 0 & 500 \\ 0 & 10 & 15 & 12 & 0 & 1 & 0 & 1800 \\ 0 & .5 & 0 & \textcircled{1} & 0 & 0 & 1 & 75 \\ \hline 1 & -1 & -1 & -2 & 0 & 0 & 0 & 0 \end{array}\right] \rightarrow \begin{array}{c} BV \\ s_1 \\ s_2 \\ x_3 \\ P \end{array} \left[\begin{array}{ccccccc|c} P & x_1 & x_2 & x_3 & s_1 & s_2 & s_3 & RHS \\ 0 & 1 & \textcircled{5} & 0 & 1 & 0 & -4 & 200 \\ 0 & 4 & 15 & 0 & 0 & 1 & -12 & 900 \\ 0 & .5 & 0 & 1 & 0 & 0 & 1 & 75 \\ \hline 1 & 0 & -1 & 0 & 0 & 0 & 2 & 150 \end{array}\right]$$

$$\rightarrow \begin{array}{c} BV \\ x_2 \\ s_2 \\ x_3 \\ P \end{array} \left[\begin{array}{ccccccc|c} P & x_1 & x_2 & x_3 & s_1 & s_2 & s_3 & RHS \\ 0 & \frac{1}{5} & 1 & 0 & \frac{1}{5} & 0 & -\frac{4}{5} & 40 \\ 0 & 1 & 0 & 0 & -3 & 1 & 0 & 300 \\ 0 & .5 & 0 & 1 & 0 & 0 & 1 & 75 \\ \hline 1 & \frac{1}{5} & 0 & 0 & \frac{1}{5} & 0 & \frac{6}{5} & 190 \end{array}\right].$$

$P = 190, x_1 = 0, x_2 = 40, x_3 = 75$.

The maximum profit of \$190 is realized when the salesperson sells none of product A, 40 units of product B, and 75 units of product C.

**29.** Let $x_1$, $x_2$, and $x_3$ represent the number of gallons of regular, premium and super premium grade gasoline, respectively. The objective is to maximize revenue $P = 1.2x_1 + 1.3x_2 + 1.4x_3$ dollars. The constraints are imposed by the availability of high and low octane gasoline and the total number of gallons that can be mixed.

$$\begin{cases} \text{high octane}: .6x_1 + .7x_2 + .8x_3 \le 140{,}000 \text{ gallons} \\ \text{low octane}: .4x_1 + .3x_2 + .2x_3 \le 120{,}000 \text{ gallons} \\ \text{total mixed}: \quad x_1 + x_2 + x_3 \le 225{,}000 \text{ gallons} \\ x_1 \ge 0, x_2 \ge 0, x_3 \ge 0 \end{cases}$$

$$\begin{array}{c} BV \\ s_1 \\ s_2 \\ s_3 \\ P \end{array} \begin{array}{c} \begin{array}{cccccccc} P & x_1 & x_2 & x_3 & s_1 & s_2 & s_3 & RHS \end{array} \\ \left[\begin{array}{ccccccc|c} 0 & .6 & .7 & \textcircled{.8} & 1 & 0 & 0 & 140{,}000 \\ 0 & .4 & .3 & .2 & 0 & 1 & 0 & 120{,}000 \\ 0 & 1 & 1 & 1 & 0 & 0 & 1 & 225{,}000 \\ \hline 1 & -1.2 & -1.3 & -1.4 & 0 & 0 & 0 & 0 \end{array}\right] \end{array} \rightarrow$$

$$\begin{array}{c} BV \\ x_3 \\ s_2 \\ s_3 \\ P \end{array} \begin{array}{c} \begin{array}{cccccccc} P & x_1 & x_2 & x_3 & s_1 & s_2 & s_3 & RHS \end{array} \\ \left[\begin{array}{ccccccc|c} 0 & .75 & .875 & 1 & 1.25 & 0 & 0 & 175{,}000 \\ 0 & .25 & .125 & 0 & -.25 & 1 & 0 & 85{,}000 \\ 0 & \textcircled{.25} & .125 & 0 & -1.25 & 0 & 1 & 50{,}000 \\ \hline 1 & -.15 & -.075 & 0 & 1.75 & 0 & 0 & 245{,}000 \end{array}\right] \end{array} \rightarrow$$

$$\begin{array}{c} BV \\ x_3 \\ s_2 \\ x_1 \\ P \end{array} \begin{array}{c} \begin{array}{cccccccc} P & x_1 & x_2 & x_3 & s_1 & s_2 & s_3 & RHS \end{array} \\ \left[\begin{array}{ccccccc|c} 0 & 0 & .5 & 1 & 5 & 0 & -3 & 25{,}000 \\ 0 & 0 & 0 & 0 & 1 & 1 & -1 & 35{,}000 \\ 0 & 1 & .5 & 0 & -5 & 0 & 4 & 200{,}000 \\ \hline 1 & 0 & 0 & 0 & 1 & 0 & .6 & 275{,}000 \end{array}\right] \end{array}.$$

$P = 275{,}000, x_1 = 200{,}000, x_2 = 0, x_3 = 25{,}000$.

The maximum revenue is \$275,000 when 200,000 gallons of regular, no gallons of premium, and 25,000 gallons of super premium gasoline are mixed.

**31.** Let $x_1$, $x_2$, and $x_3$ represent the amounts invested in stocks, corporate bonds, and municipal bonds, respectively. The objective is to maximize the return on investments: $P = .10x_1 + .08x_2 + .06x_3$. The constraints are imposed by the consultant's available funds and her investment strategy.

$$\begin{cases} x_1 + x_2 + x_3 \le \$90{,}000: \text{total funds available} \\ x_1 \le \$45{,}000: \text{stock investment not to exceed } \frac{1}{2} \text{ funds} \\ 2x_2 \le x_3: \text{twice investment in corporate bonds not to exceed investment in municipal bonds} \end{cases}$$

$$\begin{array}{c} BV \\ s_1 \\ s_2 \\ s_3 \\ P \end{array} \begin{array}{c} \begin{array}{cccccccc} P & x_1 & x_2 & x_3 & s_1 & s_2 & s_3 & RHS \end{array} \\ \left[\begin{array}{ccccccc|c} 0 & 1 & 1 & 1 & 1 & 0 & 0 & 90{,}000 \\ 0 & \textcircled{1} & 0 & 0 & 0 & 1 & 0 & 45{,}000 \\ 0 & 0 & 2 & -1 & 0 & 0 & 1 & 0 \\ \hline 1 & -.1 & -.08 & -.06 & 0 & 0 & 0 & 0 \end{array}\right] \end{array} \rightarrow$$

$$
\begin{array}{c} BV \\ s_1 \\ x_1 \\ s_3 \\ P \end{array}
\begin{array}{c} \\ \left[\begin{array}{ccccccc|c} P & x_1 & x_2 & x_3 & s_1 & s_2 & s_3 & RHS \\ 0 & 0 & 1 & 1 & 1 & -1 & 0 & 45{,}000 \\ 0 & 1 & 0 & 0 & 0 & 1 & 0 & 45{,}000 \\ 0 & 0 & \textcircled{2} & -1 & 0 & 0 & 1 & 0 \\ \hline 1 & 0 & -.08 & -.06 & 0 & .1 & 0 & 4{,}500 \end{array}\right] \rightarrow \end{array}
$$

$$
\begin{array}{c} BV \\ s_1 \\ x_1 \\ x_2 \\ P \end{array}
\left[\begin{array}{ccccccc|c} P & x_1 & x_2 & x_3 & s_1 & s_2 & s_3 & RHS \\ 0 & 0 & 0 & \textcircled{1.5} & 1 & -1 & -.5 & 45{,}000 \\ 0 & 1 & 0 & 0 & 0 & 1 & 0 & 45{,}000 \\ 0 & 0 & 1 & -.5 & 0 & 0 & .5 & 0 \\ \hline 1 & 0 & 0 & -.1 & 0 & .1 & .04 & 4{,}500 \end{array}\right] \rightarrow
$$

$$
\begin{array}{c} BV \\ x_3 \\ x_1 \\ x_2 \\ P \end{array}
\left[\begin{array}{ccccccc|c} P & x_1 & x_2 & x_3 & s_1 & s_2 & s_3 & RHS \\ 0 & 0 & 0 & 1 & \frac{2}{3} & -\frac{2}{3} & -\frac{1}{3} & 30{,}000 \\ 0 & 1 & 0 & 0 & 0 & 1 & 0 & 45{,}000 \\ 0 & 0 & 1 & 0 & \frac{1}{3} & -\frac{1}{3} & \frac{1}{3} & 15{,}000 \\ \hline 1 & 0 & 0 & 0 & \frac{1}{15} & \frac{1}{30} & \frac{1}{150} & 7{,}500 \end{array}\right].
$$

$P = 7{,}500, x_1 = 45{,}000, x_2 = 15{,}000, x_3 = 30{,}000$.

The maximum return on investments is \$7,500 when \$45,000 is invested in stocks, \$15,000 is invested in corporate bonds, and \$30,000 is invested in municipal bonds.

**33.** Let $x_1$, $x_2$, and $x_3$ represent the number of acres of crops A, B, and C, respectively. The objective is to maximize profit $P = 70x_1 + 90x_2 + 50x_3$ dollars. The constraints are imposed by the acres of land available, the cost for cultivation, and the hours of labor available.

$$
\begin{cases} \text{acres of land}: & x_1 + x_2 + x_3 \le 200 \\ \text{cost of cultivation}: & 40x_1 + 50x_2 + 30x_3 \le 18{,}000 \\ \text{hours of labor}: & 20x_1 + 30x_2 + 15x_3 \le 4200 \end{cases}
$$

$$
\begin{array}{c} BV \\ s_1 \\ s_2 \\ s_3 \\ P \end{array}
\left[\begin{array}{ccccccc|c} P & x_1 & x_2 & x_3 & s_1 & s_2 & s_3 & RHS \\ 0 & 1 & 1 & 1 & 1 & 0 & 0 & 200 \\ 0 & 40 & 50 & 30 & 0 & 1 & 0 & 18{,}000 \\ 0 & 20 & \textcircled{30} & 15 & 0 & 0 & 1 & 4{,}200 \\ \hline 1 & -70 & -90 & -50 & 0 & 0 & 0 & 0 \end{array}\right] \rightarrow
$$

$$
\begin{array}{c} BV \\ s_1 \\ s_2 \\ x_2 \\ P \end{array}
\left[\begin{array}{ccccccc|c} P & x_1 & x_2 & x_3 & s_1 & s_2 & s_3 & RHS \\ 0 & \textcircled{\frac{1}{3}} & 0 & \frac{1}{2} & 1 & 0 & -\frac{1}{30} & 60 \\ 0 & \frac{20}{3} & 0 & 5 & 0 & 1 & -\frac{5}{3} & 11{,}000 \\ 0 & \frac{2}{3} & 1 & \frac{1}{2} & 0 & 0 & \frac{1}{30} & 140 \\ \hline 1 & -10 & 0 & -5 & 0 & 0 & 3 & 12{,}600 \end{array}\right] \rightarrow
$$

| BV | P | $x_1$ | $x_2$ | $x_3$ | $s_1$ | $s_2$ | $s_3$ | RHS |
|---|---|---|---|---|---|---|---|---|
| $x_1$ | 0 | 1 | 0 | $\frac{3}{2}$ | 3 | 0 | $-\frac{1}{10}$ | 180 |
| $s_2$ | 0 | 0 | 0 | $-5$ | $-20$ | 1 | $-1$ | 9,800 |
| $x_2$ | 0 | 0 | 1 | $-\frac{1}{2}$ | $-2$ | 0 | $\frac{1}{10}$ | 20 |
| $P$ | 1 | 0 | 0 | 10 | 30 | 0 | 2 | 14,400 |

$P = 14{,}400, x_1 = 180, x_2 = 20, x_3 = 0$.

The maximum profit is \$14,400 when 180 acres of crop A, 20 acres of crop B, and no acres of crop C are planted.

**35.** Let $x_1$, $x_2$ and $x_3$ represent the number of cans of type I, II and III, respectively. The objective is to maximize revenue $P = 28x_1 + 24x_2 + 20x_3$ dollars. The constraints are imposed by the available pounds of the various nuts:

$$\begin{cases} \text{peanuts}: 3x_1 + 4x_2 + 5x_3 \le 500 \\ \text{pecans}: x_1 + 1/2x_2 \le 100 \\ \text{cashews}: x_1 + 1/2x_2 \le 50 \\ x_1 \ge 0, x_2 \ge 0, x_3 \ge 0 \end{cases}$$

| BV | P | $x_1$ | $x_2$ | $x_3$ | $s_1$ | $s_2$ | $s_3$ | RHS |
|---|---|---|---|---|---|---|---|---|
| $s_1$ | 0 | 3 | 4 | 5 | 1 | 0 | 0 | 500 |
| $s_2$ | 0 | 1 | $\frac{1}{2}$ | 0 | 0 | 1 | 0 | 100 |
| $s_3$ | 0 | (1) | $\frac{1}{2}$ | 0 | 0 | 0 | 1 | 50 |
| $P$ | 1 | $-28$ | $-24$ | $-20$ | 0 | 0 | 0 | 0 |

→

| BV | P | $x_1$ | $x_2$ | $x_3$ | $s_1$ | $s_2$ | $s_3$ | RHS |
|---|---|---|---|---|---|---|---|---|
| $s_1$ | 0 | 0 | $\frac{5}{2}$ | (5) | 1 | 0 | $-3$ | 350 |
| $s_2$ | 0 | 0 | 0 | 0 | 0 | 1 | $-1$ | 50 |
| $x_1$ | 0 | 1 | $\frac{1}{2}$ | 0 | 0 | 0 | 1 | 50 |
| $P$ | 1 | 0 | $-10$ | $-20$ | 0 | 0 | 28 | 1400 |

→

| BV | P | $x_1$ | $x_2$ | $x_3$ | $s_1$ | $s_2$ | $s_3$ | RHS |
|---|---|---|---|---|---|---|---|---|
| $x_3$ | 0 | 0 | $\frac{1}{2}$ | 1 | $\frac{1}{5}$ | 0 | $-\frac{3}{5}$ | 70 |
| $s_2$ | 0 | 0 | 0 | 0 | 0 | 1 | $-1$ | 50 |
| $x_1$ | 0 | 1 | $\frac{1}{2}$ | 0 | 0 | 0 | 1 | 50 |
| $P$ | 1 | 0 | 0 | 0 | 4 | 0 | 16 | 2800 |

$P = 2{,}800, x_1 = 50, x_2 = 0, x_3 = 70$.

A maximum revenue of \$2800 is realized when 50 cans of type I, no cans of type II and 70 cans of type III are made.

**37.** Let $x_1$, $x_2$ and $x_3$ represent the number of television, stereo and radio cabinets, respectively. The objective is to maximize profit $P = 10x_1 + 25x_2 + 3x_3$ dollars. The constraints are imposed by the hours available for assembly, decorating, and crating:

$$\begin{cases} \text{assembly}: 3x_1 + 10x_2 + x_3 \le 30{,}000 \\ \text{decorating}: 5x_1 + 8x_2 + x_3 \le 40{,}000 \\ \text{crating}: 0.1x_1 + 0.6x_2 + 1.0x_3 \le 120 \\ x_1 \ge 0, x_2 \ge 0, x_3 \ge 0 \end{cases}$$

| BV | $P$ | $x_1$ | $x_2$ | $x_3$ | $s_1$ | $s_2$ | $s_3$ | RHS | |
|---|---|---|---|---|---|---|---|---|---|
| $s_1$ | 0 | 3 | 10 | 1 | 1 | 0 | 0 | 30,000 | |
| $s_2$ | 0 | 5 | 8 | 1 | 0 | 1 | 0 | 40,000 | → |
| $s_3$ | 0 | .1 | (.6) | .1 | 0 | 0 | 1 | 120 | |
| $P$ | 1 | −10 | −25 | −3 | 0 | 0 | 0 | 0 | |

| BV | $P$ | $x_1$ | $x_2$ | $x_3$ | $s_1$ | $s_2$ | $s_3$ | RHS | |
|---|---|---|---|---|---|---|---|---|---|
| $s_1$ | 0 | $\frac{4}{3}$ | 0 | $-\frac{2}{3}$ | 1 | 0 | $-\frac{50}{3}$ | 28,000 | |
| $s_2$ | 0 | $\frac{11}{3}$ | 0 | $-\frac{1}{3}$ | 0 | 1 | $-\frac{40}{3}$ | 38,400 | → |
| $x_2$ | 0 | ($\frac{1}{6}$) | 1 | $\frac{1}{6}$ | 0 | 0 | $\frac{5}{3}$ | 200 | |
| $P$ | 1 | $-\frac{35}{6}$ | 0 | $\frac{7}{6}$ | 0 | 0 | $\frac{125}{3}$ | 5,000 | |

| BV | $P$ | $x_1$ | $x_2$ | $x_3$ | $s_1$ | $s_2$ | $s_3$ | RHS |
|---|---|---|---|---|---|---|---|---|
| $s_1$ | 0 | 0 | −8 | −2 | 1 | 0 | −30 | 26,400 |
| $s_2$ | 0 | 0 | −22 | −4 | 0 | 1 | −50 | 34,000 |
| $x_1$ | 0 | 1 | 6 | 1 | 0 | 0 | 10 | 1,200 |
| $P$ | 1 | 0 | 35 | 7 | 0 | 0 | 100 | 12,000 |

$P = 12{,}000, x_1 = 1{,}200, x_2 = 0, x_3 = 0$.

The maximum profit is $12,000 when 1200 television cabinets, no stereo or radio cabinets, are produced.

**39.** Let $x_1$, $x_2$ and $x_3$ represent the number of televisions shipped to Atlanta each month from Chicago, New York and Denver, respectively. The objective is to maximize profit $P = 50x_1 + 80x_2 + 40x_3$ dollars. The constraints are imposed by the limit on televisions shipped to Atlanta and the cost for transportation and labor.

$$\begin{cases} \text{Atlanta: } x_1 + x_2 + x_3 \le 400 \text{ televisions} \\ \text{transportation: } 20x_1 + 20x_2 + 40x_3 \le \$10{,}000 \\ \text{labor: } 6x_1 + 8x_2 + 4x_3 \le \$3{,}000 \\ x_1 \ge 0, x_2 \ge 0, x_3 \ge 0 \end{cases}$$

| BV | $P$ | $x_1$ | $x_2$ | $x_3$ | $s_1$ | $s_2$ | $s_3$ | RHS | |
|---|---|---|---|---|---|---|---|---|---|
| $s_1$ | 0 | 1 | 1 | 1 | 1 | 0 | 0 | 400 | |
| $s_2$ | 0 | 20 | 20 | 40 | 0 | 1 | 0 | 10,000 | → |
| $s_3$ | 0 | 6 | (8) | 4 | 0 | 0 | 1 | 3,000 | |
| $P$ | 1 | −50 | −80 | −40 | 0 | 0 | 0 | 0 | |

| BV | $P$ | $x_1$ | $x_2$ | $x_3$ | $s_1$ | $s_2$ | $s_3$ | RHS |
|---|---|---|---|---|---|---|---|---|
| $s_1$ | 0 | $\frac{1}{4}$ | 0 | $\frac{1}{2}$ | 1 | 0 | $-\frac{1}{8}$ | 25 |
| $s_2$ | 0 | 5 | 0 | 30 | 0 | 1 | $-\frac{5}{2}$ | 2,500 |
| $x_2$ | 0 | $\frac{3}{4}$ | 1 | $\frac{1}{2}$ | 0 | 0 | $\frac{1}{8}$ | 375 |
| $P$ | 1 | 10 | 0 | 0 | 0 | 0 | 10 | 30,000 |

$P = 30{,}000, x_1 = 0, x_2 = 375, x_3 = 0$.

The maximum profit is \$30,000 when no televisions are shipped from Chicago or Denver, and 375 are shipped from New York.

## Technology Exercises

**1.**

$$\begin{array}{c} BV \\ s_1 \\ s_2 \\ P \end{array}\begin{array}{c} \begin{array}{cccccc} P & x_1 & x_2 & s_1 & s_2 & RHS \end{array} \\ \left[\begin{array}{ccccc|c} 0 & 2 & 3 & 1 & 0 & 125 \\ 0 & \textcircled{3} & 1 & 0 & 1 & 75 \\ \hline 1 & -5 & -3 & 0 & 0 & 0 \end{array}\right] \end{array} \rightarrow \begin{array}{c} BV \\ s_1 \\ x_1 \\ P \end{array}\begin{array}{c} \begin{array}{cccccc} P & x_1 & x_2 & s_1 & s_2 & RHS \end{array} \\ \left[\begin{array}{ccccc|c} 0 & 0 & \textcircled{\tfrac{7}{3}} & 1 & -\tfrac{2}{3} & 75 \\ 0 & 1 & \tfrac{1}{3} & 0 & \tfrac{1}{3} & 25 \\ \hline 1 & 0 & -\tfrac{4}{3} & 0 & \tfrac{5}{3} & 125 \end{array}\right] \end{array} \rightarrow$$

$$\begin{array}{c} BV \\ x_2 \\ x_1 \\ P \end{array}\begin{array}{c} \begin{array}{cccccc} P & x_1 & x_2 & s_1 & s_2 & RHS \end{array} \\ \left[\begin{array}{ccccc|c} 0 & 0 & 1 & \tfrac{3}{7} & -\tfrac{2}{7} & \tfrac{225}{7} \\ 0 & 1 & 0 & -\tfrac{1}{7} & \tfrac{3}{7} & \tfrac{100}{7} \\ \hline 1 & 0 & 0 & \tfrac{4}{7} & \tfrac{9}{7} & \tfrac{1175}{7} \end{array}\right] \end{array}$$

. The maximum value for $P$ is $\frac{1175}{7} \approx 167\frac{6}{7} \approx 167.86$, obtained when $x_1 = \frac{100}{7} \approx 14.29$ and $x_2 = \frac{225}{7} \approx 32.14$.

**3.**

$$\begin{array}{c} BV \\ s_1 \\ s_2 \\ P \end{array}\begin{array}{c} \begin{array}{cccccc} P & x_1 & x_2 & s_1 & s_2 & RHS \end{array} \\ \left[\begin{array}{ccccc|c} 0 & 1 & \textcircled{2} & 1 & 0 & 2 \\ 0 & 2 & 1 & 0 & 1 & 2 \\ \hline 1 & -5 & -7 & 0 & 0 & 0 \end{array}\right] \end{array} \rightarrow \begin{array}{c} BV \\ x_2 \\ s_2 \\ P \end{array}\begin{array}{c} \begin{array}{cccccc} P & x_1 & x_2 & s_1 & s_2 & RHS \end{array} \\ \left[\begin{array}{ccccc|c} 0 & \tfrac{1}{2} & 1 & \tfrac{1}{2} & 0 & 1 \\ 0 & \textcircled{\tfrac{3}{2}} & 0 & -\tfrac{1}{2} & 1 & 1 \\ \hline 1 & -\tfrac{3}{2} & 0 & \tfrac{7}{2} & 0 & 7 \end{array}\right] \end{array} \rightarrow$$

$$\begin{array}{c} BV \\ x_2 \\ x_1 \\ P \end{array}\begin{array}{c} \begin{array}{cccccc} P & x_1 & x_2 & s_1 & s_2 & RHS \end{array} \\ \left[\begin{array}{ccccc|c} 0 & 0 & 1 & \tfrac{2}{3} & -\tfrac{1}{3} & \tfrac{2}{3} \\ 0 & 1 & 0 & -\tfrac{1}{3} & \tfrac{2}{3} & \tfrac{2}{3} \\ \hline 1 & 0 & 0 & 3 & 1 & 8 \end{array}\right] \end{array}$$

. The maximum value for $P$ is 8, obtained when $x_1 = \frac{2}{3} \approx 0.67$ and $x_2 = \frac{2}{3} \approx 0.67$.

## 4.3 Solving Minimum Problems in Standard Form; The Duality Principle

1. The minimum problem is in standard form.

3. Since the coefficient of $x_2$ in the objective function is negative, the minimum problem is not in standard form.

5. Since the first constraint is not written with a $\geq$ sign, the minimum problem is not in standard form.

**7.** The matrix is: $\begin{bmatrix} 1 & 1 & | & 2 \\ 2 & 3 & | & 6 \\ 2 & 3 & | & 0 \end{bmatrix}$. The transpose is $\begin{bmatrix} 1 & 2 & | & 2 \\ 1 & 3 & | & 3 \\ 2 & 6 & | & 0 \end{bmatrix}$. Dual problem: Maximize $P = 2y_1 + 6y_2$ subject to $y_1 + 2y_2 \le 2$, $y_1 + 3y_2 \le 3$, $y_1 \ge 0$, $y_2 \ge 0$.

**9.** The matrix is: $\begin{bmatrix} 1 & 1 & 1 & | & 5 \\ 2 & 1 & 0 & | & 4 \\ 3 & 1 & 1 & | & 0 \end{bmatrix}$; The transpose is: $\begin{bmatrix} 1 & 2 & | & 3 \\ 1 & 1 & | & 1 \\ 1 & 0 & | & 1 \\ 5 & 4 & | & 0 \end{bmatrix}$. Dual problem: Maximize $P = 5y_1 + 4y_2$ subject to $y_1 + 2y_2 \le 3$, $y_1 + y_2 \le 1$, $y_1 \le 1$, $y_1 \ge 0$, $y_2 \ge 0$.

**11.** The matrix is: $\begin{bmatrix} 1 & 1 & 1 & 2 & | & 60 \\ 3 & 2 & 1 & 2 & | & 90 \\ 3 & 4 & 1 & 2 & | & 0 \end{bmatrix}$. The transpose is: $\begin{bmatrix} 1 & 3 & | & 3 \\ 1 & 2 & | & 4 \\ 1 & 1 & | & 1 \\ 2 & 2 & | & 2 \\ 60 & 90 & | & 0 \end{bmatrix}$.

Dual problem: Maximize $P = 60y_1 + 90y_2$ subject to $y_1 + 3y_2 \le 3$, $y_1 + 2y_2 \le 4$, $y_1 + y_2 \le 1$, $2y_1 + 2y_2 \le 2$, $y_1 \ge 0$, $y_2 \ge 0$.

**13.** The matrix is: $\begin{bmatrix} 1 & 1 & | & 2 \\ 2 & 6 & | & 6 \\ 6 & 3 & | & 0 \end{bmatrix}$. The transpose is: $\begin{bmatrix} 1 & 2 & | & 6 \\ 1 & 6 & | & 3 \\ 2 & 6 & | & 0 \end{bmatrix}$.

Dual problem: Maximize $P = 2y_1 + 6y_2$ subject to $y_1 + 2y_2 \le 6$, $y_1 + 6y_2 \le 3$, $y_1 \ge 0$, $y_2 \ge 0$.

$$\begin{array}{c} BV \\ s_1 \\ s_2 \\ P \end{array} \begin{array}{c} P \quad y_1 \quad y_2 \quad s_1 \quad s_2 \quad RHS \\ \left[\begin{array}{cccccc|c} 0 & 1 & 2 & 1 & 0 & 6 \\ 0 & 1 & \textcircled{6} & 0 & 1 & 3 \\ \hline 1 & -2 & -6 & 0 & 0 & 0 \end{array}\right] \end{array} \rightarrow \begin{array}{c} BV \\ s_1 \\ y_2 \\ P \end{array} \begin{array}{c} P \quad y_1 \quad y_2 \quad s_1 \quad s_2 \quad RHS \\ \left[\begin{array}{ccccc|c} 0 & 2/3 & 0 & 1 & -1/3 & 5 \\ 0 & \textcircled{1/6} & 1 & 0 & 1/6 & 1/2 \\ \hline 1 & -1 & 0 & 0 & 1 & 3 \end{array}\right] \end{array} \rightarrow$$

$$\begin{array}{c} BV \\ s_1 \\ y_1 \\ P \end{array} \begin{array}{c} P \quad y_1 \quad y_2 \quad s_1 \quad s_2 \quad RHS \\ \left[\begin{array}{ccccc|c} 0 & 0 & -4 & 1 & -1 & 3 \\ 0 & 1 & 6 & 0 & 1 & 3 \\ \hline 1 & 0 & 6 & 0 & 2 & 6 \end{array}\right] \end{array};$$

The solution to the maximum problem is $P = 6, y_1 = 3, y_2 = 0$.

The solution to the minimum problem is $C = 6, x_1 = 0, x_2 = 2$.

**15.** The matrix is: $\left[\begin{array}{cc|c} 1 & 1 & 4 \\ 3 & 4 & 12 \\ 6 & 3 & 0 \end{array}\right]$. The transpose is: $\left[\begin{array}{cc|c} 1 & 3 & 6 \\ 1 & 4 & 3 \\ 4 & 12 & 0 \end{array}\right]$.

Dual problem: Maximize $P = 4y_1 + 12y_2$ subject to $y_1 + 3y_2 \le 6$, $y_1 + 4y_2 \le 3$, $y_1 \ge 0$, $y_2 \ge 0$.

$$\begin{array}{c} BV \\ s_1 \\ s_2 \\ P \end{array} \begin{array}{c} \begin{array}{cccccc|c} P & y_1 & y_2 & s_1 & s_2 & & RHS \end{array} \\ \left[\begin{array}{ccccc|c} 0 & 1 & 3 & 1 & 0 & 6 \\ 0 & 1 & \textcircled{4} & 0 & 1 & 3 \\ \hline 1 & -4 & -12 & 0 & 0 & 0 \end{array}\right] \end{array} \to \begin{array}{c} BV \\ s_1 \\ y_2 \\ P \end{array} \begin{array}{c} \begin{array}{cccccc} P & y_1 & y_2 & s_1 & s_2 & RHS \end{array} \\ \left[\begin{array}{ccccc|c} 0 & \frac{1}{4} & 0 & 1 & -\frac{3}{4} & \frac{15}{4} \\ 0 & \textcircled{\frac{1}{4}} & 1 & 0 & \frac{1}{4} & \frac{3}{4} \\ \hline 1 & -1 & 0 & 0 & 3 & 9 \end{array}\right] \end{array} \to$$

$$\begin{array}{c} BV \\ s_1 \\ y_1 \\ P \end{array} \begin{array}{c} \begin{array}{cccccc} P & y_1 & y_2 & s_1 & s_2 & RHS \end{array} \\ \left[\begin{array}{ccccc|c} 0 & 0 & -1 & 1 & -1 & 3 \\ 0 & 1 & 4 & 0 & 1 & 3 \\ \hline 1 & 0 & 4 & 0 & 4 & 12 \end{array}\right] \end{array};$$

The solution to the maximum problem is $P = 12, y_1 = 3, y_2 = 0$.
The solution to the minimum problem is $C = 12, x_1 = 0, x_2 = 4$.

**17.** The matrix is: $\left[\begin{array}{ccc|c} 1 & -3 & 4 & 12 \\ 3 & 1 & 2 & 10 \\ 1 & -1 & -1 & -8 \\ 1 & 2 & 1 & 0 \end{array}\right]$. The transpose is: $\left[\begin{array}{ccc|c} 1 & 3 & 1 & 1 \\ -3 & 1 & -1 & 2 \\ 4 & 2 & -1 & 1 \\ 12 & 10 & -8 & 0 \end{array}\right]$.

Dual problem: Maximize $P = 12y_1 + 10y_2 - 8y_3$ subject to $y_1 + 3y_2 + y_3 \le 1$, $-3y_1 + y_2 - y_3 \le 2$, $4y_1 + 2y_2 - y_3 \le 1, y_1 \ge 0$, $y_2 \ge 0$, $y_3 \ge 0$.

$$\begin{array}{c} BV \\ s_1 \\ s_2 \\ s_3 \\ P \end{array} \begin{array}{c} \begin{array}{cccccccc} P & y_1 & y_2 & y_3 & s_1 & s_2 & s_3 & RHS \end{array} \\ \left[\begin{array}{ccccccc|c} 0 & 1 & 3 & 1 & 1 & 0 & 0 & 1 \\ 0 & -3 & 1 & -1 & 0 & 1 & 0 & 2 \\ 0 & \textcircled{4} & 2 & -1 & 0 & 0 & 1 & 1 \\ \hline 1 & -12 & -10 & 8 & 0 & 0 & 0 & 0 \end{array}\right] \end{array} \to \begin{array}{c} BV \\ s_1 \\ s_2 \\ y_1 \\ P \end{array} \begin{array}{c} \begin{array}{cccccccc} P & y_1 & y_2 & y_3 & s_1 & s_2 & s_3 & RHS \end{array} \\ \left[\begin{array}{ccccccc|c} 0 & 0 & \textcircled{\frac{5}{2}} & \frac{5}{4} & 1 & 0 & -\frac{1}{4} & \frac{3}{4} \\ 0 & 0 & \frac{5}{2} & -\frac{7}{4} & 0 & 1 & \frac{3}{4} & \frac{11}{4} \\ 0 & 1 & \frac{1}{2} & -\frac{1}{4} & 0 & 0 & \frac{1}{4} & \frac{1}{4} \\ \hline 1 & 0 & -4 & 5 & 0 & 0 & 3 & 3 \end{array}\right] \end{array} \to$$

$$\begin{array}{c} BV \\ y_2 \\ s_2 \\ y_1 \\ P \end{array} \begin{array}{c} \begin{array}{cccccccc} P & y_1 & y_2 & y_3 & s_1 & s_2 & s_3 & RHS \end{array} \\ \left[\begin{array}{ccccccc|c} 0 & 0 & 1 & \frac{1}{2} & \frac{2}{5} & 0 & -\frac{1}{10} & \frac{3}{10} \\ 0 & 0 & 0 & -3 & -1 & 1 & 1 & 2 \\ 0 & 1 & 0 & -\frac{1}{2} & -\frac{1}{5} & 0 & \frac{3}{10} & \frac{1}{10} \\ \hline 1 & 0 & 0 & 7 & \frac{8}{5} & 0 & \frac{13}{5} & \frac{21}{5} \end{array}\right] \end{array};$$

The solution to the maximum problem is $P = \frac{21}{5}, y_1 = \frac{1}{10}, y_2 = \frac{3}{10}, y_3 = 0$.

The solution to the minimum problem is $C = \frac{21}{5}, x_1 = \frac{8}{5}, x_2 = 0, x_3 = \frac{13}{5}$.

**19.** The matrix is: $\left[\begin{array}{cccc|c} 1 & 0 & 1 & 0 & 1 \\ 0 & 1 & 0 & 1 & 1 \\ -1 & -1 & -1 & -1 & -3 \\ 1 & 4 & 2 & 4 & 0 \end{array}\right]$. The transpose is: $\left[\begin{array}{ccc|c} 1 & 0 & -1 & 1 \\ 0 & 1 & -1 & 4 \\ 1 & 0 & -1 & 2 \\ 0 & 1 & -1 & 4 \\ 1 & 1 & -3 & 0 \end{array}\right]$.

Dual problem: Maximize $P = y_1 + y_2 - 3y_3$ subject to $y_1 - y_3 \le 1$, $y_2 - y_3 \le 4$, $y_1 - y_3 \le 2$, $y_2 - y_3 \le 4, y_1 \ge 0$, $y_2 \ge 0$, $y_3 \ge 0$.

$$\begin{array}{c} BV \\ s_1 \\ s_2 \\ s_3 \\ s_4 \\ P \end{array} \begin{array}{c} \begin{array}{ccccccccc} P & y_1 & y_2 & y_3 & s_1 & s_2 & s_3 & s_4 & RHS \end{array} \\ \left[\begin{array}{cccccccc|c} 0 & \textcircled{1} & 0 & -1 & 1 & 0 & 0 & 0 & 1 \\ 0 & 0 & 1 & -1 & 0 & 1 & 0 & 0 & 4 \\ 0 & 1 & 0 & -1 & 0 & 0 & 1 & 0 & 2 \\ 0 & 0 & 1 & -1 & 0 & 0 & 0 & 1 & 4 \\ \hline 1 & -1 & -1 & 3 & 0 & 0 & 0 & 0 & 0 \end{array}\right] \end{array} \rightarrow \begin{array}{c} BV \\ y_1 \\ s_2 \\ s_3 \\ s_4 \\ P \end{array} \begin{array}{c} \begin{array}{ccccccccc} P & y_1 & y_2 & y_3 & s_1 & s_2 & s_3 & s_4 & RHS \end{array} \\ \left[\begin{array}{cccccccc|c} 0 & 1 & 0 & -1 & 1 & 0 & 0 & 0 & 1 \\ 0 & 0 & \textcircled{1} & -1 & 0 & 1 & 0 & 0 & 4 \\ 0 & 0 & 0 & 0 & -1 & 0 & 1 & 0 & 1 \\ 0 & 0 & 1 & -1 & 0 & 0 & 0 & 1 & 4 \\ \hline 1 & 0 & -1 & 2 & 1 & 0 & 0 & 0 & 1 \end{array}\right] \end{array}$$

$$\rightarrow \begin{array}{c} BV \\ y_1 \\ y_2 \\ s_3 \\ s_4 \\ P \end{array} \begin{array}{c} \begin{array}{ccccccccc} P & y_1 & y_2 & y_3 & s_1 & s_2 & s_3 & s_4 & RHS \end{array} \\ \left[\begin{array}{cccccccc|c} 0 & 1 & 0 & -1 & 1 & 0 & 0 & 0 & 1 \\ 0 & 0 & 1 & -1 & 0 & 1 & 0 & 0 & 4 \\ 0 & 0 & 0 & 0 & -1 & 0 & 1 & 0 & 1 \\ 0 & 0 & 0 & 0 & 0 & -1 & 0 & 1 & 0 \\ \hline 1 & 0 & 0 & 1 & 1 & 1 & 0 & 0 & 5 \end{array}\right] \end{array}$$

The solution to the maximum problem is $P = 5, y_1 = 1, y_2 = 4, y_3 = 0, y_4 = 0$.
The solution to the minimum problem is $C = 5, x_1 = 1, x_2 = 1, x_3 = 0, x_4 = 0$. ( An alternate choice of pivot element leads to the solution $C = 5, x_1 = 1, x_2 = 0, x_3 = 0, x_4 = 1$.)

**21.** Let $x_1$ and $x_2$ represent the number of P pills and Q pills, respectively, that Mr. Jones should take. The objective is to minimize cost $C = 3x_1 + 4x_2$ cents. The constraints are imposed by Mr. Jones' daily calcium and iron needs:

$$\begin{cases} 5x_1 + 10x_2 \ge 50 \text{ mg} \\ 2x_1 + x_2 \ge 8 \text{ mg} \\ x_1 \ge 0, x_2 \ge 0 \end{cases}$$

The matrix is $\left[\begin{array}{cc|c} 5 & 10 & 50 \\ 2 & 1 & 8 \\ 3 & 4 & 0 \end{array}\right]$. The transpose is: $\left[\begin{array}{cc|c} 5 & 2 & 3 \\ 10 & 1 & 4 \\ 50 & 8 & 0 \end{array}\right]$.

Dual problem: Maximize $P = 50y_1 + 8y_2$ subject to $5y_1 + 2y_2 \le 3$, $10y_1 + y_2 \le 4$, $y_1 \ge 0$, $y_2 \ge 0$.

| BV | P | $y_1$ | $y_2$ | $s_1$ | $s_2$ | RHS |
|---|---|---|---|---|---|---|
| $s_1$ | 0 | 5 | 2 | 1 | 0 | 3 |
| $s_2$ | 0 | 10 | 1 | 0 | 1 | 4 |
| $P$ | 1 | −50 | −8 | 0 | 0 | 0 |

→

| BV | P | $y_1$ | $y_2$ | $s_1$ | $s_2$ | RHS |
|---|---|---|---|---|---|---|
| $s_1$ | 0 | 0 | $\frac{3}{2}$ | 1 | $-\frac{1}{2}$ | 1 |
| $y_1$ | 0 | 1 | $\frac{1}{10}$ | 0 | $\frac{1}{10}$ | $\frac{2}{5}$ |
| $P$ | 1 | 0 | −3 | 0 | 5 | 20 |

→

| BV | P | $y_1$ | $y_2$ | $s_1$ | $s_2$ | RHS |
|---|---|---|---|---|---|---|
| $y_2$ | 0 | 0 | 1 | $\frac{2}{3}$ | $-\frac{1}{3}$ | $\frac{2}{3}$ |
| $y_1$ | 0 | 1 | 0 | $-\frac{1}{15}$ | $\frac{2}{15}$ | $\frac{1}{3}$ |
| $P$ | 1 | 0 | 0 | 2 | 4 | 22 |

$C = 22, x_1 = 2, x_2 = 4$. Mr. Jones' minimum cost is 22¢ when he takes 2 of pill P and 4 of pill Q.

**23.** Let $x_1$, $x_2$, and $x_3$ represent the number of units of products A, B, and C, respectively. The objective is to minimize cost $C = 4x_1 + 2x_2 + x_3$ dollars. The constraints are imposed by the minimum number of units that must be produced: $x_1 \ge 20$, $x_2 \ge 30$, $x_3 \ge 40$, $x_1 + x_2 + x_3 \ge 200$, $x_1 \ge 0$, $x_2 \ge 0$, $x_3 \ge 0$.

The matrix is: $\left[\begin{array}{ccc|c} 1 & 0 & 0 & 20 \\ 0 & 1 & 0 & 30 \\ 0 & 0 & 1 & 40 \\ 1 & 1 & 1 & 200 \\ 4 & 2 & 1 & 0 \end{array}\right]$. The transpose is: $\left[\begin{array}{cccc|c} 1 & 0 & 0 & 1 & 4 \\ 0 & 1 & 0 & 1 & 2 \\ 0 & 0 & 1 & 1 & 1 \\ 20 & 30 & 40 & 200 & 0 \end{array}\right]$.

Dual problem: Maximize $P = 20y_1 + 30y_2 + 40y_3 + 200y_4$ subject to: $y_1 + y_4 \le 4$, $y_2 + y_4 \le 2$, $y_3 + y_4 \le 1$, $y_1 \ge 0$, $y_2 \ge 0$, $y_3 \ge 0$.

| BV | P | $y_1$ | $y_2$ | $y_3$ | $y_4$ | $s_1$ | $s_2$ | $s_3$ | RHS |
|---|---|---|---|---|---|---|---|---|---|
| $s_1$ | 0 | 1 | 0 | 0 | 1 | 1 | 0 | 0 | 4 |
| $s_2$ | 0 | 0 | 1 | 0 | 1 | 0 | 1 | 0 | 2 |
| $s_3$ | 0 | 0 | 0 | 1 | 1 | 0 | 0 | 1 | 1 |
| $P$ | 1 | −20 | −30 | −40 | −200 | 0 | 0 | 0 | 0 |

→

| BV | P | $y_1$ | $y_2$ | $y_3$ | $y_4$ | $s_1$ | $s_2$ | $s_3$ | RHS |
|---|---|---|---|---|---|---|---|---|---|
| $s_1$ | 0 | 1 | 0 | −1 | 0 | 1 | 0 | −1 | 3 |
| $s_2$ | 0 | 0 | 1 | −1 | 0 | 0 | 1 | −1 | 1 |
| $y_4$ | 0 | 0 | 0 | 1 | 1 | 0 | 0 | 1 | 1 |
| $P$ | 1 | −20 | −30 | 160 | 0 | 0 | 0 | 200 | 200 |

→

| BV | P | $y_1$ | $y_2$ | $y_3$ | $y_4$ | $s_1$ | $s_2$ | $s_3$ | RHS |
|---|---|---|---|---|---|---|---|---|---|
| $s_1$ | 0 | 1 | 0 | −1 | 0 | 1 | 0 | −1 | 3 |
| $y_2$ | 0 | 0 | 1 | −1 | 0 | 0 | 1 | −1 | 1 |
| $y_4$ | 0 | 0 | 0 | 1 | 1 | 0 | 0 | 1 | 1 |
| $P$ | 1 | −20 | 0 | 130 | 0 | 0 | 30 | 170 | 230 |

→

$$\begin{array}{c} BV \\ y_1 \\ y_2 \\ y_4 \\ P \end{array}\begin{array}{c} \begin{array}{cccccccc|c} P & y_1 & y_2 & y_3 & y_4 & s_1 & s_2 & s_3 & RHS \end{array} \\ \left[\begin{array}{cccccccc|c} 0 & 1 & 0 & -1 & 0 & 1 & 0 & -1 & 3 \\ 0 & 0 & 1 & -1 & 0 & 0 & 1 & -1 & 1 \\ 0 & 0 & 0 & 1 & 1 & 0 & 0 & 1 & 1 \\ \hline 1 & 0 & 0 & 110 & 0 & 20 & 30 & 150 & 290 \end{array}\right] \end{array}.$$

$C = 290, x_1 = 20, x_2 = 30, x_3 = 150$. The minimum cost is \$290 when 20 units of product A, 30 units of product B, and 150 units of product C are produced.

**25.** Let $x_1$, $x_2$, $x_3$ represent the number of orders for Lunches 1, 2, 3, respectively. The objective is to minimize the cost $C = 6.2x_1 + 7.4x_2 + 9.1x_3$ dollars. The constraints are imposed by requirements for soup, salads, sandwiches and pasta:

$$\begin{cases} \text{soup}: x_1 \geq 4 \\ \text{salad}: x_1 + x_2 + x_3 \geq 9 \\ \text{sandwiches}: x_1 + x_3 \geq 6 \\ \text{pasta}: x_2 + x_3 \geq 5 \\ x_1 \geq 0, x_2 \geq 0 \end{cases}$$

The matrix is: $\left[\begin{array}{ccc|c} 1 & 0 & 0 & 4 \\ 1 & 1 & 1 & 9 \\ 1 & 0 & 1 & 6 \\ 0 & 1 & 1 & 5 \\ 6.2 & 7.4 & 9.1 & 0 \end{array}\right]$. The transpose is: $\left[\begin{array}{cccc|c} 1 & 1 & 1 & 0 & 6.2 \\ 0 & 1 & 0 & 1 & 7.4 \\ 0 & 1 & 1 & 1 & 9.1 \\ 4 & 9 & 6 & 5 & 0 \end{array}\right]$.

Dual problem:

Maximize $P = 4y_1 + 9y_2 + 6y_3 + 5y_4$ subject to $\begin{cases} y_1 + y_2 + y_3 \leq 6.2 \\ y_2 + y_4 \leq 7.4 \\ y_2 + y_3 + y_4 \leq 9.1 \\ y_1 \geq 0, y_2 \geq 0, y_3 \geq 0, y_4 \geq 0 \end{cases}$

$$\begin{array}{c} BV \\ s_1 \\ s_2 \\ s_3 \\ P \end{array}\begin{array}{c} \begin{array}{cccccccc|c} P & y_1 & y_2 & y_3 & y_4 & s_1 & s_2 & s_3 & RHS \end{array} \\ \left[\begin{array}{cccccccc|c} 0 & 1 & \textcircled{1} & 1 & 0 & 1 & 0 & 0 & 6.2 \\ 0 & 0 & 1 & 0 & 1 & 0 & 1 & 0 & 7.4 \\ 0 & 0 & 1 & 1 & 1 & 0 & 0 & 1 & 9.1 \\ \hline 1 & -4 & -9 & -6 & -5 & 0 & 0 & 0 & 0 \end{array}\right] \end{array} \rightarrow$$

$$\begin{array}{c} BV \\ y_2 \\ s_2 \\ s_3 \\ P \end{array}\begin{array}{c} \begin{array}{cccccccc|c} P & y_1 & y_2 & y_3 & y_4 & s_1 & s_2 & s_3 & RHS \end{array} \\ \left[\begin{array}{cccccccc|c} 0 & 1 & 1 & 1 & 0 & 1 & 0 & 0 & 6.2 \\ 0 & -1 & 0 & -1 & \textcircled{1} & -1 & 1 & 0 & 1.2 \\ 0 & -1 & 0 & 0 & 1 & -1 & 0 & 1 & 2.9 \\ \hline 1 & 5 & 0 & 3 & -5 & 9 & 0 & 0 & 55.8 \end{array}\right] \end{array} \rightarrow$$

$$\begin{array}{c} BVP \\ y_2 \\ y_4 \\ s_3 \\ P \end{array}\begin{array}{c} \begin{array}{cccccccc|c} P & y_1 & y_2 & y_3 & y_4 & s_1 & s_2 & s_3 & RHS \end{array} \\ \left[\begin{array}{cccccccc|c} 0 & 1 & 1 & 1 & 0 & 1 & 0 & 0 & 6.2 \\ 0 & -1 & 0 & -1 & 1 & -1 & 1 & 0 & 1.2 \\ 0 & 0 & 0 & \textcircled{1} & 0 & 0 & -1 & 1 & 1.7 \\ \hline 1 & 0 & 0 & -2 & 0 & 4 & 5 & 0 & 61.8 \end{array}\right] \end{array} \rightarrow$$

$$\begin{array}{c} BVP \\ y_2 \\ y_4 \\ y_3 \\ P \end{array}\begin{array}{c} \begin{array}{cccccccc|c} P & y_1 & y_2 & y_3 & y_4 & s_1 & s_2 & s_3 & RHS \end{array} \\ \left[\begin{array}{cccccccc|c} 0 & 1 & 1 & 0 & 0 & 1 & 1 & -1 & 4.5 \\ 0 & -1 & 0 & 0 & 1 & -1 & 0 & 1 & 2.9 \\ 0 & 0 & 0 & 1 & 0 & 0 & -1 & 1 & 1.7 \\ \hline 1 & 0 & 0 & 0 & 0 & 4 & 3 & 2 & 65.2 \end{array}\right] \end{array}.$$

The minimum value for $C$ is 65.2, obtained. when $x_1 = 4$, $x_2 = 3$ and $x_3 = 2$. The group's minimum cost is \$65.20 when they order 4 of Lunch #1, 3 of Lunch #2 and 2 of Lunch #3.

## Technology Exercises

**1.** Matrix: $\left[\begin{array}{cc|c} 2 & 3 & 125 \\ 3 & 1 & 75 \\ 5 & 3 & 0 \end{array}\right]$. Transpose: $\left[\begin{array}{cc|c} 2 & 3 & 5 \\ 3 & 1 & 3 \\ 125 & 75 & 0 \end{array}\right]$

Dual problem: Maximize $P = 125y_1 + 75y_2$ subject to $\begin{cases} 2y_1 + 3y_2 \le 5 \\ 3y_1 + \ \ y_2 \le 3 \\ y_1 \ge 0, y_2 \ge 0 \end{cases}$.

$$\begin{array}{c} BVP \\ s_1 \\ s_2 \\ P \end{array}\begin{array}{c} \begin{array}{ccccc|c} P & y_1 & y_2 & s_1 & s_2 & RHS \end{array} \\ \left[\begin{array}{ccccc|c} 0 & 2 & 3 & 1 & 0 & 5 \\ 0 & \textcircled{3} & 1 & 0 & 1 & 3 \\ \hline 1 & -125 & -75 & 0 & 0 & 0 \end{array}\right] \end{array} \rightarrow \begin{array}{c} BVP \\ s_1 \\ y_1 \\ P \end{array}\begin{array}{c} \begin{array}{ccccc|c} P & y_1 & y_2 & s_1 & s_2 & RHS \end{array} \\ \left[\begin{array}{ccccc|c} 0 & 0 & \textcircled{\frac{7}{3}} & 1 & -\frac{2}{3} & 3 \\ 0 & 1 & \frac{1}{3} & 0 & \frac{1}{3} & 1 \\ \hline 1 & 0 & -\frac{100}{3} & 0 & \frac{125}{3} & 125 \end{array}\right] \end{array} \rightarrow$$

$$\begin{array}{c} BVP \\ y_2 \\ y_1 \\ P \end{array}\begin{array}{c} \begin{array}{ccccc|c} P & y_1 & y_2 & s_1 & s_2 & RHS \end{array} \\ \left[\begin{array}{ccccc|c} 0 & 0 & 1 & \frac{3}{7} & -\frac{2}{7} & \frac{9}{7} \\ 0 & 1 & 0 & -\frac{1}{7} & \frac{3}{7} & \frac{4}{7} \\ \hline 1 & 0 & 0 & \frac{100}{7} & \frac{225}{7} & \frac{1175}{7} \end{array}\right] \end{array}.$$

The minimum value for $C$ is $\frac{1175}{7} \approx 167.86$, obtained when $x_1 = \frac{100}{7} \approx 14.29$ and $x_2 = \frac{225}{7} \approx 32.14$.

**3.** Matrix: $\left[\begin{array}{cc|c} 1 & 2 & 2 \\ 2 & 1 & 2 \\ 5 & 7 & 0 \end{array}\right]$. Transpose: $\left[\begin{array}{cc|c} 1 & 2 & 5 \\ 2 & 1 & 7 \\ 2 & 2 & 0 \end{array}\right]$.

Dual problem: Maximize $P = 2y_1 + 2y_2$ subject to $\begin{cases} y_1 + 2y_2 \le 5 \\ 2y_1 + y_2 \le 7 \\ y_1 \ge 0, y_2 \ge 0 \end{cases}$.

$$\begin{array}{c} BV \\ s_1 \\ s_2 \\ P \end{array}\left[\begin{array}{ccccc|c} P & y_1 & y_2 & s_1 & s_2 & RHS \\ 0 & 1 & 2 & 1 & 0 & 5 \\ 0 & \textcircled{2} & 1 & 0 & 1 & 7 \\ \hline 1 & -2 & -2 & 0 & 0 & 0 \end{array}\right] \rightarrow \begin{array}{c} BV \\ s_1 \\ y_1 \\ P \end{array}\left[\begin{array}{ccccc|c} P & y_1 & y_2 & s_1 & s_2 & RHS \\ 0 & 0 & \textcircled{\tfrac{3}{2}} & 1 & -\tfrac{1}{2} & \tfrac{3}{2} \\ 0 & 1 & \tfrac{1}{2} & 0 & \tfrac{1}{2} & \tfrac{7}{2} \\ \hline 1 & 0 & -1 & 0 & 1 & 7 \end{array}\right] \rightarrow$$

$$\begin{array}{c} BV \\ y_2 \\ y_1 \\ P \end{array}\left[\begin{array}{ccccc|c} P & y_1 & y_2 & s_1 & s_2 & RHS \\ 0 & 0 & 1 & \tfrac{2}{3} & -\tfrac{1}{3} & 1 \\ 0 & 1 & 0 & -\tfrac{1}{3} & \tfrac{2}{3} & 3 \\ \hline 1 & 0 & 0 & \tfrac{2}{3} & \tfrac{2}{3} & 8 \end{array}\right].$$

The minimum value for $C$ is 8, obtained when $x_1 = \dfrac{2}{3}$ and $x_2 = \dfrac{2}{3}$.

## 4.4 The Simplex Method with Mixed Constraints

**1.** Rewrite the constraints:

$$\begin{aligned} x_1 + x_2 &\le 12 \\ -5x_1 - 2x_2 &\le -36 \\ -7x_1 - 4x_2 &\le -14 \\ x_1 \ge 0, x_2 &\ge 0 \end{aligned}$$

Introduce slack variables:

$$\begin{aligned} x_1 + x_2 + s_1 \qquad\qquad &= 12 \\ -5x_1 - 2x_2 \qquad + s_2 \qquad &= -36 \\ -7x_1 - 4x_2 \qquad\qquad + s_3 &= -14 \\ s_1 \ge 0, s_2 \ge 0, s_3 &\ge 0 \end{aligned}$$

Initial Tableau:

$$\begin{array}{c} BV \\ s_1 \\ s_2 \\ s_3 \\ P \end{array}\left[\begin{array}{cccccc|c} P & x_1 & x_2 & s_1 & s_2 & s_3 & RHS \\ 0 & 1 & 1 & 1 & 0 & 0 & 12 \\ 0 & \textcircled{-5} & -2 & 0 & 1 & 0 & -36 \\ 0 & -7 & -4 & 0 & 0 & 1 & -14 \\ \hline 1 & -3 & -4 & 0 & 0 & 0 & 0 \end{array}\right] \xrightarrow{\text{Alternative Pivoting Strategy}} \begin{array}{c} BV \\ s_1 \\ x_1 \\ s_3 \\ P \end{array}\left[\begin{array}{cccccc|c} P & x_1 & x_2 & s_1 & s_2 & s_3 & RHS \\ 0 & 0 & \textcircled{\tfrac{3}{5}} & 1 & \tfrac{1}{5} & 0 & \tfrac{24}{5} \\ 0 & 1 & \tfrac{2}{5} & 0 & -\tfrac{1}{5} & 0 & \tfrac{36}{5} \\ 0 & 0 & -\tfrac{6}{5} & 0 & -\tfrac{7}{5} & 1 & \tfrac{182}{5} \\ \hline 1 & 0 & -\tfrac{14}{5} & 0 & -\tfrac{3}{5} & 0 & \tfrac{108}{5} \end{array}\right]$$

$$\xrightarrow{\text{Standard Pivoting Strategy}} \begin{array}{c} BV \\ x_2 \\ x_1 \\ s_3 \\ P \end{array}\left[\begin{array}{cccccc|c} P & x_1 & x_2 & s_1 & s_2 & s_3 & RHS \\ 0 & 0 & 1 & \tfrac{5}{3} & \tfrac{1}{3} & 0 & 8 \\ 0 & 1 & 0 & -\tfrac{2}{3} & -\tfrac{1}{3} & 0 & 4 \\ 0 & 0 & 0 & 2 & -1 & 1 & 46 \\ \hline 1 & 0 & 0 & \tfrac{14}{3} & \tfrac{1}{3} & 0 & 44 \end{array}\right]$$

The maximum value of $P$ is 44, when $x_1 = 4$ and $x_2 = 8$.

**3.** Rewrite the constraints:

$$\begin{aligned} x_1+3x_2+x_3 &\le 9\\ -2x_1-3x_2+x_3 &\le -2\\ -3x_1+2x_2-x_3 &\le -5\\ x_1\ge 0, x_2\ge 0, x_3 &\ge 0\end{aligned}$$

Introduce slack variables:

$$\begin{aligned} x_1+3x_2+x_3+s_1 \qquad\qquad &= 9\\ -2x_1-3x_2+x_3 \qquad +s_2 \qquad &= -2\\ -3x_1+2x_2-x_3 \qquad\qquad +s_3 &= -5\\ s_1\ge 0, s_2\ge 0, s_3 &\ge 0\end{aligned}$$

Initial Tableau:

$$\begin{array}{c} BV\\ s_1\\ s_2\\ s_3\\ P\end{array}\begin{array}{c} \begin{array}{cccccccc} P & x_1 & x_2 & x_3 & s_1 & s_2 & s_3 & RHS\end{array}\\ \left[\begin{array}{ccccccc|c} 0 & 1 & 3 & 1 & 1 & 0 & 0 & 9\\ 0 & \enclose{circle}{-2} & -3 & 1 & 0 & 1 & 0 & -2\\ 0 & -3 & 2 & -1 & 0 & 0 & 1 & -5\\ \hline 1 & -3 & -2 & 1 & 0 & 0 & 0 & 0\end{array}\right]\end{array}
\xrightarrow{\text{Alternative Pivoting Strategy}}
\begin{array}{c} BV\\ s_1\\ x_1\\ s_3\\ P\end{array}\begin{array}{c} \begin{array}{cccccccc} P & x_1 & x_2 & x_3 & s_1 & s_2 & s_3 & RHS\end{array}\\ \left[\begin{array}{ccccccc|c} 0 & 0 & \frac{3}{2} & \frac{3}{2} & 1 & \frac{1}{2} & 0 & 8\\ 0 & 1 & \frac{3}{2} & -\frac{1}{2} & 0 & -\frac{1}{2} & 0 & 1\\ 0 & 0 & \frac{13}{2} & \enclose{circle}{-\frac{5}{2}} & 0 & -\frac{3}{2} & 1 & -2\\ \hline 1 & 0 & \frac{5}{2} & -\frac{1}{2} & 0 & -\frac{3}{2} & 0 & 3\end{array}\right]\end{array}$$

$$\xrightarrow{\text{Alternative Pivoting Strategy}}
\begin{array}{c} BV\\ s_1\\ x_1\\ x_3\\ P\end{array}\begin{array}{c} \begin{array}{cccccccc} P & x_1 & x_2 & x_3 & s_1 & s_2 & s_3 & RHS\end{array}\\ \left[\begin{array}{ccccccc|c} 0 & 0 & \frac{27}{5} & 0 & 1 & -\frac{2}{5} & \frac{3}{5} & \frac{34}{5}\\ 0 & 1 & \frac{1}{5} & 0 & 0 & -\frac{1}{5} & -\frac{1}{5} & \frac{7}{5}\\ 0 & 0 & -\frac{13}{5} & 1 & 0 & \enclose{circle}{\frac{3}{5}} & -\frac{2}{5} & \frac{4}{5}\\ \hline 1 & 0 & \frac{6}{5} & 0 & 0 & -\frac{6}{5} & -\frac{1}{5} & \frac{17}{5}\end{array}\right]\end{array}$$

$$\xrightarrow{\text{Standard Pivoting Strategy}}
\begin{array}{c} BV\\ s_1\\ x_1\\ s_2\\ P\end{array}\begin{array}{c} \begin{array}{cccccccc} P & x_1 & x_2 & x_3 & s_1 & s_2 & s_3 & RHS\end{array}\\ \left[\begin{array}{ccccccc|c} 0 & 0 & \enclose{circle}{\frac{11}{3}} & \frac{2}{3} & 1 & 0 & \frac{1}{3} & \frac{22}{3}\\ 0 & 1 & -\frac{2}{3} & \frac{1}{3} & 0 & 0 & -\frac{1}{3} & \frac{5}{3}\\ 0 & 0 & -\frac{13}{3} & \frac{5}{3} & 0 & 1 & -\frac{2}{3} & \frac{4}{3}\\ \hline 1 & 0 & -4 & 2 & 0 & 0 & -1 & 5\end{array}\right]\end{array}$$

$$\rightarrow
\begin{array}{c} BV\\ x_2\\ x_1\\ s_2\\ P\end{array}\begin{array}{c} \begin{array}{cccccccc} P & x_1 & x_2 & x_3 & s_1 & s_2 & s_3 & RHS\end{array}\\ \left[\begin{array}{ccccccc|c} 0 & 0 & 1 & \frac{2}{11} & \frac{3}{11} & 0 & \enclose{circle}{\frac{1}{11}} & 2\\ 0 & 1 & 0 & \frac{5}{11} & \frac{2}{11} & 0 & -\frac{3}{11} & 3\\ 0 & 0 & 0 & \frac{27}{11} & \frac{13}{11} & 1 & -\frac{3}{11} & 10\\ \hline 1 & 0 & 0 & \frac{30}{11} & \frac{12}{11} & 0 & -\frac{7}{11} & 13\end{array}\right]\end{array}
\rightarrow
\begin{array}{c} BV\\ s_3\\ x_1\\ s_2\\ P\end{array}\begin{array}{c} \begin{array}{cccccccc} P & x_1 & x_2 & x_3 & s_1 & s_2 & s_3 & RHS\end{array}\\ \left[\begin{array}{ccccccc|c} 0 & 0 & 11 & 2 & 3 & 0 & 1 & 22\\ 0 & 1 & 3 & 1 & 1 & 0 & 0 & 9\\ 0 & 0 & 3 & 3 & 2 & 1 & 0 & 16\\ \hline 1 & 0 & 7 & 4 & 3 & 0 & 0 & 27\end{array}\right]\end{array}$$

The maximum value of $P$ is 27, when $x_1 = 9$, $x_2 = 0$ and $x_3 = 0$.

**5.** Maximize $P = -z = -6x_1 - 8x_2 - x_3$.

Rewrite the constraints:

$$\begin{aligned} -3x_1-5x_2-3x_3 &\le -20\\ -x_1-3x_2-2x_3 &\le -9\\ -6x_1-2x_2-5x_3 &\le -30\\ x_1+x_2+x_3 &\le 10\\ x_1\ge 0, x_2 &\ge 0\end{aligned}$$

Introduce slack variables:

$$\begin{aligned} -3x_1-5x_2-3x_3+s_1 \qquad\qquad\qquad &= -20\\ -x_1-3x_2-2x_3 \qquad +s_2 \qquad\qquad &= -9\\ -6x_1-2x_2-5x_3 \qquad\qquad +s_3 \qquad &= -30\\ x_1+x_2+x_3 \qquad\qquad\qquad +s_4 &= 10\\ s_1\ge 0, s_2\ge 0, s_3\ge 0, s_4 &\ge 0\end{aligned}$$

Initial Tableau:

| BV | P | $x_1$ | $x_2$ | $x_3$ | $s_1$ | $s_2$ | $s_3$ | $s_4$ | RHS |
|---|---|---|---|---|---|---|---|---|---|
| $s_1$ | 0 | $\boxed{-3}$ | $-5$ | $-3$ | 1 | 0 | 0 | 0 | $-20$ |
| $s_2$ | 0 | $-1$ | $-3$ | $-2$ | 0 | 1 | 0 | 0 | $-9$ |
| $s_3$ | 0 | $-6$ | $-2$ | $-5$ | 0 | 0 | 1 | 0 | $-30$ |
| $s_4$ | 0 | 1 | 1 | 1 | 0 | 0 | 0 | 1 | 10 |
| $P$ | 1 | 6 | 8 | 1 | 0 | 0 | 0 | 0 | 0 |

Alternative Pivoting Strategy → $s_3$

| BV | P | $x_1$ | $x_2$ | $x_3$ | $s_1$ | $s_2$ | $s_3$ | $s_4$ | RHS |
|---|---|---|---|---|---|---|---|---|---|
| $x_1$ | 0 | 1 | $\frac{5}{3}$ | 1 | $-\frac{1}{3}$ | 0 | 0 | 0 | $\frac{20}{3}$ |
| $s_2$ | 0 | 0 | $\boxed{-\frac{4}{3}}$ | $-1$ | $-\frac{1}{3}$ | 1 | 0 | 0 | $-\frac{7}{3}$ |
| $s_3$ | 0 | 0 | 8 | 1 | $-2$ | 0 | 1 | 0 | 10 |
| $s_4$ | 0 | 0 | $-\frac{2}{3}$ | 0 | $\frac{1}{3}$ | 0 | 0 | 1 | $\frac{10}{3}$ |
| $P$ | 1 | 0 | $-2$ | $-5$ | 2 | 0 | 0 | 0 | $-40$ |

Alternative Pivoting Strategy → $s_3$

| BV | P | $x_1$ | $x_2$ | $x_3$ | $s_1$ | $s_2$ | $s_3$ | $s_4$ | RHS |
|---|---|---|---|---|---|---|---|---|---|
| $x_1$ | 0 | 1 | 0 | $-\frac{1}{4}$ | $-\frac{3}{4}$ | $\frac{5}{4}$ | 0 | 0 | $\frac{15}{4}$ |
| $x_2$ | 0 | 0 | 1 | $\frac{3}{4}$ | $\frac{1}{4}$ | $-\frac{3}{4}$ | 0 | 0 | $\frac{7}{4}$ |
| $s_3$ | 0 | 0 | 0 | $\boxed{-5}$ | $-4$ | 6 | 1 | 0 | $-4$ |
| $s_4$ | 0 | 0 | 0 | $\frac{1}{2}$ | $\frac{1}{2}$ | $-\frac{1}{2}$ | 0 | 1 | $\frac{9}{2}$ |
| $P$ | 1 | 0 | 0 | $-\frac{7}{2}$ | $\frac{5}{2}$ | $-\frac{3}{2}$ | 0 | 0 | $-\frac{73}{2}$ |

Alternative Pivoting Strategy → $x_3$

| BV | P | $x_1$ | $x_2$ | $x_3$ | $s_1$ | $s_2$ | $s_3$ | $s_4$ | RHS |
|---|---|---|---|---|---|---|---|---|---|
| $x_1$ | 0 | 1 | 0 | 0 | $-\frac{11}{20}$ | $\boxed{\frac{19}{20}}$ | $-\frac{1}{20}$ | 0 | $\frac{79}{20}$ |
| $x_2$ | 0 | 0 | 1 | 0 | $-\frac{7}{20}$ | $\frac{3}{20}$ | $\frac{3}{20}$ | 0 | $\frac{23}{20}$ |
| $x_3$ | 0 | 0 | 0 | 1 | $\frac{4}{5}$ | $-\frac{6}{5}$ | $-\frac{1}{5}$ | 0 | $\frac{4}{5}$ |
| $s_4$ | 0 | 0 | 0 | 0 | $\frac{1}{10}$ | $\frac{1}{10}$ | $\frac{1}{10}$ | 1 | $\frac{41}{10}$ |
| $P$ | 1 | 0 | 0 | 0 | $\frac{53}{10}$ | $-\frac{57}{10}$ | $-\frac{7}{10}$ | 0 | $-\frac{337}{10}$ |

Standard Pivoting Strategy → $x_3$

| BV | P | $x_1$ | $x_2$ | $x_3$ | $s_1$ | $s_2$ | $s_3$ | $s_4$ | RHS |
|---|---|---|---|---|---|---|---|---|---|
| $s_2$ | 0 | $\frac{20}{19}$ | 0 | 0 | $-\frac{11}{19}$ | 1 | $-\frac{1}{19}$ | 0 | $\frac{79}{19}$ |
| $x_2$ | 0 | $-\frac{3}{19}$ | 1 | 0 | $-\frac{5}{19}$ | 0 | $\boxed{\frac{3}{19}}$ | 0 | $\frac{10}{19}$ |
| $x_3$ | 0 | $\frac{24}{19}$ | 0 | 1 | $\frac{2}{19}$ | 0 | $-\frac{5}{19}$ | 0 | $\frac{110}{19}$ |
| $s_4$ | 0 | $-\frac{2}{19}$ | 0 | 0 | $\frac{3}{19}$ | 0 | $\frac{2}{19}$ | 1 | $\frac{70}{19}$ |
| $P$ | 1 | 6 | 0 | 0 | 2 | 0 | $-1$ | 0 | $-10$ |

| BV | P | $x_1$ | $x_2$ | $x_3$ | $s_1$ | $s_2$ | $s_3$ | $s_4$ | RHS |
|---|---|---|---|---|---|---|---|---|---|
| $s_2$ | 0 | 1 | $\frac{1}{3}$ | 0 | $-\frac{2}{3}$ | 1 | 0 | 0 | $\frac{13}{3}$ |
| $s_3$ | 0 | $-1$ | $\frac{19}{3}$ | 0 | $-\frac{5}{3}$ | 0 | 1 | 0 | $\frac{10}{3}$ |
| $\to x_3$ | 0 | 1 | $\frac{5}{3}$ | 1 | $-\frac{1}{3}$ | 0 | 0 | 0 | $\frac{20}{3}$ |
| $s_4$ | 0 | 0 | $-\frac{2}{3}$ | 0 | $\frac{1}{3}$ | 0 | 0 | 1 | $\frac{10}{3}$ |
| $P$ | 1 | 5 | $\frac{19}{3}$ | 0 | $\frac{1}{3}$ | 0 | 0 | 0 | $-\frac{20}{3}$ |

The maximum value for $P$ is $-\frac{20}{3}$, so the minimum value for $z$ is $\frac{20}{3}$. This is attained when $x_1 = 0, x_2 = 0$, $x_3 = \frac{20}{3}$.

**7.** Rewrite the constraints:

$$2x_1 + x_2 \le 4$$
$$x_1 + x_2 \le 3$$
$$-x_1 - x_2 \le -3$$
$$x_1 \ge 0, x_2 \ge 0$$

Introduce slack variables:

$$2x_1 + x_2 + s_1 = 4$$
$$x_1 + x_2 + s_2 = 3$$
$$-x_1 - x_2 + s_3 = -3$$
$$s_1 \ge 0, s_2 \ge 0, s_3 \ge 0$$

Initial Tableau:

| BV | P | $x_1$ | $x_2$ | $s_1$ | $s_2$ | $s_3$ | RHS |
|---|---|---|---|---|---|---|---|
| $s_1$ | 0 | 2 | 1 | 1 | 0 | 0 | 4 |
| $s_2$ | 0 | 1 | 1 | 0 | 1 | 0 | 3 |
| $s_3$ | 0 | (−1) | −1 | 0 | 0 | 1 | −3 |
| $P$ | 1 | −3 | −2 | 0 | 0 | 0 | 0 |

Alternative Pivoting Strategy →

| BV | P | $x_1$ | $x_2$ | $s_1$ | $s_2$ | $s_3$ | RHS |
|---|---|---|---|---|---|---|---|
| $s_1$ | 0 | 0 | (−1) | 1 | 0 | 2 | −2 |
| $s_2$ | 0 | 0 | 0 | 0 | 1 | 1 | 0 |
| $x_1$ | 0 | 1 | 1 | 0 | 0 | −1 | 3 |
| $P$ | 1 | 0 | 1 | 0 | 0 | −3 | 9 |

Alternative Pivoting Strategy →

| BV | P | $x_1$ | $x_2$ | $s_1$ | $s_2$ | $s_3$ | RHS |
|---|---|---|---|---|---|---|---|
| $x_2$ | 0 | 0 | 1 | −1 | 0 | −2 | 2 |
| $s_2$ | 0 | 0 | 0 | 0 | 1 | (1) | 0 |
| $x_1$ | 0 | 1 | 0 | 1 | 0 | 1 | 1 |
| $P$ | 1 | 0 | 0 | 1 | 0 | −1 | 7 |

Standard Pivoting Strategy →

| BV | P | $x_1$ | $x_2$ | $s_1$ | $s_2$ | $s_3$ | RHS |
|---|---|---|---|---|---|---|---|
| $x_2$ | 0 | 0 | 1 | −1 | 2 | 0 | 2 |
| $s_3$ | 0 | 0 | 0 | 0 | 1 | 1 | 0 |
| $x_1$ | 0 | 1 | 0 | 1 | −1 | 0 | 1 |
| $P$ | 1 | 0 | 0 | 1 | 1 | 0 | 7 |

The maximum value for $P$ is 7, when $x_1 = 1$ and $x_2 = 2$.

**9.** As suggested by the hint, let $x_1$ = the number of units shipped from M1 to A1, $x_2$ = the number of units shipped from M1 to A2, $x_3$ = the number of units shipped from M2 to A1 and $x_4$ = the number of units shipped from M2 to A2. The objective is to minimize shipping costs $C = 400x_1 + 100x_2 + 200x_3 + 300x_4$ dollars.
Constraints are imposed by the production capacities of plants M1 and M2, and by the needs of assembly plants A1 and A2.

$$\text{minimize } C = 400x_1 + 100x_2 + 200x_3 + 300x_4,$$

where

$$\begin{cases} \text{M1 capacity}: x_1 + x_2 \leq 600 \\ \text{M2 capacity}: x_3 + x_4 \leq 400 \\ \text{A1 needs}: x_1 + x_3 \geq 500 \\ \text{A2 needs}: x_2 + x_4 \geq 300 \\ x_1 \geq 0, x_2 \geq 0, x_3 \geq 0, x_4 \geq 0 \end{cases}$$

Next transform the problem to a maximization problem:

maximize $P = -C = -400x_1 - 100x_2 - 200x_3 - 300x_4$, subject to the same constraints.

Rewrite the constraints:

$$\begin{aligned} x_1 + x_2 &\leq 600 \\ x_3 + x_4 &\leq 400 \\ -x_1 - x_3 &\leq -500 \\ -x_2 - x_4 &\leq -300 \\ x_1 \geq 0, x_2 \geq 0, x_3 &\geq 0, x_4 \geq 0 \end{aligned}$$

Introduce slack variables:

$$\begin{aligned} x_1 + x_2 + s_1 &= 600 \\ x_3 + x_4 + s_2 &= 400 \\ -x_1 - x_3 + s_3 &= -500 \\ -x_2 - x_4 + s_4 &= -300 \\ s_1 \geq 0, s_2 \geq 0, s_3 \geq 0, s_4 &\geq 0 \end{aligned}$$

Initial tableau:

| BV | P | $x_1$ | $x_2$ | $x_3$ | $x_4$ | $s_1$ | $s_2$ | $s_3$ | $s_4$ | RHS |
|---|---|---|---|---|---|---|---|---|---|---|
| $s_1$ | 0 | 1 | 1 | 0 | 0 | 1 | 0 | 0 | 0 | 600 |
| $s_2$ | 0 | 0 | 0 | 1 | 1 | 0 | 1 | 0 | 0 | 400 |
| $s_3$ | 0 | (−1) | 0 | −1 | 0 | 0 | 0 | 1 | 0 | −500 |
| $s_4$ | 0 | 0 | −1 | 0 | −1 | 0 | 0 | 0 | 1 | −300 |
| P | 1 | 400 | 100 | 200 | 300 | 0 | 0 | 0 | 0 | 0 |

Alternative Pivoting Strategy → $x_1$

| BV | P | $x_1$ | $x_2$ | $x_3$ | $x_4$ | $s_1$ | $s_2$ | $s_3$ | $s_4$ | RHS |
|---|---|---|---|---|---|---|---|---|---|---|
| $s_1$ | 0 | 0 | 1 | −1 | 0 | 1 | 0 | 1 | 0 | 100 |
| $s_2$ | 0 | 0 | 0 | 1 | 1 | 0 | 1 | 0 | 0 | 400 |
| $x_1$ | 0 | 1 | 0 | 1 | 0 | 0 | 0 | −1 | 0 | 500 |
| $s_4$ | 0 | 0 | (−1) | 0 | −1 | 0 | 0 | 0 | 1 | −300 |
| P | 1 | 0 | 100 | −200 | 300 | 0 | 0 | 400 | 0 | −200,000 |

Alternative Pivoting Strategy → $x_1$

| BV | P | $x_1$ | $x_2$ | $x_3$ | $x_4$ | $s_1$ | $s_2$ | $s_3$ | $s_4$ | RHS |
|---|---|---|---|---|---|---|---|---|---|---|
| $s_1$ | 0 | 0 | 0 | (−1) | −1 | 1 | 0 | 1 | 1 | −200 |
| $s_2$ | 0 | 0 | 0 | 1 | 1 | 0 | 1 | 0 | 0 | 400 |
| $x_1$ | 0 | 1 | 0 | 1 | 0 | 0 | 0 | −1 | 0 | 500 |
| $x_2$ | 0 | 0 | 1 | 0 | 1 | 0 | 0 | 0 | −1 | 300 |
| P | 1 | 0 | 0 | −200 | 200 | 0 | 0 | 400 | 100 | −230,000 |

Alternative Pivoting Strategy → $x_1$

| BV | P | $x_1$ | $x_2$ | $x_3$ | $x_4$ | $s_1$ | $s_2$ | $s_3$ | $s_4$ | RHS |
|---|---|---|---|---|---|---|---|---|---|---|
| $x_3$ | 0 | 0 | 0 | 1 | 1 | −1 | 0 | −1 | −1 | 200 |
| $s_2$ | 0 | 0 | 0 | 0 | 0 | (1) | 1 | 1 | 1 | 200 |
| $x_1$ | 0 | 1 | 0 | 0 | −1 | 1 | 0 | 0 | 1 | 300 |
| $x_2$ | 0 | 0 | 1 | 0 | 1 | 0 | 0 | 0 | −1 | 300 |
| $P$ | 1 | 0 | 0 | 0 | 400 | −200 | 0 | 200 | −100 | −190,000 |

Standard Pivoting Strategy → $x_1$

| BV | P | $x_1$ | $x_2$ | $x_3$ | $x_4$ | $s_1$ | $s_2$ | $s_3$ | $s_4$ | RHS |
|---|---|---|---|---|---|---|---|---|---|---|
| $x_3$ | 0 | 0 | 0 | 1 | 1 | 0 | 1 | 0 | 0 | 400 |
| $s_1$ | 0 | 0 | 0 | 0 | 0 | 1 | 1 | 1 | 1 | 200 |
| $x_1$ | 0 | 1 | 0 | 0 | −1 | 0 | −1 | −1 | 0 | 100 |
| $x_2$ | 0 | 0 | 1 | 0 | 1 | 0 | 0 | 0 | −1 | 300 |
| $P$ | 1 | 0 | 0 | 0 | 400 | 0 | 200 | 400 | 100 | −150,000 |

The maximum value for $P$ is –150,000, so the minimum shipping cost is $150,000. This is attained when $x_1 = 100$, $x_2 = 300$, $x_3 = 400$ and $x_4 = 0$. Ship 100 engines from M1 to A1, 300 engines from M1 to A2, 400 engines from M2 to A1, and no engines from M2 to A2 for a minimum total shipping charge of $150,000.

**11.** Let $x_1$, $x_2$, $x_3$ represent the number of units of foods I, II, III, respectively, in the mixture. The objective is to minimize the cost of the mixture $C = 2x_1 + x_2 + 3x_3$ dollars. Constraints are imposed by the requirements for protein and carbohydrates.

minimize $C = 2x_1 + x_2 + 3x_3$, where

$$\begin{cases} \text{protein}: 2x_1 + 3x_2 + 4x_3 \ge 20 \text{ ounces} \\ \text{carbohydrates}: 4x_1 + 2x_2 + 2x_3 \ge 15 \text{ ounces} \\ x_1 \ge 0, x_2 \ge 0, x_3 \ge 0 \end{cases}$$

Next transform the problem to a maximization problem: maximize $P = -C = -2x_1 - x_2 - 3x_3$, subject to the same constraints.

Rewrite the constraints:

$$-2x_1 - 3x_2 - 4x_3 \le -20$$
$$-4x_1 - 2x_2 - 2x_3 \le -15$$
$$x_1 \ge 0, x_2 \ge 0, x_3 \ge 0$$

Introduce slack variables:

$$-2x_1 - 3x_2 - 4x_3 + s_1 = -20$$
$$-4x_1 - 2x_2 - 2x_3 + s_2 = -15$$
$$s_1 \ge 0, s_2 \ge 0$$

Initial tableau:

| BV | P | $x_1$ | $x_2$ | $x_3$ | $s_1$ | $s_2$ | RHS |
|---|---|---|---|---|---|---|---|
| $s_1$ | 0 | (−2) | −3 | −4 | 1 | 0 | −20 |
| $s_2$ | 0 | −4 | −2 | −2 | 0 | 1 | −15 |
| $P$ | 1 | 2 | 1 | 3 | 0 | 0 | 0 |

Alternative Pivoting Strategy → $s_2$

| BV | P | $x_1$ | $x_2$ | $x_3$ | $s_1$ | $s_2$ | RHS |
|---|---|---|---|---|---|---|---|
| $x_1$ | 0 | 1 | $\frac{3}{2}$ | 2 | $-\frac{1}{2}$ | 0 | 10 |
| $s_2$ | 0 | 0 | (4) | 6 | −2 | 1 | 25 |
| $P$ | 1 | 0 | −2 | −1 | 1 | 0 | −20 |

| | BV | P | $x_1$ | $x_2$ | $x_3$ | $s_1$ | $s_2$ | RHS |
|---|---|---|---|---|---|---|---|---|
| | $x_1$ | 0 | 1 | 0 | $-\frac{1}{4}$ | $\frac{1}{4}$ | $-\frac{3}{8}$ | $\frac{5}{8}$ |
| Standard Pivoting Strategy → | $x_2$ | 0 | 0 | 1 | $\frac{3}{2}$ | $-\frac{1}{2}$ | $\frac{1}{4}$ | $\frac{25}{4}$ |
| | $P$ | 1 | 0 | 0 | 2 | 0 | $\frac{1}{2}$ | $-\frac{15}{2}$ |

Maximum $P=-C=-\dfrac{15}{2}$, so minimum $C=\dfrac{15}{2}$, when $x_1=\dfrac{5}{8}$, $x_2=\dfrac{25}{4}$ and $x_3=0$. Thus the cheapest mixture satisfying the requirements for protein and carbohydrates costs $7.50 and consists of $\dfrac{5}{8}=0.625$ units of food I, $\dfrac{25}{4}=6.25$ units of food II and no units of food III.

**13.** Let $x_1$ represent the number of television sets shipped from $W_1$ to $R_1$; $x_2$, the number shipped from $W_1$ to $R_2$; $x_3$, the number shipped from $W_2$ to $R_1$; and $x_4$, the number shipped from $W_2$ to $R_2$. The objective is to minimize cost $C=8x_1+12x_2+13x_3+7x_4$ dollars. Constraints are imposed by the number of television sets ordered by the retailers, and the number available in storage: $x_1+x_3=55$, $x_2+x_4=75$, $x_1+x_2\le 100$, $x_3+x_4\le 120$, $x_1\ge 0$, $x_2\ge 0$, $x_3\ge 0$, $x_4\ge 0$.

Transform the problem to a maximization problem:
maximize $P=-C=-8x_1-12x_2-13x_3-7x_4$ subject to the same conatraints.

| Rewrite the constraints: | Introduce slack variables: |
|---|---|
| $x_1+x_3\le 55$ | $x_1+x_3+s_1 = 55$ |
| $-x_1-x_3\le -55$ | $-x_1-x_3+s_2 = -55$ |
| $x_2+x_4\le 75$ | $x_2+x_4+s_3 = 75$ |
| $-x_2-x_4\le -75$ | $-x_2-x_4+s_4 = -75$ |
| $x_1+x_2\le 100$ | $x_1+x_2+s_5 = 100$ |
| $x_3+x_4\le 120$ | $x_3+x_4+s_6 = 120$ |
| $x_1\ge 0, x_2\ge 0, x_3\ge 0, x_4\ge 0$ | $s_1\ge 0, s_2\ge 0, s_3\ge 0, s_4\ge 0, s_5\ge 0, s_6\ge 0$ |

Initial tableau:

| BV | P | $x_1$ | $x_2$ | $x_3$ | $x_4$ | $s_1$ | $s_2$ | $s_3$ | $s_4$ | $s_5$ | $s_6$ | RHS |
|---|---|---|---|---|---|---|---|---|---|---|---|---|
| $s_1$ | 0 | 1 | 0 | 1 | 0 | 1 | 0 | 0 | 0 | 0 | 0 | 55 |
| $s_2$ | 0 | (−1) | 0 | −1 | 0 | 0 | 1 | 0 | 0 | 0 | 0 | −55 |
| $s_3$ | 0 | 0 | 1 | 0 | 1 | 0 | 0 | 1 | 0 | 0 | 0 | 75 |
| $s_4$ | 0 | 0 | −1 | 0 | −1 | 0 | 0 | 0 | 1 | 0 | 0 | −75 |
| $s_5$ | 0 | 1 | 1 | 0 | 0 | 0 | 0 | 0 | 0 | 1 | 0 | 100 |
| $s_6$ | 0 | 0 | 0 | 1 | 1 | 0 | 0 | 0 | 0 | 0 | 1 | 120 |
| $P$ | 1 | 8 | 12 | 13 | 7 | 0 | 0 | 0 | 0 | 0 | 0 | 0 |

Alternative Pivoting Strategy → $s_4$

| BV | $P$ | $x_1$ | $x_2$ | $x_3$ | $x_4$ | $s_1$ | $s_2$ | $s_3$ | $s_4$ | $s_5$ | $s_6$ | RHS |
|---|---|---|---|---|---|---|---|---|---|---|---|---|
| $s_1$ | 0 | 0 | 0 | 0 | 0 | 1 | 1 | 0 | 0 | 0 | 0 | 0 |
| $x_1$ | 0 | 1 | 0 | 1 | 0 | 0 | −1 | 0 | 0 | 0 | 0 | 55 |
| $s_3$ | 0 | 0 | 1 | 0 | 1 | 0 | 0 | 1 | 0 | 0 | 0 | 75 |
| $s_4$ | 0 | 0 | **(−1)** | 0 | −1 | 0 | 0 | 0 | 1 | 0 | 0 | −75 |
| $s_5$ | 0 | 0 | 1 | −1 | 0 | 0 | 1 | 0 | 0 | 1 | 0 | 45 |
| $s_6$ | 0 | 0 | 0 | 1 | 1 | 0 | 0 | 0 | 0 | 0 | 1 | 120 |
| $P$ | 1 | 0 | 12 | 5 | 7 | 0 | 8 | 0 | 0 | 0 | 0 | −440 |

Alternative Pivoting Strategy → $x_2$

| BV | $P$ | $x_1$ | $x_2$ | $x_3$ | $x_4$ | $s_1$ | $s_2$ | $s_3$ | $s_4$ | $s_5$ | $s_6$ | RHS |
|---|---|---|---|---|---|---|---|---|---|---|---|---|
| $s_1$ | 0 | 0 | 0 | 0 | 0 | 1 | 1 | 0 | 0 | 0 | 0 | 0 |
| $x_1$ | 0 | 1 | 0 | 1 | 0 | 0 | −1 | 0 | 0 | 0 | 0 | 55 |
| $s_3$ | 0 | 0 | 0 | 0 | 0 | 0 | 0 | 1 | 1 | 0 | 0 | 0 |
| $x_2$ | 0 | 0 | 1 | 0 | 1 | 0 | 0 | 0 | −1 | 0 | 0 | 75 |
| $s_5$ | 0 | 0 | 0 | **(−1)** | −1 | 0 | 1 | 0 | 1 | 1 | 0 | −30 |
| $s_6$ | 0 | 0 | 0 | 1 | 1 | 0 | 0 | 0 | 0 | 0 | 1 | 120 |
| $P$ | 1 | 0 | 0 | 5 | −5 | 0 | 8 | 0 | 12 | 0 | 0 | −1,340 |

Alternative Pivoting Strategy → $x_2$

| BV | $P$ | $x_1$ | $x_2$ | $x_3$ | $x_4$ | $s_1$ | $s_2$ | $s_3$ | $s_4$ | $s_5$ | $s_6$ | RHS |
|---|---|---|---|---|---|---|---|---|---|---|---|---|
| $s_1$ | 0 | 0 | 0 | 0 | 0 | 1 | 1 | 0 | 0 | 0 | 0 | 0 |
| $x_1$ | 0 | 1 | 0 | 0 | −1 | 0 | 0 | 0 | 1 | 1 | 0 | 25 |
| $s_3$ | 0 | 0 | 0 | 0 | 0 | 0 | 0 | 1 | 1 | 0 | 0 | 0 |
| $x_2$ | 0 | 0 | 1 | 0 | 1 | 0 | 0 | 0 | −1 | 0 | 0 | 75 |
| $x_3$ | 0 | 0 | 0 | 1 | **(1)** | 0 | −1 | 0 | −1 | −1 | 0 | 30 |
| $s_6$ | 0 | 0 | 0 | 0 | 0 | 0 | 1 | 0 | 1 | 1 | 1 | 90 |
| $P$ | 1 | 0 | 0 | 0 | −10 | 0 | 13 | 0 | 17 | 5 | 0 | −1,490 |

Standard Pivoting Strategy → $x_2$

| BV | $P$ | $x_1$ | $x_2$ | $x_3$ | $x_4$ | $s_1$ | $s_2$ | $s_3$ | $s_4$ | $s_5$ | $s_6$ | RHS |
|---|---|---|---|---|---|---|---|---|---|---|---|---|
| $s_1$ | 0 | 0 | 0 | 0 | 0 | 1 | 1 | 0 | 0 | 0 | 0 | 0 |
| $x_1$ | 0 | 1 | 0 | 1 | 0 | 0 | −1 | 0 | 0 | 0 | 0 | 55 |
| $s_3$ | 0 | 0 | 0 | 0 | 0 | 0 | 0 | 1 | 1 | 0 | 0 | 0 |
| $x_2$ | 0 | 0 | 1 | −1 | 0 | 0 | 1 | 0 | 0 | **(1)** | 0 | 45 |
| $x_4$ | 0 | 0 | 0 | 1 | 1 | 0 | −1 | 0 | −1 | −1 | 0 | 30 |
| $s_6$ | 0 | 0 | 0 | 0 | 0 | 0 | 1 | 0 | 1 | 1 | 1 | 90 |
| $P$ | 1 | 0 | 0 | 10 | 0 | 0 | 3 | 0 | 7 | −5 | 0 | −1,190 |

| BV | P | $x_1$ | $x_2$ | $x_3$ | $x_4$ | $s_1$ | $s_2$ | $s_3$ | $s_4$ | $s_5$ | $s_6$ | RHS |
|---|---|---|---|---|---|---|---|---|---|---|---|---|
| $s_1$ | 0 | 0 | 0 | 0 | 0 | 1 | 1 | 0 | 0 | 0 | 0 | 0 |
| $x_1$ | 0 | 1 | 0 | 1 | 0 | 0 | −1 | 0 | 0 | 0 | 0 | 55 |
| $s_3$ | 0 | 0 | 0 | 0 | 0 | 0 | 0 | 1 | 1 | 0 | 0 | 0 |
| → $s_5$ | 0 | 0 | 1 | −1 | 0 | 0 | 1 | 0 | 0 | 1 | 0 | 45 |
| $x_4$ | 0 | 0 | 1 | 0 | 1 | 0 | 0 | 0 | −1 | 0 | 0 | 75 |
| $s_6$ | 0 | 0 | −1 | 1 | 0 | 0 | 0 | 0 | 1 | 0 | 1 | 45 |
| $P$ | 1 | 0 | 5 | 5 | 0 | 0 | 8 | 0 | 7 | 0 | 0 | −965 |

The maximum value for $P$ is –965, so the minimum value for $C$ is 965. This is attained when $x_1 = 55$, $x_2 = 0$, $x_3 = 0$, and $x_4 = 75$. Ship 55 televisions from warehouse $W_1$ to retailer $R_1$, and 75 televisions from warehouse $W_2$ to retailer $R_2$ for a minimum cost of \$965.

**15.** Let $x_1$ and $x_2$ represent the number of pounds of raw materials A and B, respectively. The objective is to minimize cost $C = 4x_1 + 8x_2$ dollars. The constraints are imposed by the number of units of each raw material needed to produce the machine: $5x_1 + 10x_2 = 150$ pounds, $x_1 \le 20$, $x_2 \ge 14$, $x_1 \ge 0, x_2 \ge 0$.
Transform the problem to a maximization problem:
maximize $P = -C = -4x_1 - 8x_2$ subject to the same constraints.

Rewrite the constraints:

$$5x_1 + 10x_2 \le 150$$
$$-5x_1 - 10x_2 \le -150$$
$$x_1 \le 20$$
$$-x_2 \le -14$$
$$x_1 \ge 0, x_2 \ge 0$$

Introduce slack variables:

$$x_1 + 10x_2 + s_1 = 150$$
$$-5x_1 - 10x_2 + s_2 = -150$$
$$x_1 + s_3 = 20$$
$$-x_2 + s_4 = -14$$
$$s_1 \ge 0, s_2 \ge 0, s_3 \ge 0, s_4 \ge 0$$

Initial tableau:

| BV | P | $x_1$ | $x_2$ | $s_1$ | $s_2$ | $s_3$ | $s_4$ | RHS |
|---|---|---|---|---|---|---|---|---|
| $s_1$ | 0 | 5 | 10 | 1 | 0 | 0 | 0 | 150 |
| $s_2$ | 0 | (−5) | −10 | 0 | 1 | 0 | 0 | −150 |
| $s_3$ | 0 | 1 | 0 | 0 | 0 | 1 | 0 | 20 |
| $s_4$ | 0 | 0 | −1 | 0 | 0 | 0 | 1 | −14 |
| $P$ | 1 | 4 | 8 | 0 | 0 | 0 | 0 | 0 |

Alternative Pivoting Strategy → $s_3$

| BV | $P$ | $x_1$ | $x_2$ | $s_1$ | $s_2$ | $s_3$ | $s_4$ | RHS |
|---|---|---|---|---|---|---|---|---|
| $s_1$ | 0 | 0 | 0 | 1 | 1 | 0 | 0 | 0 |
| $x_1$ | 0 | 1 | 2 | 0 | $-\frac{1}{5}$ | 0 | 0 | 30 |
| $s_3$ | 0 | 0 | **$-2$** | 0 | $\frac{1}{5}$ | 1 | 0 | $-10$ |
| $s_4$ | 0 | 0 | $-1$ | 0 | 0 | 0 | 1 | $-14$ |
| $P$ | 1 | 0 | 0 | 0 | $\frac{4}{5}$ | 0 | 0 | $-120$ |

Alternative Pivoting Strategy → $x_2$

| BV | $P$ | $x_1$ | $x_2$ | $s_1$ | $s_2$ | $s_3$ | $s_4$ | RHS |
|---|---|---|---|---|---|---|---|---|
| $s_1$ | 0 | 0 | 0 | 1 | 1 | 0 | 0 | 0 |
| $x_1$ | 0 | 1 | 0 | 0 | 0 | 1 | 0 | 20 |
| $x_2$ | 0 | 0 | 1 | 0 | $-\frac{1}{10}$ | $-\frac{1}{2}$ | 0 | 5 |
| $s_4$ | 0 | 0 | 0 | 0 | **$-\frac{1}{10}$** | $-\frac{1}{2}$ | 1 | $-9$ |
| $P$ | 1 | 0 | 0 | 0 | $\frac{4}{5}$ | 0 | 0 | $-120$ |

Alternative Pivoting Strategy → $x_2$

| BV | $P$ | $x_1$ | $x_2$ | $s_1$ | $s_2$ | $s_3$ | $s_4$ | RHS |
|---|---|---|---|---|---|---|---|---|
| $s_1$ | 0 | 0 | 0 | 1 | 0 | **$-5$** | 10 | $-90$ |
| $x_1$ | 0 | 1 | 0 | 0 | 0 | 1 | 0 | 20 |
| $x_2$ | 0 | 0 | 1 | 0 | 0 | 0 | $-1$ | 14 |
| $s_2$ | 0 | 0 | 0 | 0 | 1 | 5 | $-10$ | 90 |
| $P$ | 1 | 0 | 0 | 0 | 0 | $-4$ | 8 | $-192$ |

Alternative Pivoting Strategy → $x_2$

| BV | $P$ | $x_1$ | $x_2$ | $s_1$ | $s_2$ | $s_3$ | $s_4$ | RHS |
|---|---|---|---|---|---|---|---|---|
| $s_3$ | 0 | 0 | 0 | $-\frac{1}{5}$ | 0 | 1 | $-2$ | 18 |
| $x_1$ | 0 | 1 | 0 | $\frac{1}{5}$ | 0 | 0 | 2 | 2 |
| $x_2$ | 0 | 0 | 1 | 0 | 0 | 0 | $-1$ | 14 |
| $s_2$ | 0 | 0 | 0 | **1** | 1 | 0 | 0 | 0 |
| $P$ | 1 | 0 | 0 | $-\frac{4}{5}$ | 0 | 0 | 0 | $-120$ |

Standard Pivoting Strategy → $x_2$

| BV | $P$ | $x_1$ | $x_2$ | $s_1$ | $s_2$ | $s_3$ | $s_4$ | RHS |
|---|---|---|---|---|---|---|---|---|
| $s_3$ | 0 | 0 | 0 | 0 | $\frac{1}{5}$ | 1 | $-2$ | 18 |
| $x_1$ | 0 | 1 | 0 | 0 | $-\frac{1}{5}$ | 0 | 2 | 2 |
| $x_2$ | 0 | 0 | 1 | 0 | 0 | 0 | $-1$ | 14 |
| $s_1$ | 0 | 0 | 0 | 1 | 1 | 0 | 0 | 0 |
| $P$ | 1 | 0 | 0 | 0 | $\frac{4}{5}$ | 0 | 0 | $-120$ |

The maximum value for $P$ is $-120$, so the minimum value for $C$ is 120. This is attained when $x_1 = 2$ and $x_2 = 14$.

The minimum cost is \$120 when 2 units of raw material A and 14 units of raw material B are used for each unit of the final product.

## Chapter 4 Review

### True or False

**1.**   True            **3.**   True            **5.**   True

### Fill in the Blank

**1.**   slack variables            **3.**   $\geq$

### Review Exercises

**1.**

$$
\begin{array}{c|ccccccc|c}
BV & P & x_1 & x_2 & x_3 & s_1 & s_2 & s_3 & RHS \\
\hline
s_1 & 0 & 5 & 5 & 10 & 1 & 0 & 0 & 1000 \\
s_2 & 0 & 10 & 8 & 5 & 0 & 1 & 0 & 2000 \\
s_3 & 0 & 10 & \boxed{5} & 0 & 0 & 0 & 1 & 500 \\
\hline
P & 1 & -100 & -200 & -50 & 0 & 0 & 0 & 0
\end{array}
$$

$$
\begin{array}{c|ccccccc|c}
BV & P & x_1 & x_2 & x_3 & s_1 & s_2 & s_3 & RHS \\
\hline
s_1 & 0 & -5 & 0 & \boxed{10} & 1 & 0 & -1 & 500 \\
\to s_2 & 0 & -6 & 0 & 5 & 0 & 1 & -\frac{8}{5} & 1200 \\
x_2 & 0 & 2 & 1 & 0 & 0 & 0 & \frac{1}{5} & 100 \\
\hline
P & 1 & 300 & 0 & -50 & 0 & 0 & 40 & 20000
\end{array}
$$

$$
\begin{array}{c|ccccccc|c}
BV & P & x_1 & x_2 & x_3 & s_1 & s_2 & s_3 & RHS \\
\hline
x_3 & 0 & -\frac{1}{2} & 0 & 1 & \frac{1}{10} & 0 & -\frac{1}{10} & 50 \\
\to s_2 & 0 & -\frac{7}{2} & 0 & 0 & -\frac{1}{2} & 1 & -\frac{11}{10} & 950 \\
x_2 & 0 & 2 & 1 & 0 & 0 & 0 & \frac{1}{5} & 100 \\
\hline
P & 1 & 275 & 0 & 0 & 5 & 0 & 35 & 22500
\end{array}
$$

The maximum value for $P$ is 22,500, obtained when $x_1 = 0$, $x_2 = 100$ and $x_3 = 50$.

**3.**

$$
\begin{array}{c|cccccc|c}
BV & P & x_1 & x_2 & x_3 & s_1 & s_2 & RHS \\
\hline
s_1 & 0 & 2 & \boxed{2} & 1 & 1 & 0 & 8 \\
s_2 & 0 & 1 & -4 & 3 & 0 & 1 & 12 \\
\hline
P & 1 & -40 & -60 & -50 & 0 & 0 & 0
\end{array}
\to
\begin{array}{c|cccccc|c}
BV & P & x_1 & x_2 & x_3 & s_1 & s_2 & RHS \\
\hline
x_2 & 0 & 1 & 1 & \frac{1}{2} & \frac{1}{2} & 0 & 4 \\
s_2 & 0 & 5 & 0 & \boxed{5} & 2 & 1 & 28 \\
\hline
P & 1 & 20 & 0 & -20 & 30 & 0 & 240
\end{array}
$$

$$
\begin{array}{c|cccccc|c}
BV & P & x_1 & x_2 & x_3 & s_1 & s_2 & RHS \\
\hline
x_2 & 0 & \frac{1}{2} & 1 & 0 & \frac{3}{10} & -\frac{1}{10} & \frac{6}{5} \\
\to x_3 & 0 & 1 & 0 & 1 & \frac{2}{5} & \frac{1}{5} & \frac{28}{5} \\
\hline
P & 1 & 40 & 0 & 0 & 38 & 4 & 352
\end{array}
$$

The maximum value for $P$ is 352, obtained when $x_1 = 0$, $x_2 = \dfrac{6}{5}$ and $x_3 = \dfrac{28}{5}$.

**5.** This is a minimum problem in standard form.

Using Von Neumann duality: Matrix: $\left[\begin{array}{cc|c} 2 & 2 & 8 \\ 1 & -1 & 2 \\ 2 & 1 & 0 \end{array}\right]$. Transpose: $\left[\begin{array}{cc|c} 2 & 1 & 2 \\ 2 & -1 & 1 \\ 8 & 2 & 0 \end{array}\right]$.

Dual problem: Maximize $P = 8y_1 + 2y_2$ subject to $2y_1 + y_2 \le 2$, $2y_1 - y_2 \le 1$, $y_1 \ge 0$, $y_2 \ge 0$.

$$\begin{array}{c} \\ s_1 \\ s_2 \\ P \end{array}\begin{array}{c} BV\ P\quad y_1\quad y_2\quad s_1\quad s_2\quad RHS \\ \left[\begin{array}{ccccc|c} 0 & 2 & 1 & 1 & 0 & 2 \\ 0 & \textcircled{2} & -1 & 0 & 1 & 1 \\ \hline 1 & -8 & -2 & 0 & 0 & 0 \end{array}\right] \end{array} \rightarrow \begin{array}{c} \\ s_1 \\ y_1 \\ P \end{array}\begin{array}{c} BV\ P\quad y_1\quad y_2\quad s_1\quad s_2\quad RHS \\ \left[\begin{array}{ccccc|c} 0 & 0 & \textcircled{2} & 1 & -1 & 1 \\ 0 & 1 & -\frac{1}{2} & 0 & \frac{1}{2} & \frac{1}{2} \\ \hline 1 & 0 & -6 & 0 & 4 & 4 \end{array}\right] \end{array} \rightarrow$$

$$\begin{array}{c} \\ y_2 \\ y_1 \\ P \end{array}\begin{array}{c} BV\ P\quad y_1\quad y_2\quad s_1\quad s_2\quad RHS \\ \left[\begin{array}{ccccc|c} 0 & 0 & 1 & \frac{1}{2} & -\frac{1}{2} & \frac{1}{2} \\ 0 & 1 & 0 & \frac{1}{4} & \frac{1}{4} & \frac{3}{4} \\ \hline 1 & 0 & 0 & 3 & 1 & 7 \end{array}\right] \end{array}$$

The maximum value for $P$ is 7, when $y_1 = \frac{3}{4}$ and $y_2 = \frac{1}{2}$. The minimum value for $C$ is 7, when $x_1 = 3$ and $x_2 = 1$.

**7.** This is a minimum problem in standard form.

Using Von Neumann duality: Matrix: $\left[\begin{array}{ccc|c} 1 & 1 & 1 & 100 \\ 2 & 1 & 0 & 50 \\ 5 & 4 & 3 & 0 \end{array}\right]$. Transpose: $\left[\begin{array}{cc|c} 1 & 2 & 5 \\ 1 & 1 & 4 \\ 1 & 0 & 3 \\ 100 & 50 & 0 \end{array}\right]$.

Dual problem: Maximize $P = 100y_1 + 50y_2$ subject to $y_1 + 2y_2 \le 5$, $y_1 + y_2 \le 4$, $y_1 \le 3$, $y_1 \ge 0$, $y_2 \ge 0$.

$$\begin{array}{c} \\ s_1 \\ s_2 \\ s_3 \\ P \end{array}\begin{array}{c} BV\ P\quad y_1\quad y_2\quad s_1\quad s_2\quad s_3\quad RHS \\ \left[\begin{array}{cccccc|c} 0 & 1 & 2 & 1 & 0 & 0 & 5 \\ 0 & 1 & 1 & 0 & 1 & 0 & 4 \\ 0 & \textcircled{1} & 0 & 0 & 0 & 1 & 3 \\ \hline 1 & -100 & -50 & 0 & 0 & 0 & 0 \end{array}\right] \end{array} \begin{array}{c} \\ s_1 \\ \rightarrow s_2 \\ y_1 \\ P \end{array}\begin{array}{c} BV\ P\quad y_1\quad y_2\quad s_1\quad s_2\quad s_3\quad RHS \\ \left[\begin{array}{cccccc|c} 0 & 0 & 2 & 1 & 0 & -1 & 2 \\ 0 & 0 & \textcircled{1} & 0 & 1 & -1 & 1 \\ 0 & 1 & 0 & 0 & 0 & 1 & 3 \\ \hline 1 & 0 & -50 & 0 & 0 & 100 & 300 \end{array}\right] \end{array}$$

$$\begin{array}{c} \\ s_1 \\ \rightarrow y_2 \\ y_1 \\ P \end{array}\begin{array}{c} BV\ P\quad y_1\quad y_2\quad s_1\quad s_2\quad s_3\quad RHS \\ \left[\begin{array}{cccccc|c} 0 & 0 & 0 & 1 & -2 & -1 & 0 \\ 0 & 0 & 1 & 0 & 1 & -1 & 1 \\ 0 & 1 & 0 & 0 & 0 & 1 & 3 \\ \hline 1 & 0 & 0 & 0 & 50 & 50 & 350 \end{array}\right] \end{array}$$

The maximum value for $P$ is 350 when $y_1 = 3$ and $y_2 = 1$. The minimum value for $C$ is 350 when $x_1 = 0$, $x_2 = 50$ and $x_3 = 50$.
(In the last step, choosing the entry "2" as the pivot leads to the answer $C = 350$ when $x_1 = 25$, $x_2 = 0$ and $x_3 = 75$.)

**11.** Let $x_1$ and $x_2$ represent the number of pounds of hamburger and picnic, respectively, patties made. The objective is to maximize the total amount of meat used for the patties, $P = x_1 + x_2$ pounds. The constraints are imposed by the amounts of round steak, chuck steak and pork available to be ground into the patties.

$$\begin{cases} \text{round steak}: 0.2x_1 \le 160/2 = 80 \\ \text{chuck steak}: 0.6x_1 + 0.5x_2 \le 600/2 = 300 \\ \text{pork}: 0.3x_2 \le 300/2 = 150 \\ x_1 \ge 0, x_2 \ge 0 \end{cases}.$$

This is a maximum problem in standard form.

$$\begin{array}{c} BV \\ s_1 \\ s_2 \\ s_3 \\ P \end{array} \begin{array}{cccccc|c} P & x_1 & x_2 & s_1 & s_2 & s_3 & RHS \\ 0 & \textcircled{\tfrac{1}{5}} & 0 & 1 & 0 & 0 & 80 \\ 0 & \frac{3}{5} & \frac{1}{2} & 0 & 1 & 0 & 300 \\ 0 & 0 & \frac{3}{10} & 0 & 0 & 1 & 150 \\ \hline 1 & -1 & -1 & 0 & 0 & 0 & 0 \end{array} \rightarrow \begin{array}{c} BV \\ x_1 \\ s_2 \\ s_3 \\ P \end{array} \begin{array}{cccccc|c} P & x_1 & x_2 & s_1 & s_2 & s_3 & RHS \\ 0 & 1 & 0 & 5 & 0 & 0 & 400 \\ 0 & 0 & \textcircled{\tfrac{1}{2}} & -3 & 1 & 0 & 60 \\ 0 & 0 & \frac{3}{10} & 0 & 0 & 1 & 150 \\ \hline 1 & 0 & -1 & 5 & 0 & 0 & 400 \end{array}$$

$$\rightarrow \begin{array}{c} BV \\ x_1 \\ x_2 \\ s_3 \\ P \end{array} \begin{array}{cccccc|c} P & x_1 & x_2 & s_1 & s_2 & s_3 & RHS \\ 0 & 1 & 0 & 5 & 0 & 0 & 400 \\ 0 & 0 & 1 & -6 & 2 & 0 & 120 \\ 0 & 0 & 0 & \textcircled{\tfrac{9}{5}} & -\frac{3}{5} & 1 & 114 \\ \hline 1 & 0 & 0 & -1 & 2 & 0 & 520 \end{array} \rightarrow \begin{array}{c} BV \\ x_1 \\ x_2 \\ s_1 \\ P \end{array} \begin{array}{cccccc|c} P & x_1 & x_2 & s_1 & s_2 & s_3 & RHS \\ 0 & 1 & 0 & 0 & \frac{5}{3} & -\frac{25}{9} & \frac{250}{3} \\ 0 & 0 & 1 & 0 & 0 & \frac{10}{3} & 500 \\ 0 & 0 & 0 & 1 & -\frac{1}{3} & \frac{5}{9} & \frac{190}{3} \\ \hline 1 & 0 & 0 & 0 & \frac{5}{3} & \frac{5}{9} & \frac{1750}{3} \end{array}$$

The maximum value for $P$ is $\frac{1750}{3} = 583\frac{1}{3}$, obtained when $x_1 = \frac{250}{3} = 83\frac{1}{3}$ and $x_2 = 500$.

Make $83\frac{1}{3}$ pounds of hamburger patties and 500 pounds of picnic patties, to use a maximum of $583\frac{1}{3}$ pounds of meat.

**13.** Let $x_1$, $x_2$, and $x_3$ represent the acres of corn, wheat, soybeans, respectively, planted. The objective is to maximize the profit, $P = 30x_1 + 40x_2 + 40x_3$ dollars. The constraints are imposed by the available land, preparation costs and available labor.

$$\begin{cases} \text{land}: x_1 + x_2 + x_3 \le 1000 \text{ acres} \\ \text{preparation}: 100x_1 + 120x_2 + 70x_3 \le 10000 \text{ dollars} \\ \text{labor}: 7x_1 + 10x_2 + 8x_3 \le 8000 \text{ days} \\ x_1 \ge 0, x_2 \ge 0, x_3 \ge 0 \end{cases}.$$

This is a maximization problem in standard form.

| BV | P | $x_1$ | $x_2$ | $x_3$ | $s_1$ | $s_2$ | $s_3$ | RHS |
|---|---|---|---|---|---|---|---|---|
| $s_1$ | 0 | 1 | 1 | 1 | 1 | 0 | 0 | 1000 |
| $s_2$ | 0 | 100 | 120 | ⑦⓪ | 0 | 1 | 0 | 10000 |
| $s_3$ | 0 | 7 | 10 | 8 | 0 | 0 | 1 | 8000 |
| P | 1 | $-30$ | $-40$ | $-40$ | 0 | 0 | 0 | 0 |

$\rightarrow$

| BV | P | $x_1$ | $x_2$ | $x_3$ | $s_1$ | $s_2$ | $s_3$ | RHS |
|---|---|---|---|---|---|---|---|---|
| $s_1$ | 0 | $-\frac{3}{7}$ | $-\frac{5}{7}$ | 0 | 1 | $-\frac{1}{70}$ | 0 | $\frac{6000}{7}$ |
| $x_3$ | 0 | $\frac{10}{7}$ | $\frac{12}{7}$ | 1 | 0 | $\frac{1}{70}$ | 0 | $\frac{1000}{7}$ |
| $s_3$ | 0 | $-\frac{31}{7}$ | $-\frac{26}{7}$ | 0 | 0 | $-\frac{4}{35}$ | 1 | $\frac{48000}{7}$ |
| P | 1 | $\frac{190}{7}$ | $\frac{200}{7}$ | 0 | 0 | $\frac{4}{7}$ | 0 | $\frac{40000}{7}$ |

The maximum value for $P$ is $\dfrac{40000}{7}$, obtained when $x_1 = 0$, $x_2 = 0$ and $x_3 = \dfrac{1000}{7}$. Plant no acres of corn, no acres of wheat and $\dfrac{1000}{7} = 142\dfrac{6}{7}$ acres of soybeans for a maximum profit of $\dfrac{40000}{7} \approx \$5714.29$.

## Professional Exam Questions

1.  $c$           3. $c$           5. $b$           7. $c$

9. $d$: gross profit $=$ (revenue) $-$ (cost)

$$= (4A + 2B + 3.5C) - (A + 0.5B + 1.5C) = 3A + 1.5B + 2C$$

11. $d$

# Chapter 5

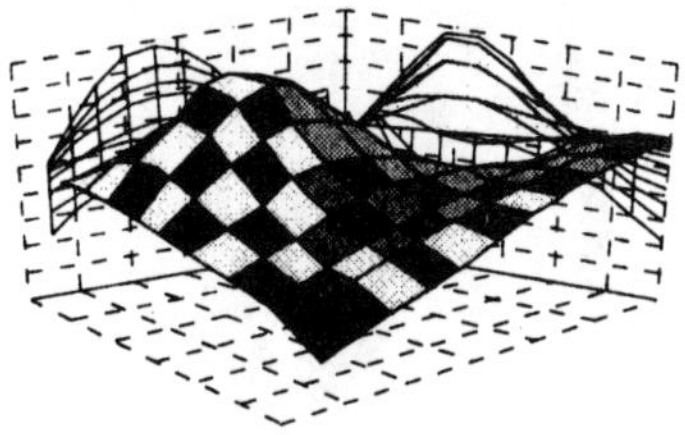

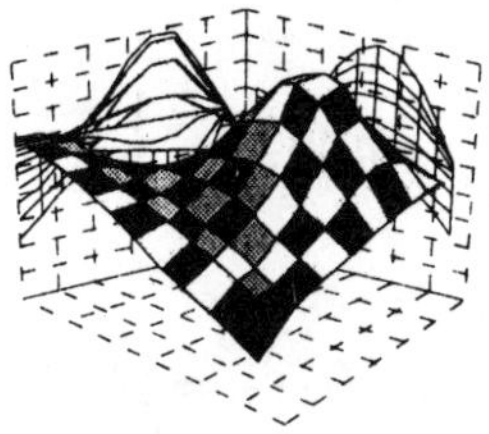

# Finance

## 5.1 Interest

**1.** $0.60 = \frac{60}{100} = 60\%$

**3.** $1.1 = \frac{110}{100} = 110\%$

**5.** $0.06 = \frac{6}{100} = 6\%$

**7.** $0.0025 = \frac{.25}{100} = 0.25\%$

**9.** $25\% = \frac{25}{100} = 0.25$

**11.** $100\% = \frac{100}{100} = 1.0$

**13.** $6.5\% = \frac{6.5}{100} = .065$

**15.** $73.4\% = \frac{73.4}{100} = 0.734$

**17.** $0.15 \cdot 1000 = 150$

**19.** $0.18 \cdot 100 = 18$

**21.** $2.10 \cdot 50 = 105$

**23.** Solve $\frac{x}{100} \cdot 80 = 4$; $80x = 400$; $x = 5$; 5%

**25.** Solve $\frac{x}{100} \cdot 5 = 8$; $5x = 800$; $x = 160$; 160%

**27.** Solve $0.08 \cdot x = 20$; $x = \frac{20}{0.08}$; $x = 250$

**29.** Solve $0.15 \cdot x = 50$; $x = \frac{50}{0.15}$; $x = 333\frac{1}{3}$

**31.** $\$1000 \cdot 0.04 \cdot \frac{3}{12} = \$10$

**33.** $\$500 \cdot 0.12 \cdot \frac{9}{12} = \$45$

**35.** $\$1000 \cdot 0.10 \cdot \frac{18}{12} = \$150$

**37.** $1000 + 1000 \cdot r \cdot \frac{6}{12} = 1050$, $r = 0.10$, or 10%

**39.** $300 + 300 \cdot r \cdot \frac{12}{12} = 400$, $r = \frac{1}{3}$, or $33\frac{1}{3}\%$

**41.** $900 + 900 \cdot r \cdot \frac{10}{12} = 1000$, $r = \frac{2}{15}$, or $13\frac{1}{3}\%$

**43.** $\$1200 - \$1200 \cdot 0.10 \cdot \frac{6}{12} = \$1200 - \$60 = \$1140$

**45.** $\$2000 - \$2000 \cdot 0.08 \cdot \frac{24}{12} = \$2000 - \$320 = \$1680$

**47.** $1200 = A\left(1 - 0.10 \cdot \frac{6}{12}\right)$, $A = \$1263.16$

**49.** $2000 = A\left(1 - 0.08 \cdot \frac{24}{12}\right)$, $A = \$2380.95$

**51.** $\$500 = P\left(1 + 0.08 \cdot \frac{9}{12}\right)$, $P = \frac{1}{1.06} \cdot \$500 = \$471.70$. Invest \$471.70.

**53.** If $t$ is the duration of the loan, then solve $600(0.08)t = 156$, or $48t = 156$; $t = 3.25$. The loan was for $3\frac{1}{4}$ years.

**55.** The amount $A$ repaid for a 6 month discounted loan at 9% satisfies $\$1000 = A\left(1 - 0.09 \cdot \frac{6}{12}\right)$, or $A = \frac{1}{0.955}\$1000 \approx \$1047.12$. The amount paid on a 6 month loan using a simple interest rate of 10% is $\$1000 \cdot \left(1 + 0.10 \cdot \frac{6}{12}\right) = \$1050$. Thus, the interest paid on the discounted loan is approximately \$47, the interest paid on the simple interest loan is \$50, so less interest is paid on the discounted loan.

**57.** The amount $A$ repaid for a one year discounted loan at 6% satisfies $\$4000 = A(1 - 0.06 \cdot 1)$ or $A = \frac{1}{.94} \cdot \$4000 = \$4255.32$. The amount paid on a one year loan using a simple interest rate of 6.3% is $\$4000 \cdot (1 + (0.063) \cdot 1) = \$4252$. Thus, less interest is paid on the simple interest loan.

**59.** For the simple interest loan, Ruth would repay $\$2000(1+0.123)=\$2246$. For the discounted loan, Ruth would repay $\dfrac{\$2000}{1-0.121}\approx\$2275.31$. The simple interest loan at 12.3% is a better deal.

**61.** $P=\$1{,}000{,}000\left(1-0.085\cdot\frac{3}{12}\right)=\$1{,}000{,}000\cdot(.97875)=\$978{,}750$. Bid \$978,750.

**63.** $P=\$3{,}000{,}000\left(1-0.07715\cdot\frac{6}{12}\right)=\$3{,}000{,}000(0.961425)=\$2{,}884{,}275$. Bid \$2,884,275.

## 5.2 Compound Interest

**1.** $A=P(1+i)^n=1000\left(1+\dfrac{0.08}{12}\right)^{36}=\$1270.24$

**3.** $A=P(1+i)^n=500(1+.09)^3=\$647.51$

**5.** $A=P(1+i)^n=800\left(1+\dfrac{0.12}{365}\right)^{200}=\$854.36$

**7.** $P=A(1+i)^{-n}=100\left(1+\dfrac{0.10}{12}\right)^{-6}=\$95.14$

**9.** $P=A(1+i)^{-n}=500\left(1+\dfrac{0.09}{365}\right)^{-365}=\$456.97$

**11.** (a) $A=P(1+i)^n=\$1000(1+.09)^3=\$1295.03$; $I=A-P=\$295.03$

(b) $A=\$1000\left(1+\dfrac{.09}{2}\right)^6=\$1302.26$; $I=A-P=\$302.26$

(c) $A=\$1000\left(1+\dfrac{.09}{4}\right)^{12}=\$1306.05$; $I=A-P=\$306.05$

(d) $A=\$1000\left(1+\dfrac{.09}{12}\right)^{36}=\$1308.65$; $I=A-P=\$308.65$

**13.** (a) $A=P(1+i)^n=1000\left(1+\dfrac{0.12}{4}\right)^8=\$1266.77$

(b) $A=1000\left(1+\dfrac{0.12}{4}\right)^{12}=\$1425.76$ (c) $A=1000\left(1+\dfrac{0.12}{4}\right)^{16}=\$1604.71$

**15.** (a) $P = A(1+i)^{-n} = \$5000\left(1+\frac{0.06}{2}\right)^{-8} = \$3947.05$

(b) $P = \$5000\left(1+\frac{0.06}{2}\right)^{-16} = \$3115.83$

**17.** Using a principal of \$100 and a one year time period:

(a) $A = P(1+i)^n = \$100\left(1+\frac{0.08}{2}\right)^2 = \$108.16$. The interest earned is $\$108.16 - \$100 = \$8.16$. The effective rate of interest is 8.16%.

(b) $A = \$100\left(1+\frac{0.12}{12}\right)^{12} = \$112.68$. The interest earned is $\$112.68 - \$100 = \$12.68$. The effective rate of interest is 12.68%.

**19.**

$$2P = P(1+i)^3$$
$$2 = (1+i)^3$$
$$\sqrt[3]{2} = 1+i$$
$$i = \sqrt[3]{2} - 1 \approx 0.25992 \approx 25.99\%$$

**21.** $3P = P(1+0.1)^n$, $3 = (1.1)^n$, $n = \log_{1.1} 3 = \frac{\log 3}{\log 1.1} \approx 11.5267$ years. The investment triples in just over 11.5 years.

**23.** On the 12% simple interest loan he will repay $\$1000(1+0.12\cdot 2) = \$1240$, or \$240 in interest. On the 10% loan compounded monthly he will repay $\$1000\left(1+\frac{0.10}{12}\right)^{2\cdot 12} \approx \$1220.39$, or \$220.39 in interest. Mr. Nielsen would owe less interest on the 10% loan compounded monthly.

**25.** $P = A(1+i)^{-n} = \$1000(1+0.09)^{-1} + \$1000(1+0.09)^{-2} \approx \$1759.11$

**27.** Using a principal of \$100 and a one year time period:
$A = P(1+i)^n = \$100\left(1+\frac{0.0525}{4}\right)^4 = \$105.35$. The interest earned is $\$105.35 - \$100 = \$5.35$. The effective rate of interest is 5.35%.

**29.** Using a principal of \$100 and a one year time period, with a 7% effective interest rate:
$A = P(1+rt) = \$100(1+0.07) = \$107$. With quarterly compounding:
$A = P\left(1+\frac{r}{4}\right)^4 = 100\left(1+\frac{r}{4}\right)^4$. Solve $\$100\left(1+\frac{r}{4}\right)^4 = \$107$ for $r$: $\left(1+\frac{r}{4}\right)^4 = 1.07$;
$1+\frac{r}{4} = \sqrt[4]{1.07}$; $\frac{r}{4} = \sqrt[4]{1.07} - 1$; $r = 4\left(\sqrt[4]{1.07} - 1\right) = 0.0682$. The interest rate compounded

quarterly is 6.82%.

**31.** 6% compounded quarterly: $\$10{,}000\left(1+\frac{0.06}{4}\right)^4 \approx \$10{,}613.64$; $6\frac{1}{4}\%$ compounded annually: $\$10{,}000(1+0.0625)=\$10{,}625$. $6\frac{1}{4}\%$ compounded annually yields the larger amount.

**33.** 9% compounded monthly: $\$10{,}000\left(1+\frac{0.09}{12}\right)^{12} \approx \$10{,}938.07$; 8.8% compounded daily: $\$10{,}000\left(1+\frac{0.088}{365}\right)^{365} \approx \$10{,}919.77$. 9% compounded monthly yields the larger amount.

**35.** $P=(1+i)^n=\$90{,}000(1+0.05)^4 \approx \$109{,}395.56$. To the nearest hundred dollars, the selling price will be \$109,400.

**37.** $A=P(1+i)^n=600(1+0.015)^6=\$656.07$

**39.** $P=A(1+i)^{-n}=40{,}000\left(1+\frac{0.08}{4}\right)^{-16}=\$29{,}137.83$

**41.** $A=P(1+i)^n=6000\left(1+\frac{0.08}{2}\right)^{50}=\$42{,}640.10$

**43.** $A=P(1+i)^n=17{,}000(1+0.02)^8=19{,}918$

**45.** $A=\$15\cdot 1000=\$15{,}000$. $P=\$20\cdot 1000=\$20{,}000$. Solve:

$20{,}000=15{,}000\left(1+\frac{r}{4}\right)^{16}$ for $r$.

$\frac{4}{3}=\left(1+\frac{r}{4}\right)^{16}$; $\sqrt[16]{\left(\frac{4}{3}\right)}=1+\frac{r}{4}$; $r=4\left(\sqrt[16]{\frac{4}{3}}-1\right)$; $r\approx 0.072571$

The return is approximately 7.26%; Jack should buy the stock.

**47.** $A=P(1+i)^n=2000\left(1+\frac{0.09}{4}\right)^{100}=\$18{,}508.09$

**49.** $P=A(1+i)^{-n}=40{,}000(1+0.08)^{-17}=\$10{,}810.76$

**51.** Solve $25{,}000=12{,}485.52(1+i)^8$ for $i$: $i=\sqrt[8]{\frac{25000}{12485.52}}-1$; $i\approx 0.0906657$, or approximately 9.07%.

**53.** Solve for $t$: $\$25{,}000 = \$10{,}000\left(1+\frac{0.06}{365}\right)^{365t}$; $2.5 = \left(1+\frac{0.06}{365}\right)^{365t}$; $365t = \log_{\left(1+\frac{.06}{365}\right)}(2.5)$;

$t = \dfrac{\log(2.5)}{365\cdot\log\left(1+\frac{.06}{365}\right)} = 15.27$. It will take approximately $15\frac{1}{4}$ years.

**55.** $A = \$1000(1-0.03)^2 = 1000(0.97)^2 = \$940.90$

**57.** $A = \$1000(1-0.03)^5 = 1000(0.97)^5 = \$858.73$

**59.** Solve for $n$: $\frac{1}{2}P = P(1-0.03)^n$; $\frac{1}{2} = (.97)^n$; $n = \log_{(.97)}\left(\frac{1}{2}\right) = \frac{\log(.5)}{\log(.97)} = 22.76$. It will take approximately $22\frac{3}{4}$ years.

**61.** Solve for $n$: $\frac{1}{2}P = P(1-0.06)^n$; $\frac{1}{2} = (.94)^n$; $n = \log_{(.94)}\left(\frac{1}{2}\right) = \frac{\log(.5)}{\log(.94)} = 11.20$. It will take approximately 11.2 years.

## Technology Exercises

**1.** Answers will vary.

**3.** Answers will vary

## 5.3 Annuities; Sinking Funds

**1.** $A = \$100\left[\frac{(1+0.10)^{10}-1}{0.10}\right] = \$1593.74$

**3.** $A = \$400\left[\frac{\left(1+\frac{0.12}{12}\right)^{12}-1}{\frac{0.12}{12}}\right] = \$5073$

**5.** $A = \$200\left[\frac{\left(1+\frac{.06}{12}\right)^{36}-1}{\frac{0.06}{12}}\right] \approx \$7867.22$

**7.** $A = \$100\left[\frac{\left(1+\frac{.06}{12}\right)^{60}-1}{\frac{.06}{12}}\right] = \$6977$

**9.** $A = \$9000\left[\frac{(1+0.05)^{10}-1}{0.05}\right] = \$113{,}201.03$

**11.** $P = \$10{,}000\left[\frac{\left(1+\frac{.05}{12}\right)^{60}-1}{\frac{.05}{12}}\right]^{-1} = \$147.05$

**13.** $P = \$20{,}000\left[\frac{\left(1+\frac{0.06}{4}\right)^{10}-1}{\frac{0.06}{4}}\right]^{-1} = \$1868.68$

**15.** $P = \$25{,}000\left[\frac{\left(1+\frac{0.055}{12}\right)^{6}-1}{\frac{0.055}{12}}\right]^{-1} = \$4119.18$

**17.** $P = \$5{,}000\left[\frac{\left(1+\frac{0.04}{12}\right)^{24}-1}{\frac{0.04}{12}}\right]^{-1} = \$200.46$

**19.** $P = \$9{,}000\left[\frac{(1+0.05)^{4}-1}{0.05}\right]^{-1} = \$2088.11$

**21.** $A = \$2500\left[\frac{(1+0.07)^{15}-1}{0.07}\right] = \$62{,}822.56$

**23.** $A = \$300\left[\frac{\left(1+\frac{0.08}{4}\right)^{24}-1}{\frac{0.08}{4}}\right] = \$9126.56$

**25.** $P = \$350{,}000\left[\frac{\left(1+\frac{0.09}{12}\right)^{240}-1}{\frac{0.09}{12}}\right]^{-1} = \$524.04$

**27.** $P = \$100{,}000\left[\frac{(1+0.08)^{4}-1}{0.08}\right]^{-1} = \$22{,}192.08$

**29.** Solve $30{,}000 = 0.14p + p\left(\frac{(1+0.1)^{30}-1}{0.1}\right)^{-1}$; $30{,}000 = 0.146079248p; p = \$205{,}367.98$

**31.** $P = \$1{,}000{,}000\left[\frac{\left(1+\frac{0.08}{4}\right)^{40}-1}{\frac{0.08}{4}}\right]^{-1} = \$16{,}555.75$

**33.** (a) $A_{20} = P(1+i)^{20} = \$100{,}000(1+0.03)^{20} = \$180{,}611.12$

(b) $P = A\left[\frac{(1+i)^n - 1}{i}\right]^{-1} = \$180{,}611.12\left[\frac{\left(1+\frac{0.06}{2}\right)^{40} - 1}{\frac{0.06}{2}}\right]^{-1} = \$2395.33$

**35.** Solve for $t$: $\$1{,}000{,}000 = \$600\left[\frac{\left(1+\frac{0.07}{12}\right)^{12t} - 1}{\frac{0.07}{12}}\right]$; $\frac{175}{18} = \left(1+\frac{0.07}{12}\right)^{12t} - 1$;

$\frac{193}{18} = \left(1+\frac{0.07}{12}\right)^{12t}$; $12t = \log_{\left(1+\frac{0.07}{12}\right)}\left(\frac{193}{18}\right)$; $t = \frac{\log\left(\frac{193}{18}\right)}{12 \cdot \log\left(1+\frac{0.07}{12}\right)} \approx 33.99$. It will take approximately 34 years.

## Technology Exercises

**1.** Answers will vary

**3.** Answer will vary

## 5.4 Present Value of an Annuity; Amortization

**1.** $V = \$500 \cdot \frac{1-\left(1+\frac{0.10}{12}\right)^{-36}}{\left(\frac{0.10}{12}\right)} \approx \$15{,}495.62$

**3.** $V = \$100 \cdot \left[\frac{1-\left(1+\frac{0.12}{12}\right)^{-9}}{\frac{0.12}{12}}\right] \approx \$856.60$

**5.** $V = \$10{,}000 \cdot \left[\frac{1-(1+0.1)^{-20}}{0.1}\right] \approx \$85135.64$

**7.** $A = P\left[\frac{(1+i)^n - 1}{i}\right] = \$4000\left[\frac{(1+0.10)^{20} - 1}{0.10}\right] = \$229{,}100$. After 20 years, the value of their IRA is \$229,100. $P = V\left[\frac{1-(1+i)^{-n}}{i}\right]^{-1} = \$229{,}100\left[\frac{1-(1+0.10)^{-25}}{0.10}\right]^{-1} = \$25{,}239.51$. The couple can then withdraw \$25,239.51 each year for 25 years.

**9.** $P = \$10{,}000 \cdot \left[ \dfrac{1 - \left(1 + \frac{0.12}{12}\right)^{-24}}{\frac{0.12}{12}} \right]^{-1} \approx \$470.73$

**11.** $P = \$55{,}000 \cdot \left[ \dfrac{1 - \left(1 + \frac{0.10}{12}\right)^{-240}}{\frac{0.10}{12}} \right]^{-1} \approx \$530.76$

**13.** $P = \$15{,}000 \cdot \left[ \dfrac{1 - (1 + 0.12)^{-20}}{0.12} \right]^{-1} \approx \$2008.18$

**15.** $V = \$250 \cdot \left[ \dfrac{1 - \left(1 + \frac{0.10}{12}\right)^{-240}}{\left(\frac{0.10}{12}\right)} \right] \approx \$25{,}906.15$

**17.** The principal for either loan is $\$120{,}000 - \$25{,}000 = \$95{,}000$. The 9% loan for 25 years has a larger monthly payment. Clearly, the interest is greater on the 9% loan for 25 years, since the monthly payment is larger and there are more monthly payments.

| Loan | Monthly Payment |
|---|---|
| 8% for 20 years | $\$95{,}000 \cdot \left[ \dfrac{1 - \left(1 + \frac{0.08}{12}\right)^{-240}}{\left(\frac{0.08}{12}\right)} \right]^{-1} \approx \$794.62$ |
| 9% for 25 years | $\$95{,}000 \cdot \left[ \dfrac{1 - \left(1 + \frac{0.09}{12}\right)^{-300}}{\left(\frac{0.09}{12}\right)} \right]^{-1} \approx \$797.24$ |

After 10 years, the equity from the 8%, 20 year, loan is greater.

| Loan | Equity after 10 years |
|---|---|
| 8% for 20 years | $\$120{,}000 - \$794.62 \cdot \left[ \dfrac{1 - \left(1 + \frac{0.08}{12}\right)^{-120}}{\left(\frac{0.08}{12}\right)} \right] \approx \$54{,}506.24$ |
| 9% for 25 years | $\$120{,}000 - \$797.24 \cdot \left[ \dfrac{1 - \left(1 + \frac{0.09}{12}\right)^{-180}}{\left(\frac{0.09}{12}\right)} \right] \approx \$41{,}397.39$ |

**19.** The amount that John needs to accumulate is $V = \$300 \cdot \left[\frac{1-\left(1+\frac{0.09}{12}\right)^{-360}}{\left(\frac{0.09}{12}\right)}\right] \approx \$37{,}284.56$.

Thus, the monthly payment $P$ is given by $P = A\left[\frac{(1+i)^n - 1}{i}\right]^{-1}$

$= \$37{,}284.56\left[\frac{\left(1+\frac{0.09}{12}\right)^{-240} - 1}{\left(\frac{0.09}{12}\right)}\right]^{-1} = \$55.82$. John should deposit \$55.82 each month.

**21.** (a) $A = P \cdot \left[\frac{(1+i)^n - 1}{i}\right] = \$100\left[\frac{\left(1+\frac{0.06}{52}\right)^{12} - 1}{\frac{0.06}{52}}\right] \approx \$1207.64$. Dan has saved \$1207.64 after 12 weeks.

(b) The monthly withdrawal $P$ is given by $P = V\left[\frac{1-(1+i)^{-n}}{i}\right]^{-1}$

$= \$1207.64\left[\frac{1-\left(1+\frac{0.06}{52}\right)^{-34}}{\frac{0.06}{52}}\right]^{-1} = \$36.24$. The most Dan can withdraw each week for 34 weeks is \$36.24.

**23.** (a) The down payment is 20% of \$76,000, or \$15,200.

(b) The loan amount is $\$76{,}000 - (\text{down payment}) = \$76{,}000 - \$15{,}200 = \$60{,}800$.

(c) The monthly payment $P = \$60{,}800\left[\frac{1-\left(1+\frac{0.09}{12}\right)^{-360}}{\frac{0.09}{12}}\right]^{-1} \approx \$489.21$.

(d) The total amount that they pay will be $30 \cdot 12 \cdot (\$489.21) = \$176{,}115.60$, so the total interest paid is $\$176{,}115.60 - \$60{,}800 = \$115{,}315.60$.

(e) Using $V = P \cdot \frac{1-(1+i)^{-n}}{i}$, solve $\$60{,}800 = \$589.21 \cdot \left[\frac{1-\left(1+\frac{0.09}{12}\right)^{-n}}{\left(\frac{0.09}{12}\right)}\right]$ for $n$:

$$60{,}800 \cdot \frac{0.0075}{589.21} = 1 - 1.0075^{-n}$$

$$1.0075^{-n} = 1 - 60{,}800 \cdot \frac{0.0075}{589.21}$$

$$-n \cdot \log(1.0075) = \log\left(1 - 60{,}800 \cdot \frac{0.0075}{589.21}\right)$$

$$n = \frac{-\log\left(1 - 60{,}800 \cdot \frac{0.0075}{589.21}\right)}{\log(1.0075)} \approx 198.99$$

The loan will be paid in 199 months, i.e., 16 years and 7 months.

(f) The total amount that they pay would be $199\cdot(\$589.21)=\$177,252.79$, so the total interest paid is $\$117,252.79-\$60,800=\$56,452.79$.

**25.** The down payment is 20% of \$12,000 or \$2400. The loan amount is \$12,000 – \$2400 = \$9600. The monthly payment is given by $P=V\left[\frac{1-(1+i)^{-n}}{i}\right]^{-1}$

$$=\$9600\left[\frac{1-\left(1+\frac{0.15}{12}\right)^{-36}}{\frac{0.15}{12}}\right]^{-1}=\$332.79.$$

**27.** The down payment is 10% of \$20,000 or \$2000. The loan amount is \$20,000 – \$2000 = \$18,000. The monthly payment is given by $P=V\left[\frac{1-(1+i)^{-n}}{i}\right]^{-1}$

$=\$18,000\left[\frac{1-\left(1+\frac{0.12}{12}\right)^{-48}}{\frac{0.12}{12}}\right]^{-1}=\$474.01$. The total amount paid is $4\cdot12\cdot(\$474.01)=\$22,752.48$. The total interest paid is $\$22,757.48-\$18,000=\$4752.48$.

**29.** $V=140,000-140,000(0.2)=112,000$

For 30-year mortgage, $P=\$112,000\left[\frac{1-\left(1+\frac{0.098}{12}\right)^{-360}}{\frac{0.098}{12}}\right]^{-1}=\$966.37$;

$I=966.37(360)-112,000=\$235,893.20$

For 15-year mortgage, $P=\$112,000\left[\frac{1-\left(1+\frac{0.098}{12}\right)^{-180}}{\frac{0.098}{12}}\right]^{-1}=\$1189.89$;

$I=1189.89(180)-112,000=\$102,180.20$

**31.** Using $V=P\left[\frac{1-(1+i)^{-n}}{i}\right]$, solve $\$100,000=\$2,000\left[\frac{1-\left(1+\frac{0.05}{12}\right)^{-n}}{\frac{0.05}{12}}\right]$ for $n$:

$\frac{100,000}{2,000}\cdot\frac{(0.05)}{12}=1-\left(1+\frac{.05}{12}\right)^{-n}$; $\left(1+\frac{.05}{12}\right)^{-n}=1-\frac{5}{24}=\frac{19}{24}$; $-n\log\left(1+\frac{.05}{12}\right)=\log\left(\frac{19}{24}\right)$;

$n=\frac{-\log\left(\frac{19}{24}\right)}{\log\left(1+\frac{.05}{12}\right)}\approx56.18$. It will take approximately 56 months (4 years, 8 months) to exhaust the IRA.

## Technology Exercises

**1.** Answer will vary

**3.** Answer will vary

## 5.5 Applications: Leasing; Capital Expenditure; Bonds

1. The present value of an annuity of \$2000 for 5 years at 10% is $\$2000\cdot\left[\frac{1-(1+.10)^{-5}}{.10}\right]\approx\$7581.57$, which is less than the purchase price of \$8100. The corporation should lease the machine.

3. The annual cost of machine A is $\frac{\$10{,}000}{\frac{1-(1+.10)^{-8}}{.10}}\approx\$1874.44$, which is \$125.56 *less* than its annual labor savings. The annual cost of machine B is $\frac{\$8{,}000}{\frac{1-(1+.10)^{-6}}{.10}}\approx\$1836.86$, which is \$36.86 *more* than its annual labor savings. Thus, machine A is preferable.

5. Step 1: Each semiannual interest payment is $\$1000\cdot\left(\frac{0.09}{2}\right)=\$45$.

   Step 2: The present value of these payments is $\$45\cdot\left[\frac{1-\left(1+\frac{.08}{2}\right)^{-30}}{\frac{.08}{2}}\right]\approx\$778.14$.

   Step 3: The present value of the amount payable at maturity is
   $\$1000\cdot\left(1+\frac{0.08}{2}\right)^{-30}\approx\$308.32$.

   Step 4: The price of the bond is the sum of the present values of the interest payment and the maturity payment: $\$778.14+\$308.32=\$1086.46$.

   Step 3: The present value of the amount payable at maturity is
   $\$1000\cdot\left(1+\frac{0.10}{2}\right)^{-30}\approx\$231.38$.

   Step 4: The price of the bond is the sum of the present values of the interest payment and the maturity payment: $\$691.76+\$231.38=\$923.14$.

## Chapter 5 Review

### True or False

1. True
3. False

### Fill in the Blank

1. proceeds
3. annuity

## Review Exercises

**1.** $I = P \cdot r \cdot t = \$400 \cdot 0.12 \cdot \left(\frac{9}{12}\right) = \$36$. $A = P + I = \$400 + \$36 = \$436$.

**3.** $\$100\left(1+\frac{0.1}{12}\right)^{(2\cdot 12+3)} = \$125.12$.

**5.** The simple interest loan costs $\$3000(3 \cdot 0.12) = \$1080$ in interest. The interest on the compound interest loan is $\$3000\left(1+\frac{0.10}{12}\right)^{3\cdot 12} - \$3000 \approx \$1044.55$. The 10% loan compounded monthly costs Mike less.

**7.** $\$75 = P \cdot \left(1+\frac{0.10}{12}\right)^{6}$, so $P = \$75 \cdot \left(1+\frac{0.10}{12}\right)^{-6} \approx \$71.36$.

**9.** Solve $\$10{,}000 = P \cdot \left[\frac{\left(1+\frac{0.03}{12}\right)^{2\cdot 12} - 1}{\left(\frac{0.03}{12}\right)}\right]$. $P = \$10{,}000 \cdot \left[\frac{\left(1+\frac{0.03}{12}\right)^{2\cdot 12} - 1}{\left(\frac{0.03}{12}\right)}\right]^{-1} \approx \$404.81$.

**11.** The down payment is 25% of \$80,000, or \$20,000, so the loan amount is \$60,000.

(a) The monthly payment $P = \$60{,}000 \cdot \left[\frac{1-\left(1+\frac{.10}{12}\right)^{-300}}{\frac{.10}{12}}\right]^{-1} \approx \$545.22$.

(b) The total interest paid is $\$545.22 \cdot (25 \cdot 12) - \$60{,}000 = \$103{,}566$.

(c) Their equity after 5 years is

$$\$80{,}000 - \$545.22 \cdot \left[\frac{1-\left(1+\frac{.10}{12}\right)^{-240}}{\frac{.10}{12}}\right] = \$80{,}000 - \$56{,}498.21 \approx \$23{,}501.79.$$

**13.** The monthly payment $P = \$125{,}000 \cdot \left[ \dfrac{1 - \left(1 + \frac{.09}{12}\right)^{-300}}{\frac{.09}{12}} \right]^{-1} = \$1049.00$. The equity after 10 years is $\$125{,}000 - \$1049 \cdot \left[ \dfrac{1 - \left(1 + \frac{.09}{12}\right)^{-180}}{\frac{.09}{12}} \right] = \$125{,}000 - \$103{,}424.49 \approx \$21{,}575.51$

**15.** Solve $\$20{,}000 = 0.15p + p \cdot \left[ \dfrac{(1 + 0.10)^{20} - 1}{0.1} \right]^{-1}$; $p = \$119{,}431.77$. A purchase price of $119,431.77 will achieve Mr. Graff's goals.

**17.** Solve $\$25{,}000 = 0.20p + p \cdot \left[ \dfrac{(1 + 0.10)^{15} - 1}{0.1} \right]^{-1}$; $p = \$108{,}003.59$.

**19.** $A = R\left[ \dfrac{(1+i)^n - 1}{i} \right] = 500\left[ \dfrac{\left(1 + \frac{0.06}{2}\right)^{16} - 1}{\frac{0.06}{2}} \right] = \$10{,}078.44$

**21.** After 7 years Bill will owe $A = 1000(1 + 0.05)^7 = \$1407.10$ on his loan. Quarterly payments into the sinking fund should be: $P = \$1407.10\left[ \dfrac{\left(1 + \frac{0.08}{4}\right)^{28} - 1}{\frac{0.08}{4}} \right]^{-1} = \$37.98$

**23.** $P = V\left[ \dfrac{1 - (1+i)^{-n}}{i} \right]^{-1} = \$3{,}000\left[ \dfrac{1 - \left(1 + \frac{.12}{12}\right)^{-24}}{\frac{.12}{12}} \right]^{-1} = \$141.22$

**25.** Using a principal of $100 and a one year time period:

$A = P(1+i)^n = \$100\left(1 + \dfrac{0.09}{12}\right)^{12} = \$109.38$. The interest earned is $109.38 – $100 = $9.38.

The effective interest rate is 9.38%.

**27.** $P = V\left[\frac{1-(1+i)^{-n}}{i}\right]^{-1} = \$20{,}000 \cdot \left[\frac{1-\left(1+\frac{.08}{2}\right)^{-30}}{\frac{.08}{2}}\right]^{-1} = \$1156.60$

**29.** $A = P\frac{(1+i)^{n}-1}{1} = 60\left[\frac{\left(1+\frac{0.12}{12}\right)^{30}-1}{\frac{0.12}{12}}\right] = \$2087.09$

**31.** $P = V\left[\frac{1-(1+i)^{-n}}{i}\right]^{-1} = \$4000 \cdot \left[\frac{1-\left(1+\frac{.14}{4}\right)^{-16}}{\frac{.14}{4}}\right]^{-1} = \$330.74$

## Professional Exam Questions

**1.** b **3.** b **5.** d **7.** c

# Chapter 6

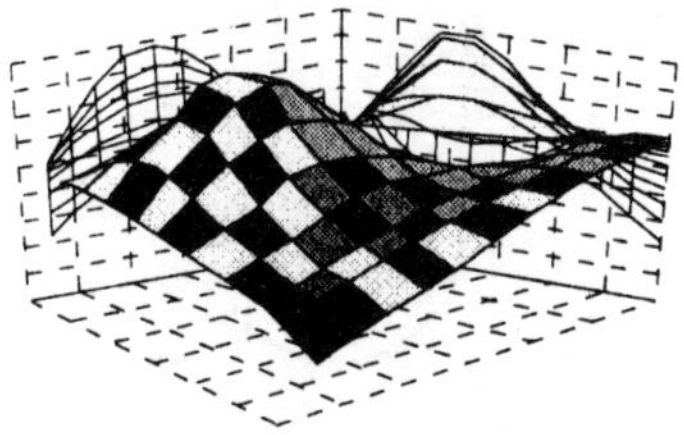

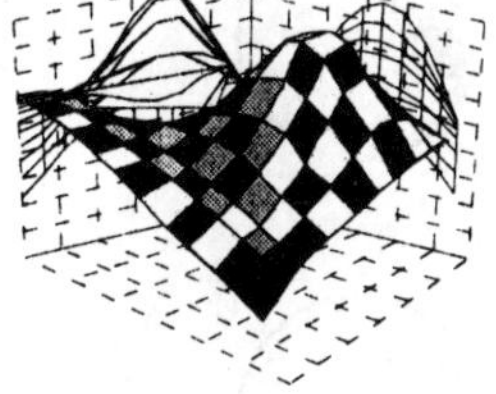

## Sets; Counting Techniques

### 6.1 Sets

**1.** True **3.** False **5.** False **7.** True

**9.** True **11.** $\{2, 3\}$ **13.** $\{1, 2, 3, 4, 5\}$ **15.** $\varnothing$

**17.** $\{a, b, d, e, f, q\}$

**19.** (a) $\{0, 1, 2, 3, 5, 7, 8\}$ (b) $\{5\}$
(c) $\{5\}$ (d) $\overline{\{5\}} = \{0, 1, 2, 3, 4, 6, 7, 8, 9\}$
(e) $\{2, 3, 4, 6, 8, 9\} \cap \{0, 1, 4, 6, 7, 9\} = \{4, 6, 9\}$
(f) $A \cup \{5\} = \{0, 1, 5, 7\}$
(g) $\{5\} \cap \{2, 3, 4, 6, 8, 9\} = \varnothing$ (h) $\{5\} \cup \{5\} = \{5\}$

**21.** (a) $\{b, c, d, e, f, g\}$ (b) $\{c\}$
(c) $\{a, h, i, j, \ldots, x, y, z\}$ (d) $\{a, b, d, e, f, g, h, i, j, \ldots, x, y, z\}$

**23.** (a) $\overline{A} \cap B$

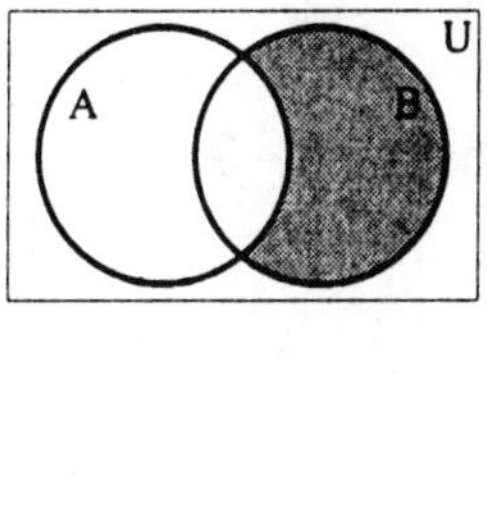


(b) $(\overline{A} \cap \overline{B}) \cup C$

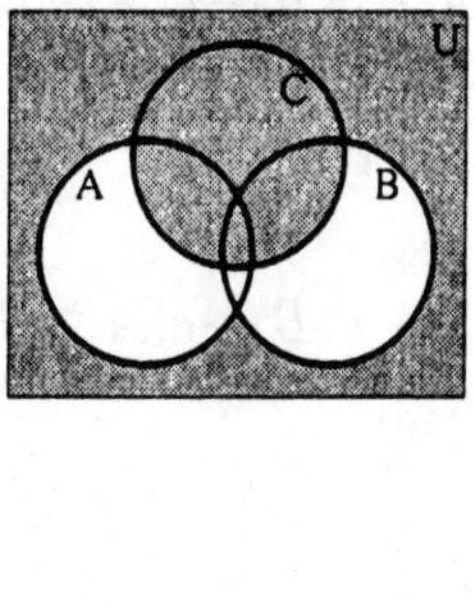

(c & d): $A \cap (A \cup B) = A \cup (A \cap B) = A$

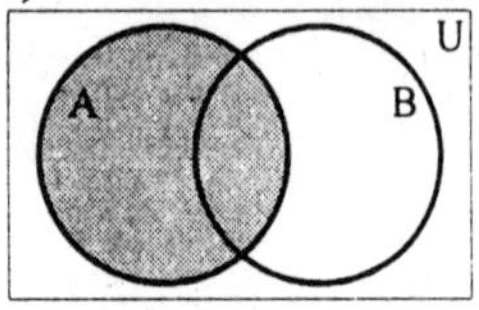


(e & f): $(A \cup B) \cap (A \cup C) = A \cup (B \cap C)$

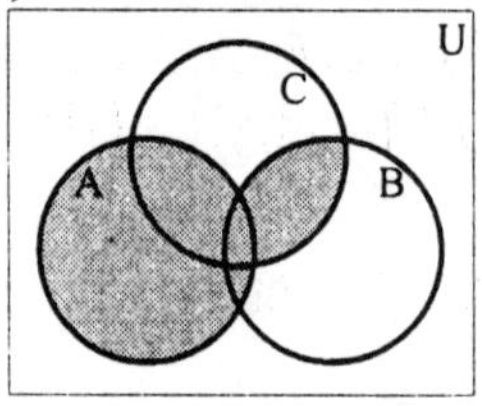


(g) $(A \cap B) \quad \cup \quad (A \cap \overline{B}) \quad = \quad A$

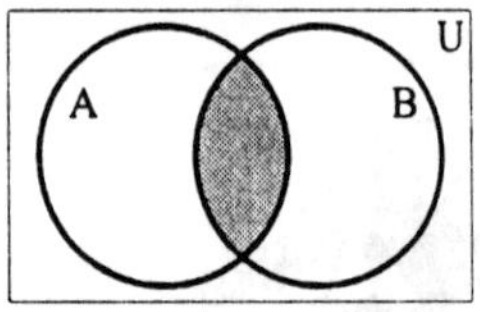


$\cup$

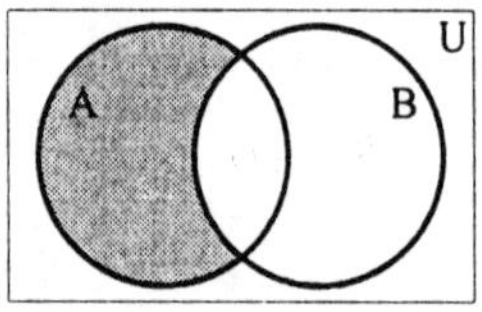


$=$

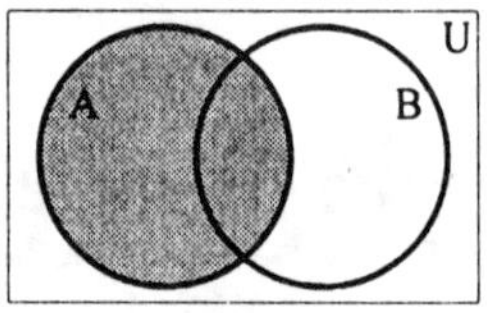


(h) $(A \cap B) \quad \cup \quad (\overline{A} \cap B) \quad = \quad B$

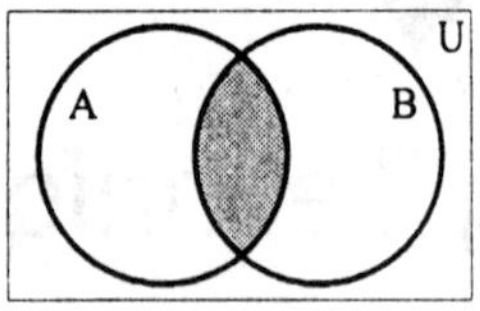


$\cup$

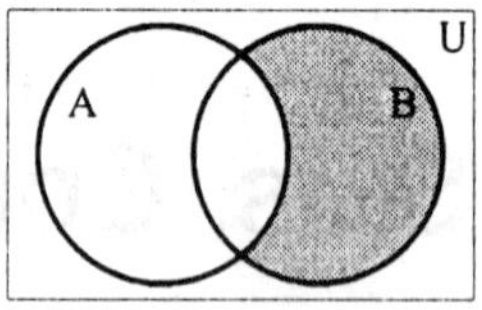


$=$

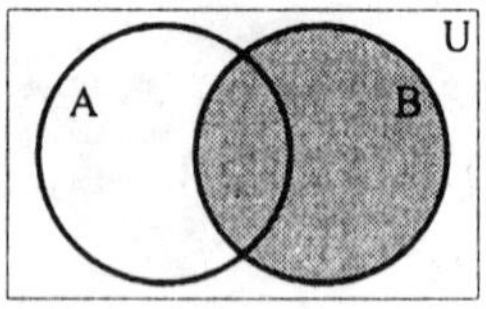


**25.** $A \cap E$ is the set of customers of IBM who are also members of the Board of Directors of IBM.

**27.** $A \cup D$ is the set of people who either are customers of IBM or are stockholders of IBM.

**29.** $\overline{A} \cap D$ is the set of people who are not customers of IBM but do hold stock in IBM.

**31.** $M \cap S$ is the set of male college students who smoke.

**33.** $\overline{M} \cup \overline{F}$ is the set of college students who are female or not freshmen.

**35.** $F \cap S \cap M$ is the set of male freshmen college students who smoke.

**37.** ∅, {a}, {b}, {c}, {a, b}, {a, c}, {b, c}, {a, b, c}

## 6.2 The Number of Elements in a Set

**1.** 6

**3.** $c(\{2, 4, 6\}) = 3$

**5.** $c(\{1,2,3,4,5,6\}) = 6$

**7.** $c(A \cup B) = c(A) + c(B) - c(A \cap B) = 4 + 3 - 2 = 5$

**9.** $c(A \cup B) = c(A) + c(B) - c(A \cap B); 7 = 5 + 4 - c(A \cap B); c(A \cap B) = 9 - 7 = 2$

**11.** $c(A \cup B) = c(A) + c(B) - c(A \cap B)$: $14 = c(A) + 8 - 4$; $c(A) = 10$

**13.** Let $A = \{\text{cars with automatic transmissions}\}$ and $P = \{\text{cars with power steering}\}$. Then $c(A) = 325$, $c(P) = 216$, and $c(A \cap P) = 89$. The number of cars with at least one of these options is $c(A \cup P) = c(A) + c(P) - c(A \cap P) = 325 + 216 - 89 = 452$.

**15.** $c(A) = 10 + 6 + 3 + 5 = 24$

**17.** $c(A \cup B) = 10 + 6 + 3 + 5 + 8 + 2 = 34$

**19.** $c(A \cap \overline{B}) = 10 + 5 = 15$

**21.** $c(A \cup B \cup C)' = 10 + 6 + 3 + 5 + 8 + 2 + 20 = 54$

**23.** $c(A \cap B \cap C) = 3$

**25.** (a) $(27 + 33 + 7) + (44 + 47 + 33) + (82 + 152 + 111) = 67 + 124 + 345 = 536$
(b) $(27 + 33 + 7) + (44 + 47 + 33) + 111 + 15 = 67 + 124 + 111 + 15 = 317$
(c) $42 + 44 + 15 + 33 = 134$

**27.** (a) $31 + 45 + 87 + 96 = 259$
(b) $31 + 62 + 87 + 275 = 455$
(c) $31 + 62 + 45 + 89 = 227$
(d) $31 + 45 = 76$
(e) $31 + 87 = 118$
(f) $31 + 62 = 93$
(g) $31 + 62 + 45 + 87 + 96 + 275 + 89 + 227 = 912$

**29.** Let AC = {cars with air conditioning}, AT = {cars with automatic transmission}, PS = {cars with power steering}.
(a) $35 + 5 = 40$
(b) $30 + 5 = 35$
(c) $20 + 20 = 40$
(d) $60 + 30 + 5 + 5 + 20 + 35 + 30 + 20 = 205$ or
$[90 + 100 + 75 - (35 + 40 + 10) + 5] + 20 = 205$
(e) $90 + 100 - 35 = 155$

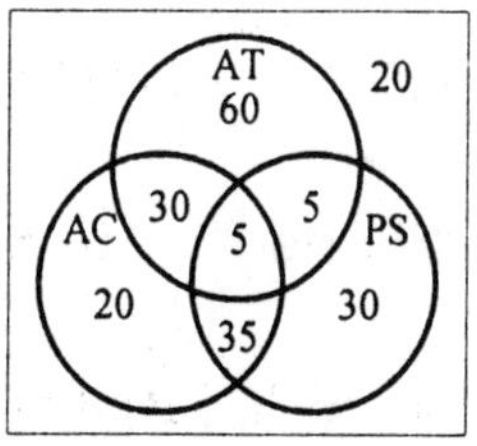


**31.** There are 8 possible blood types: A+, A–, AB+, AB–, B+, B–, O+ and O–.

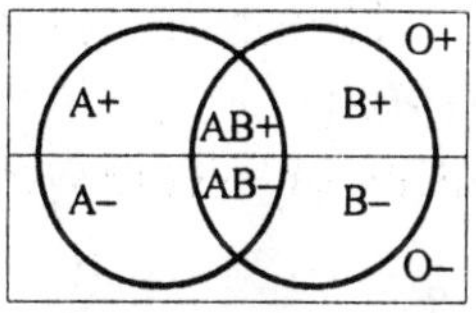


**33.** $10 + 18 + 18 = 46$

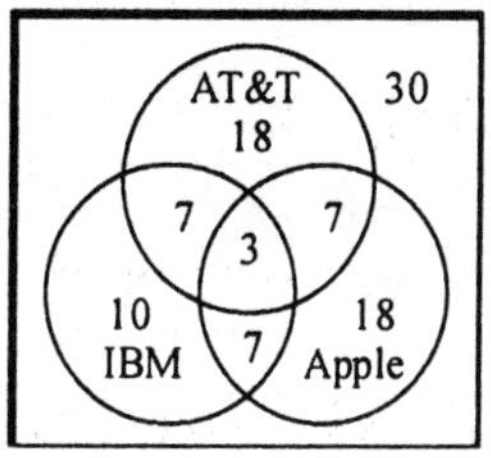

**35.** The subsets of $\{a, b, c, d\}$ are $\varnothing$, $\{a\}$, $\{b\}$, $\{c\}$, $\{d\}$, $\{a, b\}$, $\{a, c\}$, $\{a, d\}$, $\{b, c\}$, $\{b, d\}$, $\{c, d\}$, $\{a, b, c\}$, $\{a, b, d\}$, $\{a, c, d\}$, $\{b, c, d\}$, $\{a, b, c, d\}$. There are $16(=2^4)$ subsets of $\{a, b, c, d\}$.

## 6.3 The Multiplication Principle

**1.** $2 \cdot 4 = 8$

**3.** $3 \cdot 2 \cdot 4 = 24$

**5.** $3 \cdot 8 \cdot 4 \cdot 9 = 864$

**7.** $3 \cdot 12 = 36$

**9.** $10 \cdot 9 \cdot 8 = 720$

**11.** $3 \cdot 8 \cdot 10 \cdot 4 = 960$

**13.** 6 people in 6 seats: $6 \cdot 5 \cdot 4 \cdot 3 \cdot 2 \cdot 1 = 720$, 8 people in 8 seats: $8 \cdot 7 \cdot 6 \cdot 5 \cdot 4 \cdot 3 \cdot 2 \cdot 1 = 40{,}320$

**15.** no repeated letters: $6 \cdot 5 \cdot 4 \cdot 3 = 360$, allowing repeated letters: $6 \cdot 6 \cdot 6 \cdot 6 = 1296$

**17.** $7 \cdot 6 \cdot 5 \cdot 4 \cdot 3 \cdot 2 \cdot 1 = 5040$

**19.** $(4^{10}) \cdot (2^{15}) = 2^{35} = 34{,}359{,}738{,}368$

**21.** (a) $(26^2) \cdot (10^4) = 6{,}760{,}000$ (b) $(26^2) \cdot (10 \cdot 9 \cdot 8 \cdot 7) = 3{,}407{,}040$
(c) $(26 \cdot 25) \cdot (10 \cdot 9 \cdot 8 \cdot 7) = 3{,}276{,}000$

**23.** $(3 \cdot 2)$ model A's and $(3 \cdot 2)$ model B's and $(2 \cdot 2)$ model C's: $6 + 6 + 4 = 16$ distinguishable types.

**25.** $5 \cdot 3 \cdot 4 = 60$

**27.** $2^8 = 256$

**29.** $50^3 = 125{,}000$

**31.** $4 \cdot 2 \cdot 1 = 8$

## 6.4 Permutations

**1.** $\dfrac{5 \cdot 4 \cdot 3 \cdot 2!}{2!} = 5 \cdot 4 \cdot 3 = 60$

**3.** $\dfrac{6 \cdot 5 \cdot 4 \cdot 3!}{3!} = 6 \cdot 5 \cdot 4 = 120$

**5.** $\dfrac{10 \cdot 9 \cdot 8!}{8!} = 90$

**7.** $\dfrac{9 \cdot 8!}{8!} = 9$

**9.** $\dfrac{8 \cdot 7 \cdot 6!}{2 \cdot 6!} = \dfrac{8 \cdot 7}{2} = 28$

**11.** $\dfrac{7!}{(7-2)!} = \dfrac{7!}{5!} = \dfrac{7 \cdot 6 \cdot 5!}{5!} = 7 \cdot 6 = 42$

**13.** $\dfrac{8!}{(8-7)!} = \dfrac{8!}{1!} = 8! = 8 \cdot 7 \cdot 6 \cdot 5 \cdot 4 \cdot 3 \cdot 2 \cdot 1 = 40{,}320$

**15.** $\dfrac{6!}{(6-0)!} = \dfrac{6!}{6!} = 1$

**17.** $\dfrac{8!}{5! \cdot 3!} = \dfrac{8 \cdot 7 \cdot 6 \cdot 5!}{5! \cdot 3 \cdot 2} = 8 \cdot 7 = 56$

**19.** $\frac{6!}{0!\cdot 6!} = \frac{6!}{6!} = 1$

**21.** (a) $6! = 720$ (b) $1 \cdot 5! = 120$ (c) $1 \cdot 4! \cdot 1 = 24$

**23.** $P(10,4) = \frac{10!}{(10-4)!} = \frac{10!}{6!} = \frac{10 \cdot 9 \cdot 8 \cdot 7 \cdot 6!}{6!} = 10 \cdot 9 \cdot 8 \cdot 7 = 5040$

**25.** $P(9,5) = \frac{9!}{(9-5)!} = \frac{9!}{4!} = \frac{9 \cdot 8 \cdot 7 \cdot 6 \cdot 5 \cdot 4!}{4!} = 9 \cdot 8 \cdot 7 \cdot 6 \cdot 5 = 15{,}120$

**27.** $P(12,8) = \frac{12!}{(12-8)!} = \frac{12!}{4!} = \frac{12 \cdot 11 \cdot 10 \cdot 9 \cdot 8 \cdot 7 \cdot 6 \cdot 5 \cdot 4!}{4!} = 19{,}958{,}400$

**29.** $P(1500,3) = \frac{1500!}{(1500-3)!} = \frac{1500!}{1497!} = 1500 \cdot 1499 \cdot 1498 = 3{,}368{,}253{,}000$

**31.** $P(15,4) = \frac{15!}{(15-4)!} = \frac{15!}{11!} = \frac{15 \cdot 14 \cdot 13 \cdot 12 \cdot 11!}{11!} = 32{,}760$

**33.** $5! = 120$

## 6.5 Combinations

**1.** $\frac{6!}{4!\cdot 2!} = \frac{6 \cdot 5 \cdot 4!}{4!\cdot 2!} = \frac{6 \cdot 5}{2} = 15$

**3.** $\frac{7!}{2!\cdot 5!} = \frac{7 \cdot 6 \cdot 5!}{2 \cdot 5!} = 21$

**5.** $\frac{5!}{1!\cdot 4!} = \frac{5 \cdot 4!}{1 \cdot 4!} = 5$

**7.** $\frac{8!}{6!\cdot 2!} = \frac{8 \cdot 7 \cdot 6!}{6!\cdot 2} = 28$

**9.** $C(8,5) = \frac{8!}{5!\cdot 3!} = \frac{8 \cdot 7 \cdot 6 \cdot 5!}{5!\cdot 3 \cdot 2} = 56$

**11.** $C(17,4) = \frac{17!}{4!\cdot 13!} = \frac{17 \cdot 16 \cdot 15 \cdot 14 \cdot 13!}{4 \cdot 3 \cdot 2 \cdot 13!} = 2380$

**13.** There are $C(20,3) = \frac{20!}{3!\cdot 17!} = 1140$ ways to select a subcommittee of 3 members.

**15.** $C(6,\ 2) = \frac{6!}{2!\cdot 4!} = 15$

**17.** (3 choices for the faculty member)$\cdot$(2 choices for the administrator)

$\cdot$($C(5,\ 2)$ choices for the 2 students) $= 3 \cdot 2 \cdot \frac{5!}{2!\cdot 3!} = 60$

**19.** $(\text{quarterback choices})\cdot(\text{halfback choices})\cdot(\text{fullback choices})\cdot(\text{lineman choices})$
$= C(3,1)\cdot C(8,2)\cdot C(4,1)\cdot C(20,7) = 3\cdot 28\cdot 4\cdot 77{,}520 = 26{,}046{,}720$

**21.** Note that a different choice of 3 coins produces a different sum. (For example if her coins sum to 31¢, then the little girl must have pulled out a penny, a nickel and a quarter.) Thus, the number of different sums is the number of different ways to select 3 coins from a purse containing 5 different coins, or $C(5,\ 3) = 10$.

**23.** A lottery ticket corresponds to a choice of 6 distinct numbers from a collection of 100 numbers (00 through 99); there are $C(100,\ 6) = 1{,}192{,}052{,}400$ such choices.

**25.** There are 100 (voting) members of the U.S. Senate; there are $C(100,\ 5) = 75{,}287{,}520$ ways to select a committee of 5 from this group.

**27.** $C(55,8) = \dfrac{55!}{8!47!} = 1{,}217{,}566{,}350$

**29.** $P(5,3) = \dfrac{5!}{(5-3)!} = \dfrac{5!}{2!} = \dfrac{5\cdot 4\cdot 3\cdot 2!}{2!} = 60$

**31.** $C(24,4) = \dfrac{24!}{4!\cdot 20!} = 10{,}626$

**33.** $P(50,15) = \dfrac{50!}{(50-15)!} = \dfrac{50!}{35!} \approx 2.9\times 10^{24}$

**35.** $P(10,4) = \dfrac{10!}{(10-4)!} = \dfrac{10!}{6!} = \dfrac{10\cdot 9\cdot 8\cdot 7\cdot 6!}{6!} = 5040$

## 6.6 More Counting Problems

**1.** (a) $2^{10} = 1024$

(b) $C(10,4) = \dfrac{10!}{4!6!} = \dfrac{10\cdot 9\cdot 8\cdot 7\cdot 6!}{4\cdot 3\cdot 2\cdot 1\cdot 6!} = 210$

(c) $C(10,0) + C(10,1) + C(10,2) = 1 + 10 + 45 = 56$

(d) $1024 - [C(10,0) + C(10,1) + C(10,2)] = 1024 - 56 = 968$

**3.** (a) $C(7,2)\cdot C(3,1) = 21\cdot 3 = 63$

(b) $C(7,3) = 35$

(c) $C(3,3) = 1$

**5.** $\dfrac{8!}{4!4!} = 70$ (This is equal to the number of sequences of 8 letters in which A and N each appear 4 times. Consider AAAANNNN to be equivalent to AAAA, NAAAANNN TO NAAAA, etc.)

**7.** $\dfrac{9!}{4!3!2!} = 1260$

**9.** $\dfrac{11!}{2!2!2!} = 4,989,600$ (There are 11 letters, including 2 M's, 2 A's and 2 T's.)

**11.** $\dfrac{12!}{3!5!4!} = 27,720$

**13.** (a) $C(5,1)\cdot C(8,3) = 280$ (b) $C(8,2)\cdot C(5,2) = 280$

(c) $C(5,1)\cdot C(8,3) + C(5,2)\cdot C(8,2) + C(5,3)\cdot C(8,1) = 280 + 280 + 80 = 640$

**15.** $\dfrac{30!}{(6!)^5} \simeq 1.37\times 10^{18}$

**17.** $C(8,5) + C(8,6) + C(8,7) + C(8,8) = 56 + 28 + 8 + 1 = 93$

**19.** $C(6,2)\cdot C(8,2) + C(6,1)\cdot C(8,3) + C(6,0)\cdot C(8,4)$

$= \dfrac{6!}{2!4!}\cdot\dfrac{8!}{2!6!} + \dfrac{6!}{1!5!}\cdot\dfrac{8!}{3!5!} + \dfrac{6!}{0!6!}\cdot\dfrac{8!}{4!4!} = 826$

**21.** Even number of 1's: $C(8,0) + C(8,2) + C(8,4) + C(8,6) + C(8,8) = 1 + 28 + 70 + 28 + 1 = 128$;
Odd number of 1's = Total number of 8-bit strings – Even number of 1's = $2^8 - 128 = 128$.

## 6.7 The Binomial Theorem

**1.**

$$(x+y)^5 = \binom{5}{0}x^5 + \binom{5}{1}x^4y + \binom{5}{2}x^3y^2 + \binom{5}{3}x^2y^3 + \binom{5}{4}xy^4 + \binom{5}{5}y^5$$
$$= x^5 + 5x^4y + 10x^3y^2 + 10x^2y^3 + 5xy^4 + y^5$$

**3.**

$$(x+3y)^3 = \binom{3}{3}x^3 + \binom{3}{2}x^2(3y) + \binom{3}{1}x(3y)^2 + \binom{3}{0}(3y)^3$$
$$= x^3 + 3x^2\cdot 3y + 3x\cdot 9y^2 + 27y^3$$
$$= x^3 + 9x^2y + 27xy^2 + 27y^3$$

**5.**

$$(2x-y)^4 = \binom{4}{0}(2x)^4 + \binom{4}{1}(2x)^3(-y) + \binom{4}{2}(2x)^2(-y)^2 + \binom{4}{3}(2x)(-y)^3 + \binom{4}{4}(-y)^4$$
$$= 16x^4 + 4\cdot 8x^3(-y) + 6\cdot 4x^2y^2 + 4\cdot 2x(-y^3) + y^4$$
$$= 16x^4 - 32x^3y + 24x^2y^2 - 8xy^3 + y^4$$

**7.** $\binom{5}{3} = 10$

**9.** $\binom{10}{2}\cdot 3^2 = 45\cdot 9 = 405$

**11.** There are $2^5 = 32$ different subsets of a set with 5 elements.

**13.** There are $2^{10} - 1 = 1023$ *nonempty* subsets of a set with 10 elements.

**15.** $\binom{10}{1}+\binom{10}{3}+\binom{10}{5}+\binom{10}{7}+\binom{10}{9} = 10+120+252+120+10 = 512$

**17.** Using $\binom{n}{k}=\binom{n-1}{k}+\binom{n-1}{k-1}$: $\binom{10}{7}=\binom{9}{7}+\binom{9}{6}$ and $\binom{9}{7}=\binom{8}{7}+\binom{8}{6}$, so $\binom{10}{7}=\binom{8}{7}+\binom{8}{6}+\binom{9}{6}$. Also, $\binom{8}{7}=\binom{7}{7}+\binom{7}{6}$ and $\binom{7}{7}=1=\binom{6}{6}$, so $\binom{10}{7}=\binom{6}{6}+\binom{7}{6}+\binom{8}{6}+\binom{9}{6}$.

**19.** Apply $\binom{n}{k}=\binom{n-1}{k}+\binom{n-1}{k-1}$ with $n = 12$ and $k = 6$: $\binom{12}{6}=\binom{11}{6}+\binom{11}{5}$.

**21.** Suppose we wish to select a team of $k$ people from a class of $n$ students, and designate one of the team members as "team leader." We could do this by selecting the team $\left(\text{in } \binom{n}{k} \text{ ways}\right)$ and then designate the leader ($k$ choices), so there are $k \cdot \binom{n}{k}$ possible results. *Or*, we could first choose the leader ($n$ choices) and then choose the remaining $k-1$ team members $\left(\text{in } \binom{n-1}{k-1} \text{ ways}\right)$, so there are $n \cdot \binom{n-1}{k-1}$ possible results. Since these two methods count the same set of results, the two answers must be the same, i.e., $k \cdot \binom{n}{k} = n \cdot \binom{n-1}{k-1}$.

$$\left[\text{Alternatively, } k \cdot \binom{n}{k} = k \cdot \frac{n!}{k!(n-k)!} = k \cdot \frac{n!}{k \cdot (k-1)!(n-k)!} = \frac{n \cdot (n-1)!}{(k-1)!(n-k)!}\right.$$
$$\left. = n \cdot \frac{(n-1)!}{(k-1)!((n-1)-(k-1))!} = n \cdot \binom{n-1}{k-1}.\right]$$

## Chapter 6 Review

### True or False

**1.** True **3.** False **5.** True **7.** False

## Fill in the Blank

**1.** disjoint **3.** combination **5.** binomial coefficients **7.** $\binom{5}{2} \cdot 2^2 = 40$

## Review Exercises

**1.** None of these. **3.** None of these. **5.** None of these.

**7.** $\subset, \subseteq$ **9.** $\subseteq, =$ **11.** None of these.

**13.** $\subset, \subseteq$ **15.** None of these.

**17.** (a) $(A \cap B) \cup C = \{3, 6\} \cup \{6, 8, 9\} = \{3, 6, 8, 9\}$
(b) $(A \cap B) \cap C = \{3, 6\} \cap \{6, 8, 9\} = \{6\}$
(c) $(A \cup B) \cap B = \{1, 2, 3, 5, 6, 7, 8\} \cap \{2, 3, 6, 7\} = \{2, 3, 6, 7\}$

**19.** $c(A \cup B) = c(A) + c(B) - c(A \cap B)$, so $33 = 24 + 12 - c(A \cap B)$, and $c(A \cap B) = 3$.

**21.** In the Venn diagram, circle CT represents *Chicago Tribune* readers, circle NYT represents *New York Times* readers, and circle WSJ represents *Wall Street Journal* readers.

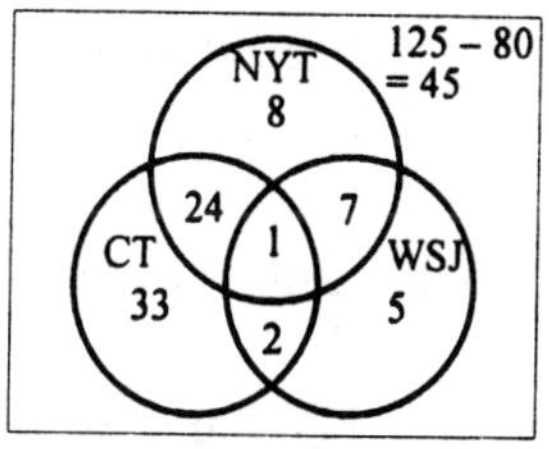


(a) $125 - (33 + 24 + 8 + 7 + 5 + 2 + 1) = 45$ students read none of these papers.
(b) $60 - (25 + 3 - 1) = 33$ read only the *Chicago Tribune.*
(c) $45 + 5 = 50$ read neither the *Chicago Tribune* nor the *New York Times.*

**23.** $P(6, 3) = \frac{6!}{3!} = 120$

**25.** $C(5,3) = \frac{5!}{3!2!} = \frac{5 \cdot 4}{2} = 10$ ways

**27.** The number of different orderings of 3 books on a shelf is $3! = 6$.

**29.** Assuming the customer may only choose one roof design, one window design, and one of the choices for brick, there are $3 \cdot 4 \cdot 6 = 72$ possible combinations (or house styles.)

**31.** The maximum number of 2 digit words is $P(4,2) = \frac{4!}{2!} = 4 \cdot 3 = 12$. In the latter case, each word is a subset of $\{1, 2, 3, 4\}$ of size 2, which there are $C(4,2) = \frac{4!}{2!2!} = 6$ possibilities.

**33.** $\left(\binom{16}{4}\text{choices for the 4 north side houses}\right)\cdot\left(\binom{10}{3}\text{choices for the 3 south side houses}\right)$

$= 1820 \cdot 120 = 218{,}400$ ways

**35.** (a) $\left(\binom{7}{3}\text{choices for the 3 boys}\right)\cdot\left(\binom{6}{4}\text{choices for the 4 girls}\right) = 35 \cdot 15 = 525$ ways

(b) There are $\binom{13}{7} = 1716$ ways to choose 7 from the 13 boys and girls. There is only one possible committee which does not have at least one member of each sex, namely to choose all seven of the boys for the committee. Thus there are $1716 - 1 = 1715$ ways to select the committee with at least one member of each sex.

**37.** ((4! arrangements of the history books) $\cdot$ (5! arrangements of the English books) $\cdot$(6! arrangements of the mathematics books)) $\cdot$ (3! orderings of the three subjects) $= 12{,}441{,}600$ arrangements.

**39.** $\left(\binom{10}{8}\text{choices for the 8 boys}\right)\cdot\left(\binom{11}{5}\text{choices for the 5 girls}\right) = 45 \cdot 462 = 20{,}790$ ways.

**41.** $5 \cdot 6 \cdot 8 = 240$ ways

**43.** With 3 boys and each team including at least one boy, one team has one boy and the other has the remaining two boys. Thus, the boys and girls can be divided into teams by selecting the singleton boy and then selecting his 3 female teammates. There are $3 \cdot \binom{5}{3} = 30$ ways to do this.

**45.** The number of possible orderings of {AB, C, D, E} is $4! = 24$.

**47.** (a) $\binom{20}{4} = 4845$ (b) $\binom{20}{3}\binom{5}{1} = 5700$

(c) $\binom{25}{4} - \binom{20}{4} = 12650 - 4845 = 7805$

**49.** One path from A to B is RRRRRRUUUUUU. All paths from A to B are exactly 12 moves, of which exactly 6 are to the right (R). Thus, in all, there are $\binom{12}{6} = 924$ paths from A to B.

**51.** $(x+2)^4 = \binom{4}{0}x^4 + \binom{4}{1}x^3(2) + \binom{4}{2}x^2(2^2) + \binom{4}{3}x(2^3) + \binom{4}{4}(2^4) = x^4 + 8x^3 + 24x^2 + 32x + 16$

**53.** $\binom{7}{4}(2^4) = 35 \cdot 16 = 560$.

# Chapter 7

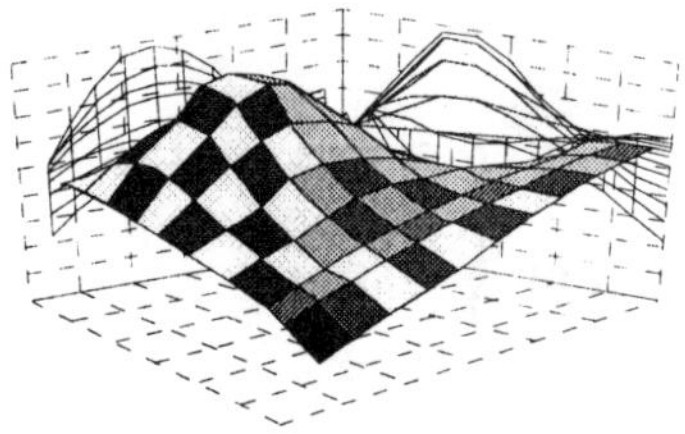

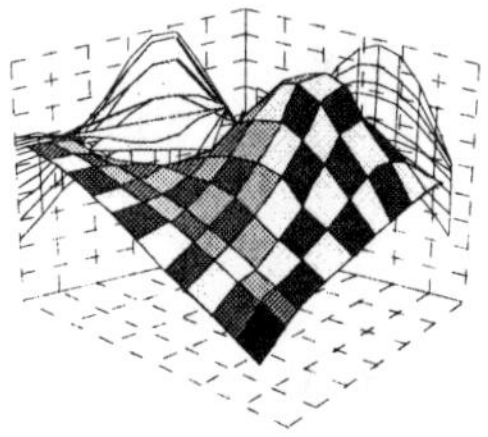

# Probability

## 7.1 Sample Spaces and Assignment of Probabilities

**1.** (a) $\{H, T\}$ (b) $\{0,1,2\}$ (c) $\{M, D\}$

**3.** $S = \{HH, HT, TH, TT\}$

**5.** $S = \{HHH, HHT, HTH, HTT, THH, THT, TTH, TTT\}$

**7.** $S = \{HH1, HH2, HH3, HH4, HH5, HH6, HT1, HT2, HT3, HT4, HT5, HT6$
$TH1, TH2, TH3, TH4, TH5, TH6, TT1, TT2, TT3, TT4, TT5, TT6\}$

**9.** $S = \{RA, RB, RC, GA, GB, GC\}$

**11.** $S = \{AA, AB, AC, BA, BB, BC, CA, CB, CC\}$

**13.** $S = \{AA1, AA2, AA3, AA4, AB1, AB2, AB3, AB4, AC1, AC2, AC3, AC4, BA1, BA2,$
$BA3, BA4, BB1, BB2, BB3, BB4, BC1, BC2, BC3, BC4, CA1, CA2, CA3, CA4, CB1,$
$CB2, CB3, CB4, CC1, CC2, CC3, CC4\}$

**15.** $S = \{RA1, RA2, RA3, RA4, RB1, RB2, RB3, RB4, RC1, RC2, RC3, RC4, GA1, GA2,$
$GA3, GA4, GB1, GB2, GB3, GB4, GC1, GC2, GC3, GC4\}$

**17.** $2^4 = 16$

**19.** $6^3 = 216$

**21.** $\binom{52}{2} = \frac{52 \cdot 51}{2!} = 1326$

**23.** $26 \cdot 26 = 676$

**25.** Probability assignments 1, 2, 3 and 6 are consistent with the definition of the probability of an outcome. (Assignment 4 fails since one of the probabilities is negative; assignment 5 fails since the sum of the probabilities does not equal 1.)

**27.** Assignment 2: $P(H) = 0$ and $P(T) = 1$, so $P(HH) = 0^2$, $P(HT) = P(TH) = (0)(1) = 0$, and $P(TT) = 1^2 = 1$.

**29.** Let $p =$ probability of tails then $3p =$ probability of heads. $p + 3p = 1$; $4p = 1$; $p = \frac{1}{4}; 3p = \frac{3}{4}$. We should assign probabilities of $\frac{3}{4}$ to heads and $\frac{1}{4}$ to tails.

**31.** $P(1) + P(3) + P(5) = \frac{2}{3}$; $P(1) = P(3) = P(5) = \frac{1}{3}$ of $\frac{2}{3} = \frac{2}{9}$, $P(2) + P(4) + P(6) = \frac{1}{3}$, $P(2) = P(4) = P(6) = \frac{1}{3}$ of $\frac{1}{3} = \frac{1}{9}$

In Problems 33-38, the sample space is $S = \{(x, y) | x = 1, \ldots, 6, y = 1, \ldots, 6\}$ for which the probability of each event is $\frac{1}{36}$.

**33.** $P(A) = \frac{2}{36} = \frac{1}{18}$

**35.** $P(C) = \frac{4}{36} = \frac{1}{9}$

**37.** $P(E) = \frac{6}{36} = \frac{1}{6}$

In Problems 39-44, the sample space is $S = \{(x, y) | x = 1, \ldots, 6, y = H \text{ or } T\}$, for which the probability of each event is $\frac{1}{12}$.

**39.** $A = \{(1, H), (2, H), (3, H), (4, H), (5, H), (6, H)\}$; $P(A) = \frac{6}{12} = \frac{1}{2}$

**41.** $C = \{(4, H), (4, T)\}$; $P(C) = \frac{2}{12} = \frac{1}{6}$

**43.** $E = \{(1, H), (1, T), (2, H), (2, T), (3, H), (3, T), (5, H), (5, T), (6, H), (6, T)\}$; $P(E) = \frac{10}{12} = \frac{5}{6}$

**45.** $G = \{(5, H), (5, T), (6, H), (6, T)\}$; $P(G) = \frac{4}{12} = \frac{1}{3}$

**47.** $K=\{(3,H),(4,H),(5,H)\}$; $P(K)=\frac{3}{12}=\frac{1}{4}$

**49.**

| Outcome: | $HH$ | $HT$ | $TH$ | $TT$ |
|---|---|---|---|---|
| Probability: | $\frac{1}{4}$ | $\frac{1}{4}$ | $\frac{1}{4}$ | $\frac{1}{4}$ |

**51.**

| Outcome: | $1H$ | $1T$ | $2H$ | $2T$ | $3H$ | $3T$ | $4H$ | $4T$ | $5H$ | $5T$ | $6H$ | $6T$ |
|---|---|---|---|---|---|---|---|---|---|---|---|---|
| Probability: | $\frac{1}{12}$ | $\frac{1}{12}$ | $\frac{1}{12}$ | $\frac{1}{12}$ | $\frac{1}{12}$ | $\frac{1}{12}$ | $\frac{1}{12}$ | $\frac{1}{12}$ | $\frac{1}{12}$ | $\frac{1}{12}$ | $\frac{1}{12}$ | $\frac{1}{12}$ |

**53.**

| Outcome: | $HHHH$ | $HHHT$ | $HHTH$ | $HHTT$ | $HTHH$ | $HTHT$ | $HTTH$ | $HTTT$ |
|---|---|---|---|---|---|---|---|---|
| Probability: | $\frac{1}{16}$ | $\frac{1}{16}$ | $\frac{1}{16}$ | $\frac{1}{16}$ | $\frac{1}{16}$ | $\frac{1}{16}$ | $\frac{1}{16}$ | $\frac{1}{16}$ |

| Outcome: | $THHH$ | $THHT$ | $THTH$ | $THTT$ | $TTHH$ | $TTHT$ | $TTTH$ | $TTTT$ |
|---|---|---|---|---|---|---|---|---|
| Probability: | $\frac{1}{16}$ | $\frac{1}{16}$ | $\frac{1}{16}$ | $\frac{1}{16}$ | $\frac{1}{16}$ | $\frac{1}{16}$ | $\frac{1}{16}$ | $\frac{1}{16}$ |

**55.** "last 3 tosses are tails" = $\{HTTT, TTTT\}$

**57.** "number of heads exceeds 1 but is fewer than 4" = "number of heads is 2 or 3"
$=\{HHHT, HHTH, HHTT, HTHH, HTHT, HTTH, THHH, THHT, THTH, TTHH\}$

**59.** The outcomes of the first 2 runs, $RR$, $RL$, $LR$, $LL$, are equally likely, so each has probability $\frac{1}{4}$. The probability of $R$ on the 3rd run is $\frac{1}{3}$, and the probability of $L$ on the 3rd run is $\frac{2}{3}$.
Thus, the probabilities for the outcomes of the 3 runs are :

| Outcomes: | $RRR$ | $RRL$ | $RLR$ | $RLL$ | $LRR$ | $LRL$ | $LLR$ | $LLL$ |
|---|---|---|---|---|---|---|---|---|
| Probability: | $\frac{1}{12}$ | $\frac{1}{6}$ | $\frac{1}{12}$ | $\frac{1}{6}$ | $\frac{1}{12}$ | $\frac{1}{6}$ | $\frac{1}{12}$ | $\frac{1}{6}$ |

(a) $P(\{RRR, RRL, LRR\})=\frac{1}{12}+\frac{1}{6}+\frac{1}{12}=\frac{1}{3}$

(b) $P(\{LLL\})=\frac{1}{6}$

(c) $P(\{LRR, LRL, LLR, LLL\})=\frac{1}{12}+\frac{1}{6}+\frac{1}{12}+\frac{1}{6}=\frac{1}{2}$

(d) $P(\{RRR, RRL, LRR, LRL\})=\frac{1}{12}+\frac{1}{6}+\frac{1}{12}+\frac{1}{6}=\frac{1}{2}$

**61.** $P(C_1) = p$ and $P(C_1) = \frac{1}{2}P(C_2) = P(C_3)$, so $P(C_2) = 2p$ and $P(C_3) = p$. Also, $P(C_1) + P(C_2) + P(C_3) = 1$, so $p + 2p + p = 4p = 1$, or $p = \frac{1}{4}$. Thus, $P(C_1) = \frac{1}{4}$, $P(C_2) = \frac{1}{2}$ and $P(C_3) = \frac{1}{4}$.

## 7.2 Properties of the Probability of an Event

**1.** $P(\overline{A}) = 1 - .25 = .75$

**3.** $P(A \cup B) = P(A) + P(B) = .25 + .40 = .65$

**5.** $P(A \cup B) = P(A) + P(B) - P(A \cap B)$
$= .25 + .40 - .15 = .5$

**7.** $\frac{1}{52}$

**9.** $\frac{13}{52} = \frac{1}{4}$

**11.** $\frac{3 \cdot 4}{52} = \frac{12}{52} = \frac{3}{13}$

**13.** $\frac{5 \cdot 4}{52} = \frac{5}{13}$

**15.** $\frac{52 - 4}{52} = \frac{12}{13}$

**17.** The events "Sum is 2" and "Sum is 12" are mutually exclusive. The probability of obtaining a 2 or a 12 is $\frac{1}{36} + \frac{1}{36} = \frac{1}{18}$.

**19.** $\frac{3}{3+5+8+7} = \frac{3}{23}$

**21.** $\frac{7}{23}$

**23.** $\frac{3+5}{23} = \frac{8}{23}$

**25.** $\frac{23-(5+7)}{23} = \frac{11}{23}$

**27.** "The number on one die is double the number on the other" $= \{(1,2),(2,4),(3,6),(2,1),(4,2),(6,3)\}$, so the probability that the number on one die is double the number on the other is $\frac{6}{36} = \frac{1}{6}$.

**29.** $P(\text{losing}) = 1 - P(\text{win}) - P(\text{tie}) = 1 - .65 - .05 = 0.3$

**31.** According to Jenny's estimates, $P(\text{"pass mathematics"}) = 0.4$, $P(\text{"pass English"}) = 0.6$ and $P(\text{"pass mathematics" or "pass English"}) = 0.8$. Since $P(\text{"pass mathematics" or "pass English"}) = P(\text{"pass mathematics"}) + P(\text{"pass English"}) - P(\text{"pass mathematics" and "pass English"})$, we have $0.8 = 0.4 + 0.6 - P(\text{"pass mathematics" and "pass English"})$ so $P(\text{"pass mathematics" and "pass English"}) = 0.2$. Thus, by Jenny's own estimation, the probability that she passes both courses is 0.2.

**33.** (a) $P(A \text{ or } B) = P(A) + P(B) - P(A \cap B) = 0.5 + 0.4 - 0.2 = 0.7$
(b) $P(A \text{ but not } B) = P(A) - P(A \cap B) = 0.5 - 0.2 = 0.3$
(c) $P(B \text{ but not } A) = P(B) - P(A \cap B) = 0.4 - 0.2 = 0.2$
(d) $P(\text{neither } A \text{ nor } B) = P(\text{not}(A \text{ or } B)) = 1 - P(A \text{ or } B) = 1 - 0.7$ (from part (a)) $= 0.3$

**35.** Let $T$ be the event "a car requires a tune-up", and $B$ be the event "a car requires a brake job," so that $P(T) = 0.6$, $P(B) = 0.1$, and $P(T \cap B) = 0.02$.
(a) $P(T \cup B) = P(T) + P(B) - P(T \cap B) = 0.6 + 0.1 - 0.02 = 0.68$
(b) $P(T \text{ and not } B) = P(T) - P(T \cap B) = 0.6 - 0.02 = 0.58$
(c) $P(\text{neither } T \text{ nor } B) = 1 - P(T \cup B) = 1 - 0.68 = 0.32$

**37.** (a) $0.24 + 0.33 = 0.57$ (b) $1 - 0.05 = 0.95$
(c) $1 - 0.17 = 0.83$ (d) $0.21 + 0.17 = 0.38$
(e) $0.05 + 0.24 = 0.29$ (f) $0.05$
(g) $0.24 + 0.33 + 0.21 = 0.78$ (h) $0.33 + 0.21 + 0.17 = 0.71$

**39.** $P(\text{"single"}) = \frac{52}{150}$, $P(\text{"college graduate"}) = \frac{72}{150}$,
$P(\text{"single" and "college graduate"}) = \frac{3/4 \cdot 52}{150} = \frac{39}{150}$, so
$P(\text{"single" or "college graduate"}) = \frac{52}{150} + \frac{72}{150} - \frac{39}{150} = \frac{85}{150}$. The probability that a salesperson will be neither single nor a college graduate is $1 - \frac{85}{150} - \frac{65}{150} = \frac{13}{30}$.

**41.** $P(E) = \frac{3}{3+1} = \frac{3}{4}$ **43.** $P(E) = \frac{5}{5+7} = \frac{5}{12}$ **45.** $P(E) = \frac{1}{2}$

**47.** $P(E) = .6$, $P(\overline{E}) = 1 - P(E) = 1 - .6 = .4$. The odds for $E$ are $\frac{P(E)}{P(\overline{E})} = \frac{.6}{.4}$ or 3 to 2. The odds against $E$ are 2 to 3.

**49.** $P(F) = \frac{3}{4}$, $P(\overline{F}) = 1 - P(F) = 1 - \frac{3}{4} = \frac{1}{4}$. The odds for $F$ are $\frac{P(F)}{P(\overline{F})} = \frac{3/4}{1/4}$ or 3 to 1; the odds against $F$ are 1 to 3.

**51.** $P(7) = \frac{1}{6}$, $P(\text{not } 7) = \frac{5}{6}$; so the odds for a 7 are 1 to 5,
$P(11) = \frac{1}{18}$, $P(\text{not } 11) = \frac{17}{18}$; so the odds for a 11 are 1 to 17;
$P(7 \text{ or } 11) = \frac{1}{6} + \frac{1}{18} = \frac{2}{9}$, $P(\text{neither 7 nor } 11) = \frac{7}{9}$; so the odds for 7 or 11 are 2 to 7.

**53.** $P(\text{gets interview}) = 0.54$; $P(\text{does not get interview}) = 1 - 0.54 = 0.46$. The odds against the interview are .46 to .54 or 23 to 27.

**55.** $P(A \text{ wins}) = \frac{1}{1+2} = \frac{1}{3}$. $P(B \text{ wins}) = \frac{2}{2+3} = \frac{2}{5}$. Since ties are impossible, "A wins" and "B wins" are mutually exclusive events, so:

$P(A \text{ or } B \text{ wins}) = P(A \text{ wins}) + P(B \text{ wins}) = \frac{1}{3} + \frac{2}{5} = \frac{11}{15}$.

The odds that A or B wins the race are 11/15 to 4/15 or 11 to 4.

**57.** Using the hint,

$$\begin{aligned} P(E \cup F) &= P((E \cap \overline{F}) \cup (E \cap F) \cup (\overline{E} \cap F)) \\ &= P(E \cap \overline{F}) + P(E \cap F) + P(\overline{E} \cap F) \text{ (using part (a) of the hint)} \\ &= P(E \cap \overline{F}) + P(E \cap F) + P(\overline{E} \cap F) + P(E \cap F) - P(E \cap F) \\ &= P(E) + P(F) - P(E \cap F) \text{ (using part (b) of the hint)} \end{aligned}$$

**59.** We have $\frac{P(E)}{P(\overline{E})} = \frac{a}{b}$ (since "the odds for $E$ are $a$ to $b$"), so $P(E) = \frac{a}{b} P(\overline{E})$. Also, $P(\overline{E}) = 1 - P(E)$ from Equation (3). Combining these two results, $P(E) = \frac{a}{b}(1 - P(E))$ or

$$P(E) = \frac{a}{b} - \frac{a}{b}P(E) \text{ or } P(E) + \frac{a}{b}P(E) = \frac{a}{b}$$

$$\left(1 + \frac{a}{b}\right)P(E) = \frac{a}{b}$$

$$\frac{b+a}{b}P(E) = \frac{a}{b}$$

$$P(E) = \frac{a}{b}\frac{b}{a+b} = \frac{a}{a+b}$$

## Technology Exercises

**1.** Actual probabilities: $P(H) = P(T) = \frac{1}{2}$

**3.** Actual probabilities: $P(T) = 0.25$, $P(H) = 1 - 0.25 = 0.75$

**5.** Actual probabilities: $P(\text{red}) = \frac{5}{5+2+8} = \frac{1}{3}$, $P(\text{yellow}) = \frac{2}{5+2+8} = \frac{2}{15}$,

$P(\text{white}) = \frac{8}{5+2+8} = \frac{8}{15}$

## 7.3 Probability Problems Using Counting Techniques

**1.** The probability of no digit occurring more than once is $\frac{10\cdot 9\cdot 8\cdot 7\cdot 6\cdot 5\cdot 4}{10^7}$, so the probability of at least one repeated digit is $1-\frac{10\cdot 9\cdot 8\cdot 7\cdot 6\cdot 5\cdot 4}{10^7}$, approximately 0.9395.

**3.** $\frac{26\cdot 25\cdot 24\cdot 23\cdot 22}{26^5}$, approximately 0.6644.

**5.** (a) $\frac{C(5,3)}{2^5}=\frac{10}{32}=\frac{5}{16}=0.3125$ (b) $\frac{C(5,0)}{2^5}=\frac{1}{32}=0.03125$

**7.** (a) $7=1+6=2+5=3+4=4+3=5+2=6+1$, so the probability of a 7 in a single throw of the pair of dice is $\frac{6}{36}=\frac{1}{6}$. The probability that this will occur three times in three tosses of the pair of dice is $\left(\frac{1}{6}\right)^3=\frac{1}{216}$, approximately 0.00463.

(b) $11=5+6=6+5$, so the probability of an 11 in a single throw of the pair of dice is $\frac{2}{36}=\frac{1}{18}$. The probability of a 7 or an 11 in a single throw of the pair of dice is $\frac{1}{6}+\frac{1}{18}=\frac{2}{9}$. The probability of a 7 or 11 in 2 or 3 of three tosses of the pair of dice is $C(3,2)\left(\frac{2}{9}\right)^2\left(\frac{7}{9}\right)+C(3,3)\left(\frac{2}{9}\right)^3=\frac{84+8}{729}=\frac{92}{729}$, approximately 0.126.

**9.** The number of ways of selecting 5 refrigerators from the 50 which were shipped is $\binom{50}{5}=2{,}118{,}760$. The number of ways of selecting 5 refrigerators from the 6 defective ones is $\binom{6}{5}=6$. Thus the probability that 5 refrigerators randomly selected from the 50 that were shipped will all be defective is $\frac{6}{2{,}118{,}760}\approx 0.00000283$. The number of ways that either zero or one of the five refrigerators selected is defective is $\binom{6}{0}\binom{44}{5}+\binom{6}{1}\binom{44}{4}=1{,}900{,}514$, so the probability that at least two of the 5 selected will be defective is $1-\frac{1{,}900{,}514}{2{,}118{,}760}=\frac{2227}{21{,}620}\approx 0.103$.

**11.** $P(\text{at least 2 in same month})=1-P(\text{all different months})$

$$=1-\frac{(12)(11)(10)}{(12)(12)(12)}=.236$$

**13.** The sample space $S$ is the set of ordered triples of numbers in the range 1 to 100, so $c(S) = 100^3$. The number of such triples which consist of distinct numbers is $P(100,3) = 100 \cdot 99 \cdot 98$. Thus, the probability that no two of the three slips have the same number is $\frac{100 \cdot 99 \cdot 98}{100^3} = \frac{4851}{5000}$, and the probability that at least two of the slips have the same number is $1 - \frac{4851}{5000} = \frac{149}{5000} = 0.0298$.

**15.** Assuming 365 possible birthdays (ignoring February 29), the probability that each of the 100 senators has a different birthday is $\frac{P(365,100)}{365^{100}}$, and the probability that 2 or more senators have the same birthday is $1 - \frac{365 \cdot 364 \cdot 363 \cdots 266}{365^{100}} \approx 0.999999692751$.

**17.** Since the events "*E* precedes *L*" and "*L* precedes *E*" are complementary and equally likely, each has a probability $\frac{1}{2}$. Thus, the probability that $L$ will precede $E$ is $\frac{1}{2}$.

**19.** The first letter is equally likely to be V, O, W, E or L, so the probability that it is L is $\frac{1}{5} = 0.2$.

**21.** The probability of landing on an even integer is $\frac{13}{26}$, the probability of landing on one of the last 19 integers (8,9,…,26) is $\frac{19}{26}$, and the probability of landing on an even integer which is in the range 8, …, 26 is $\frac{10}{26}$. Thus, the probability of landing on an even integer or on any of the last 19 integers is $\frac{13}{26} + \frac{19}{26} - \frac{10}{26} = \frac{22}{26} \approx 0.846$.

**23.** $$\frac{\binom{13}{3} + \binom{13}{3} + \binom{13}{3}}{\binom{52}{3}} = \frac{33}{850} \approx 0.0388$$

**25.** The number of 13-card bridge hands is $\binom{52}{13} = 635{,}013{,}559{,}600$. The number of ways of selecting 5 of the 13 spades, 4 of the 13 hearts, 3 of the 13 diamonds, and 1 of the 13 clubs is $\binom{13}{5}\binom{13}{4}\binom{13}{3}\binom{13}{1} = 3{,}421{,}322{,}190$. Thus, the probability that a bridge hand of 13 cards contains 5 spades, 4 hearts, 3 diamonds and 1 club is $\dfrac{3{,}421{,}322{,}190}{635{,}013{,}559{,}600} \approx 0.00539$.

**27.** There are $8^5 = 32{,}768$ ways for the 5 passengers to each select the one of the eight floors at which he or she will be discharged. There are $P(8,5) = 8 \cdot 7 \cdot 6 \cdot 5 \cdot 4 = 6720$ ways for the passengers to select 5 different floors. Thus, the probability that no two of the passengers leaves on the same floor is $\dfrac{6720}{32768} = \dfrac{105}{512} \approx .205$.

## 7.4 Conditional Probability

**1.** $P(E) = 0.2 + 0.3 = 0.5$

**3.** $P(E|F) = \dfrac{P(E \cap F)}{P(F)} = \dfrac{.3}{.7} = \dfrac{3}{7}$, approximately 0.4286

**5.** $P(E \cap F) = 0.3$

**7.** $P(\overline{E}) = 1 - P(E) = 1 - 0.5 = 0.5$

**9.** $P(E|F) = \dfrac{P(E \cap F)}{P(F)} = \dfrac{0.1}{0.4} = \dfrac{1}{4}$; $P(F|E) = \dfrac{P(F \cap E)}{P(E)} = \dfrac{0.1}{0.2} = \dfrac{1}{2}$.

**11.** $P(E|F) = \dfrac{P(E \cap F)}{P(F)}$, so $P(F) = \dfrac{P(E \cap F)}{P(E|F)} = \dfrac{0.2}{0.4} = \dfrac{1}{2}$.

**13.** $P(E|F) = \dfrac{P(E \cap F)}{P(F)}$, so $P(E \cap F) = P(E|F) \cdot P(F) = \left(\dfrac{4}{5}\right)\left(\dfrac{5}{13}\right) = \dfrac{4}{13}$

**15.** (a) $P(E) = \dfrac{P(E \cap F)}{P(F|E)} = \dfrac{1/3}{2/3} = \dfrac{1}{2}$ (b) $P(F) = \dfrac{P(E \cap F)}{P(E|F)} = \dfrac{1/3}{1/2} = \dfrac{2}{3}$

**17.** $P(C) = P(A) \cdot P(C \mid A) + P(B) \cdot P(C \mid B) = (.7)(.9) + (.3)(.2) = 0.69$

**19.** $P(C \mid A) = 0.9$

**21.** $P(C|B) = 0.2$

**23.** $P(E \cap F) = P(E) + P(F) - P(E \cup F) = 0.5 + 0.4 - 0.8 = 0.1$

**25.** $P(F|E) = \dfrac{P(F \cap E)}{P(E)} = \dfrac{0.1}{0.5} = 0.2$

**27.** $P(E|\overline{F}) = \dfrac{P(E \cap \overline{F})}{P(\overline{F})} = \dfrac{0.5 - 0.1}{0.6} = \dfrac{2}{3} \approx 0.67$

**29.** $\dfrac{P(\{GGB,GBG\})}{P(\{GBB,GBG,GGB,GGG\})}=\dfrac{2/8}{4/8}=\dfrac{1}{2}$

**31.** The probability of 4 heads in 4 tosses of a fair coin is $\left(\frac{1}{2}\right)^4=\frac{1}{16}$. If we know that the second throw was a head, the probability that the other 3 tosses also resulted in heads is $\left(\frac{1}{2}\right)^3=\frac{1}{8}$.

**33.** $P(\text{"first card is a heart"})\cdot P(\text{"second card is red given the first card is a heart"})$
$=\dfrac{13\text{ hearts}}{52\text{ cards}}\cdot\dfrac{25\text{ red cards}}{51\text{ remaining cards}}=\dfrac{25}{204}\approx 12.25\%$.

**35.** $P(\{WY,YW\})=P(WY)+P(YW)=\dfrac{3}{6}\cdot\dfrac{1}{5}+\dfrac{1}{6}\cdot\dfrac{3}{5}=\dfrac{6}{30}=\dfrac{1}{5}$

**37.** (a) $P(\text{red ace})=\dfrac{2\text{ red aces}}{52\text{ cards}}=\dfrac{1}{26}$ (b) $P(\text{red ace}\mid\text{ace})\dfrac{2\text{ red aces}}{4\text{ aces}}=\dfrac{1}{2}$

(c) $P(\text{red ace}\mid\text{red card})=\dfrac{2\text{ red aces}}{26\text{ red cards}}=\dfrac{1}{13}$

**39.** $16\%+8\%+6\%=30\%$ of the families have more than two children, and $100\%-20\%=80\%$ of the families have at least one child. The probability that a family has more than two children if it is known that they have at least one child is $\dfrac{30}{80}=0.375\text{ or }37.5\%$.

**41.** $P(E)=0.40$

**43.** $P(H)=0.24$

**45.** $P(E\cap H)=0.10$

**47.** $P(G\cap H)=0.08$

**49.** $P(E\mid H)=\dfrac{P(E\cap H)}{P(H)}=\dfrac{0.10}{0.24}=\dfrac{5}{12}$

**51.** $P(G\mid H)=\dfrac{P(G\cap H)}{P(H)}=\dfrac{0.08}{0.24}=\dfrac{1}{3}$

**53.** Note that there are a total of 110 Democrats, 70 Republicans, 55 Independents, 120 males, and 115 females.

(a) $P(F|I)=\dfrac{25}{55}=\dfrac{5}{11}$ (b) $P(R|F)=\dfrac{30}{115}=\dfrac{6}{23}$

(c) $P(M|D)=\dfrac{50}{110}=\dfrac{5}{11}$ (d) $P(D|M)=\dfrac{50}{120}=\dfrac{5}{12}$

(e) $P(M|R\cup I)=\dfrac{40+30}{70+55}=\dfrac{14}{25}$ (f) $P(I|M)=\dfrac{30}{120}=\dfrac{1}{4}$

**55.** (a) $\frac{1448}{2018} = \frac{724}{1009} = 0.7175$ (b) $\frac{666}{2018} = \frac{333}{1009} = 0.3300$

(c) $\frac{144}{2018} = \frac{72}{1009} = 0.0714$

(d) $P(F|E) = \frac{c(E \cap F)}{c(E)} = \frac{102}{526} = \frac{51}{263} = 0.1939$

(e) $P(A \& S|M) = \frac{c(A \& S \cap M)}{c(M)} = \frac{342}{1448} = \frac{171}{724} = 0.2362$

(f) $P(F \mid A \& S \cup E) = \frac{c(F \cap A \& S) + c(F \cap E)}{c(A \& S) + c(E)} = \frac{324 + 102}{666 + 526} = \frac{213}{596} = 0.3574$

(g) $P(\overline{B} \cap M) = P(\overline{B} \mid M) \cdot P(M) = \frac{342 + 424}{1448} \cdot \frac{1448}{2018} = \frac{766}{2018} = \frac{383}{1009} = 0.3796$

(h) $P(F \mid \overline{E}) = \frac{c(F \cap \overline{E})}{c(\overline{E})} = \frac{570 - 102}{2018 - 526} = \frac{468}{1492} = \frac{117}{373} = 0.3137$

**57.** $P(\text{"woman"}|\text{"weight under 160 pounds"}) = \frac{P(\text{"woman who weighs under 160 lbs."})}{P(\text{"weight under 160 lbs."})}$

$= \frac{(0.5)(0.7)}{(0.5)(0.7) + (0.5)(0.35)} = \frac{0.35}{0.525} = \frac{2}{3} \approx 66.7\%$

See tree diagram below.

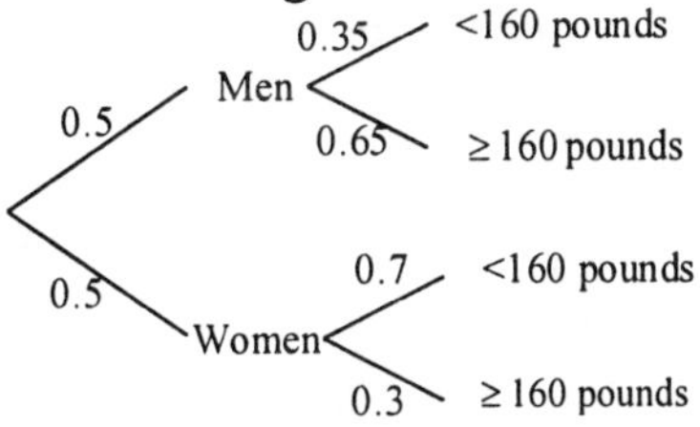


**59.** (See tree diagram to the right.)

P ("attended private school given *A* average")

$= \frac{P(\text{"private school and A average"})}{P(\text{"A average"})}$

$= \frac{(0.4)(0.3)}{0.24}$

$= \frac{1}{2} = 0.5$ or 50%

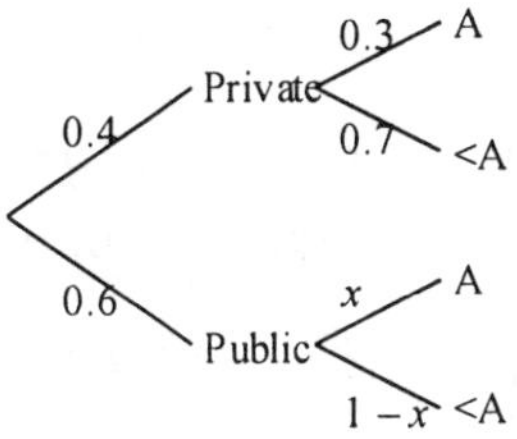

**61.** $P(\text{successful}) = P(\text{pass}) \cdot P(\text{successful} \mid \text{pass}) + P(\text{fail}) \cdot P(\text{successful} \mid \text{fail})$
$= 0.7 \cdot 0.85 + 0.3 \cdot 0.4 = 0.715$

71.5% of applicants would complete the training successfully.

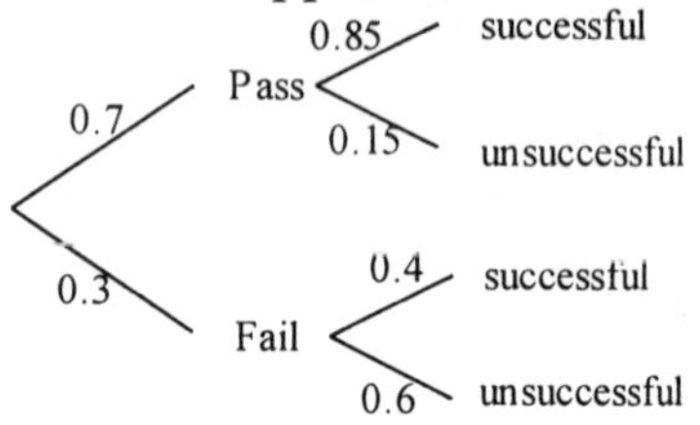


**63.** If $P(E) \neq 0$ then $P(E|E) = \dfrac{P(E \cap E)}{P(E)} = \dfrac{P(E)}{P(E)} = 1.$

**65.** If $S$ is the sample space, then $E \cap S = E$ and $P(S) = 1$. Thus,
$P(E|S) = \dfrac{P(E \cap S)}{P(S)} = \dfrac{P(E)}{1} = P(E).$

## 7.5 Independent Events

**1.** $P(E \cap F) = P(E) \cdot P(F) = (0.4)(0.6) = 0.24$

**3.** $P(E \cup F) = P(E) + P(F) - P(E \cap F) = P(E) + P(F) - P(E) \cdot P(F)$, so
$0.3 = 0.2 + P(F) - 0.2P(F);\ 0.8P(F) = 0.1;\ P(F) = \dfrac{1}{8} = 0.125$

**5.** $P(E) \cdot P(F) = \dfrac{4}{21}\dfrac{7}{12} = \dfrac{1}{9} \neq \dfrac{2}{9} = P(E \cap F)$; no

**7.** (a) $P(E|F) = P(E) = 0.2$
(b) $P(F|E) = P(F) = 0.4$
(c) $P(E \cap F) = P(E) \cdot P(F) = (0.2)(0.4) = 0.08$
(d) $P(E \cup F) = P(E) + P(F) - P(E \cap F) = 0.2 + 0.4 - 0.08 = 0.52$

**9.** $P(E \cap F \cap G) = P(E) \cdot P(F) \cdot P(G) = \dfrac{2}{3} \cdot \dfrac{3}{7} \cdot \dfrac{2}{21} = \dfrac{4}{147}$

**11.** $P(E|F) = \dfrac{P(E \cap F)}{P(F)} = \dfrac{P(E \cap F)}{0.2}$, so we need to find $P(E \cap F)$.
$P(E \cup F) = P(E) + P(F) - P(E \cap F)$, so $0.4 = 0.3 + 0.2 - P(E \cap F)$,
and $P(E \cap F) = 0.3 + 0.2 - 0.4 = 0.1$. Therefore, $P(E|F) = \dfrac{0.1}{0.2} = 0.5$. Since
$P(E|F) \neq P(E)$, $E$ and $F$ are *not* independent.

**13.** $P(E \cap F) = P(\text{"3 is rolled"}) = \frac{1}{6}$; $P(E)\cdot P(F) = \left(\frac{1}{2}\right)\left(\frac{1}{2}\right) = \frac{1}{4}$. Since $P(E \cap F) \neq P(E)\cdot P(F)$, $E$ and $F$ are *not* independent.

**15.** (See probability tree on the right.)

$P(E \cap F) = P(\text{"the family has 1 boy and 2 girls"})$
$= P(\text{"}BGG \text{ or } GBG \text{ or } GGB\text{"}) = \frac{3}{8}$.

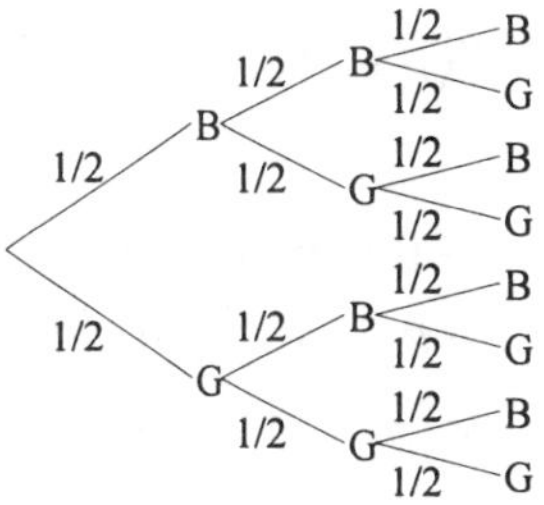


$P(E) = P(\text{"}BGG \text{ or } GBG \text{ or } GGB \text{ or } GGG\text{"}) = \frac{1}{2}$ and

$P(F) = P(\text{not "}BBB \text{ or } GGG\text{"}) = \frac{6}{8} = \frac{3}{4}$, so

$P(E)\cdot P(F) = \left(\frac{1}{2}\right)\left(\frac{3}{4}\right) = \frac{3}{8} = P(E \cap F)$. Yes, $E$ and $F$ *are* independent events.

**17.** (a) $P(\text{"1}^{\text{st}} \text{ is club"}) = \frac{13 \text{ clubs}}{52 \text{ cards}} = \frac{1}{4}$.

(b) $P(\text{"2}^{\text{nd}} \text{ is heart} \mid 1^{\text{st}} \text{ is club"}) = P(\text{"2}^{\text{nd}} \text{ is heart"}) = \frac{13 \text{ hearts}}{52 \text{ cards}} = \frac{1}{4}$.

(c) $P(\text{"first card is a club and the second is a club"})$
$= P(\text{"1}^{\text{st}} \text{ is club"})\cdot P(\text{"2}^{\text{nd}} \text{ is club"}) = \left(\frac{13}{52}\right)\left(\frac{13}{52}\right) = \left(\frac{1}{4}\right)\left(\frac{1}{4}\right) = \frac{1}{16}$.

**19.** $E \cap H = E$; $P(E \cap H) = P(E) = \frac{3}{8} \neq P(E)\cdot P(H) = \left(\frac{3}{8}\right)\left(\frac{1}{2}\right) = \frac{3}{16}$. $E$ and $H$ are *not* independent.

**21.** $P(A) = \frac{1}{4} + \frac{1}{4} = \frac{1}{2}$; $P(B) = \frac{1}{4} + \frac{1}{4} = \frac{1}{2}$; $P(C) = \frac{1}{4} + \frac{1}{4} = \frac{1}{2}$.

$P(A \cap B) = P(\{2\}) = \frac{1}{4} = \left(\frac{1}{2}\right)\left(\frac{1}{2}\right) = P(A)\cdot P(B)$, so $A$ and $B$ are independent.

$P(A \cap C) = P(\{1\}) = \frac{1}{4} = \left(\frac{1}{2}\right)\left(\frac{1}{2}\right) = P(A)\cdot P(C)$, so $A$ and $C$ are independent.

$P(B \cap C) = P(\{3\}) = \frac{1}{4} = \left(\frac{1}{2}\right)\left(\frac{1}{2}\right) = P(B)\cdot P(C)$, so $B$ and $C$ are independent.

**23.** (a) $P(\text{1st has heart disease} \cap \text{2nd has heart disease})$
$= P(\text{1st has heart disease})\cdot P(\text{2nd has heart disease})$
$= \frac{3}{4}\cdot\frac{3}{4} = \frac{9}{16}$

(b) $P(\overline{\text{1st}} \cap \overline{\text{2nd}}) = P(\overline{\text{1st}})\cdot P(\overline{\text{2nd}}) = \frac{1}{4}\frac{1}{4} = \frac{1}{16}$

(c) $P(\text{1st} \cap \overline{\text{2nd}}) + P(\overline{\text{1st}} \cap \text{2nd}) = P(\text{1st})\cdot P(\overline{\text{2nd}}) + P(\overline{\text{1st}})\cdot P(\text{2nd}) = \frac{3}{4}\frac{1}{4} + \frac{1}{4}\frac{3}{4} = \frac{3}{8}$

**25.** $P(T)=3P(H);\ P(T)+P(H)=1=4P(H);\ P(H)=\frac{1}{4};\ P(T)=\frac{3}{4}$

(a) $P(\{TTT\})=P(T)\cdot P(T)\cdot P(T)=\frac{3}{4}\frac{3}{4}\frac{3}{4}=\frac{27}{64}$

(b) $P(\{THH,HTH,HHT\})=3P(T)\cdot P(H)\cdot P(H)=3\frac{3}{4}\frac{1}{4}\frac{1}{4}=\frac{9}{64}$

**27.** Let $R$ be the event "recover", and $\overline{R}$ the event "not recover".

(a) $P(RRRR)=[P(R)]^4=(0.9)^4=0.6561$

(b) $P(\{RR\overline{R}\overline{R},R\overline{R}R\overline{R},R\overline{R}\overline{R}R,\overline{R}RR\overline{R},\overline{R}R\overline{R}R,\overline{R}\overline{R}RR\})=6P(R)\cdot P(R)\cdot P(\overline{R})\cdot P(\overline{R})$
$=6(0.9)^2(0.1)^2=0.0486$

(c) $1-P(R<2)=1-[P(R=0)+P(R=1)]=1-[(0.1)^4+4(0.9)(0.1)^3]=0.9963$

**29.** (a) $P(RR)=P(R)\cdot P(R)=\frac{6}{10}\cdot\frac{6}{10}=\frac{9}{25}$

(b) $P(\{RW,WR\})=2\cdot\frac{6}{10}\cdot\frac{4}{10}=\frac{12}{25}$

**31.** (a) $P(A|U)=\frac{c(A\cap U)}{c(U)}=\frac{40}{325}=0.1231$

(b) $P(A|\overline{U})=\frac{c(A\cap\overline{U})}{c(\overline{U})}=\frac{5}{515}=0.0097$

(c) $P(U\cap A)=\frac{40}{840}=0.0476$, $P(U)\cdot P(A)=\frac{325}{840}\frac{45}{840}=0.0207$. Since $P(U\cap A)\neq P(U)\cdot P(A)$, not independent.

(d) $P(U\cap\overline{A})=\frac{285}{840}=0.3393$, $P(U)\cdot P(\overline{A})=\frac{325}{840}\frac{795}{840}=0.3662$. Since $P(U\cap\overline{A})\neq P(U)\cdot P(\overline{A})$, not independent.

(e) $P(\overline{U}\cap A)=\frac{5}{840}=0.0060$, $P(\overline{U})\cdot P(A)=\frac{515}{840}\frac{45}{840}=0.0328$. Since $P(\overline{U}\cap A)\neq P(\overline{U})\cdot P(A)$, not independent.

(f) $P(\overline{U}\cap\overline{A})=\frac{510}{840}=0.6071$, $P(\overline{U})\cdot P(\overline{A})=\frac{515}{840}\frac{795}{840}=0.5803$. Since $P(\overline{U}\cap\overline{A})\neq P(\overline{U})\cdot P(\overline{A})$, not independent.

**33.** Let $V$ be the event "voted for her"

(a) $P(V)\cdot P(V)=\frac{2}{3}\cdot\frac{2}{3}=\frac{4}{9}$ (b) $P(\overline{V})\cdot P(\overline{V})=\frac{1}{3}\cdot\frac{1}{3}=\frac{1}{9}$

(c) $P(V)\cdot P(\overline{V})+P(\overline{V})\cdot P(V)=\frac{2}{3}\frac{1}{3}+\frac{1}{3}\frac{2}{3}=\frac{4}{9}$

**35.** $P(\text{event(a)})=1-P(\text{"no 1's in four throws"})=1-\left(\frac{5}{6}\right)^4\approx 0.5177$

$$P(\text{event(b)}) = 1 - P(\text{"no pair of 1's in 24 throws of a pair of dice"}) = 1 - \left(\frac{35}{36}\right)^{24} \approx 0.4914$$

Event (a) is more likely to occur.

**37.** If $E$ and $F$ are both independent and mutually exclusive then $P(E \cap F) = P(E) \cdot P(F)$ and $P(E \cap F) = 0$. Thus $P(E) \cdot P(F) = 0$ either $P(E) = 0$ or $P(F) = 0$, i.e., at least one of $E$ or $F$ is impossible.

**39.** Assume that $E$ and $F$ are independent, so $P(E)P(F) = P(E \cap F)$. Then,

$$\begin{aligned} P(\overline{E})P(\overline{F}) &= [1 - P(E)][1 - P(F)] \\ &= 1 - P(E) - P(F) + P(E)P(F) \\ &= 1 - P(E) - P(F) + P(E \cap F) \\ &= 1 - [P(E) + P(F) - P(E \cap F)] \\ &= 1 - P(E \cup F) \\ &= P(\overline{(E \cup F)}) = P(\overline{E} \cap \overline{F}) \end{aligned}$$

Hence, $\overline{E}$ and $\overline{F}$ are independent.

**41.** Assume that $P(E) > 0$, $P(F) > 0$ and $E$ is independent of $F$. Then $P(E|F) = P(E)$ and (by the hint) $P(E \cap F) = P(F) \cdot P(E|F) = P(F) \cdot P(E)$. Then

$$P(F|E) = \frac{P(E \cap F)}{P(E)} = \frac{P(F) \cdot P(E)}{P(E)} = P(F).$$

## Chapter 7 Review

### True or False

**1.** True **3.** False **5.** True **7.** True

### Fill in the Blank

**1.** $\frac{1}{2}$ **3.** 1; 0 **5.** for **7.** mutually exclusive

### Review Exercises

**1.** $S = \{MM, MF, FM, FF\}$

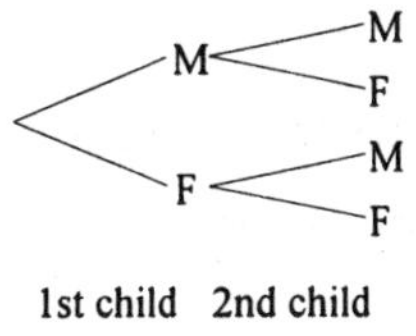

**3.** (a) $\dfrac{\binom{5}{2}}{\binom{14}{2}} = \dfrac{10}{91}$ (b) $\dfrac{\binom{5}{1}\binom{9}{1}}{\binom{14}{2}} = \dfrac{45}{91}$

(c) At least one blue means exactly one blue or both are blue (see parts (a) and (b))
$\dfrac{10}{91} + \dfrac{45}{91} = \dfrac{55}{91}$

**5.** (a) $P(3) = \dfrac{84}{400} = \dfrac{21}{100}$ (b) $P(5) = \dfrac{92}{400} = \dfrac{23}{100}$ (c) $P(6) = \dfrac{73}{400}$

**7.** (a) $P(\overline{E}) = 1 - P(E) = 1 - \dfrac{1}{2} = \dfrac{1}{2}$

(b) $P(E \cup F) = P(E) + P(F) - P(E \cap F)$, so
$P(F) = P(E \cup F) - P(E) + P(E \cap F) = \dfrac{5}{8} - \dfrac{1}{2} + \dfrac{1}{3} = \dfrac{11}{24}$

(c) $P(\overline{F}) = 1 - P(F) = 1 - \dfrac{11}{24} = \dfrac{13}{24}$

**9.**

| Outcome | 000 | 001 | 002 | 010 | 011 | 012 | 020 | 021 | 022 | 100 |
|---|---|---|---|---|---|---|---|---|---|---|
| Prob. | $\frac{1}{8}$ | $\frac{1}{32}$ | $\frac{3}{32}$ | $\frac{1}{32}$ | $\frac{1}{128}$ | $\frac{3}{128}$ | $\frac{3}{32}$ | $\frac{3}{128}$ | $\frac{9}{128}$ | $\frac{1}{32}$ |

| Outcome | 101 | 102 | 110 | 111 | 112 | 120 | 121 | 122 | 200 | 201 |
|---|---|---|---|---|---|---|---|---|---|---|
| Prob. | $\frac{1}{128}$ | $\frac{3}{128}$ | $\frac{1}{128}$ | $\frac{1}{512}$ | $\frac{3}{512}$ | $\frac{3}{128}$ | $\frac{3}{512}$ | $\frac{9}{512}$ | $\frac{3}{32}$ | $\frac{3}{128}$ |

| Outcome | 202 | 210 | 211 | 212 | 220 | 221 | 222 |
|---|---|---|---|---|---|---|---|
| Prob. | $\frac{9}{128}$ | $\frac{3}{128}$ | $\frac{3}{512}$ | $\frac{9}{512}$ | $\frac{9}{128}$ | $\frac{9}{512}$ | $\frac{27}{512}$ |

(a) No, not all outcomes are equally likely.

(b) Outcome "000" has highest probability $\left(\dfrac{1}{8}\right)$.

(c) $P(F) = P(\{012,021,102,120,201,210\}) = 6 \cdot \dfrac{1}{2} \cdot \dfrac{1}{8} \cdot \dfrac{3}{8} = \dfrac{9}{64}$

**11.** $P(\text{"The Bears win the NFL title"}) = \dfrac{7}{7+6} = \dfrac{7}{13}$.

**13.** Let $FM =$ "failed mathematics" and $FP =$ "failed physics". Then $P(FM) = 0.38$, $P(FP) = 0.27$ and $P(FM \cap FP) = 0.09$.

(a) $P(FM|FP) = \dfrac{P(FM \cap FP)}{P(FP)} = \dfrac{0.09}{0.27} = \dfrac{1}{3} \approx 0.333.$

(b) $P(FP|FM) = \dfrac{P(FP \cap FM)}{P(FM)} = \dfrac{0.09}{0.38} = \dfrac{9}{38} \approx 0.237$.

(c) $P(FM \cup FP) = P(FM) + P(FP) - P(FM \cap FP) = 0.38 + 0.27 - 0.09 = 0.56$.

**15.** Let *Blue* = "blue-eyed", *Brown* = "brown-eyed", and *Left* = "left-handed." Then $P(Blue) = 0.25$, $P(Brown) = 0.75$, $P(Left|Blue) = 0.1$, and $P(Left|Brown) = 0.05$.

(a) $P(Blue \cap Left) = P(Left|Blue) \cdot P(Blue) = (0.1)(0.25) = 0.025$.

(b)
$$\begin{aligned} P(Left) &= P(Blue \cap Left) + P(Brown \cap Left) \\ &= 0.025 + P(Left|Brown) \cdot P(Brown) \\ &= 0.025 + (0.05)(0.75) = 0.0625 \end{aligned}$$

(c) $P(Blue|Left) = \dfrac{P(Blue \cap Left)}{P(Left)} = \dfrac{0.025}{0.0625} = 0.4$

**17.** Let $X$ = ACT below 21, $Y$ = ACT 22-27, $Z$ = ACT above 28, $A$ = 3.6-4.0 GPA, $B$ = 3.0-3.5 GPA, $C$ = below 3.0 GPA

(a) $P(Z) = \dfrac{138}{400} = 0.345$

(b) $P(A) = \dfrac{168}{400} = 0.42$

(c) $P(Z \cap A) = \dfrac{104}{400} = 0.26$

(d) $P(Y) = \dfrac{160}{400} = 0.4$

(e) $P(B) = \dfrac{147}{400} = 0.3675$

(f) $P(Z \cap C) = \dfrac{4}{400} = 0.01$; $P(Z) \cdot P(C) = \dfrac{138}{400}\dfrac{85}{400} = 0.0733$. Since $P(Z \cap C) \neq P(Z) \cdot P(C)$, events $Z$ and $C$ are not independent.

**19.** Since gas station $G$ is 3 blocks north and 1 block east of his home, the probability that he will come to station $G$ is the probability of achieving exactly 3 heads in 4 flips of a fair coin
$$\frac{\binom{4}{3}}{2^4} = \frac{4}{16} = \frac{1}{4}.$$

## Profession Exam Questions

**1.** b: $P(P \cap Q) = 0 \neq P(P) \cdot P(Q) > 0$

**3.** b: $\dfrac{2}{17} = \left(\dfrac{26}{52}\right)\left(\dfrac{25}{51}\right)\left(\dfrac{24}{50}\right)$

**5.** c: 4 is the least value of $k$ for which $1 - \left(\dfrac{1}{6}\right)^k > 0.999$

**7.** b: $P(2\&6 \text{ or } 4\&4 \text{ or } 6\&2) = \left(\frac{1}{3}\right)\left(\frac{1}{6}\right) + \left(\frac{1}{3}\right)\left(\frac{1}{6}\right) + \left(\frac{1}{3}\right)\left(\frac{1}{6}\right) = \frac{1}{6}$

**9.** c: $P(S \cup T) = P(S \cap \overline{T}) + P(S \cap T) + P(T \cap \overline{S}) = 3p$

# Chapter 8

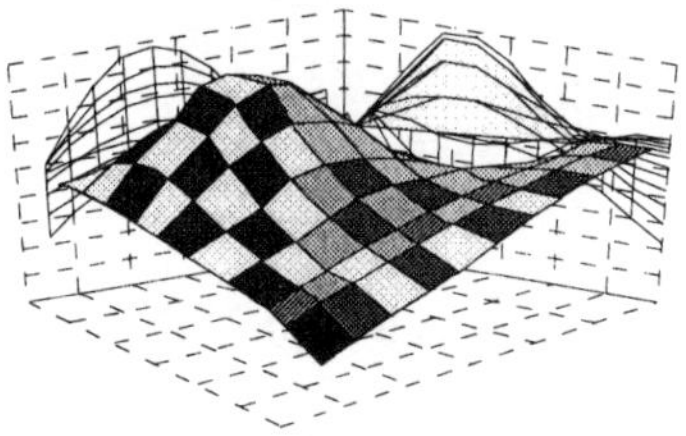

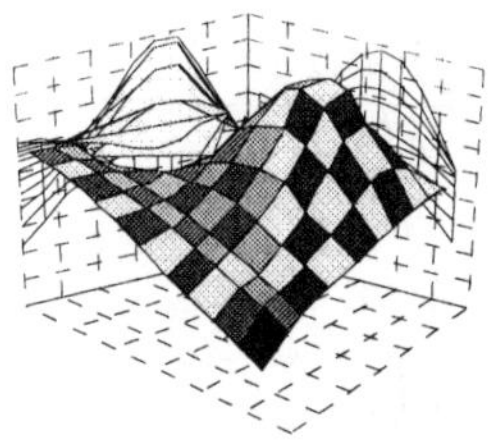

# Additional Probability Topics

## 8.1 Bayes' Formula

**1.** $P(E|A)=0.4$ **3.** $P(E|B)=0.2$ **5.** $P(E|C)=0.7$

**7.**
$$\begin{aligned} P(E) &= P(E|A)\cdot P(A)+P(E|B)\cdot P(B)+P(E|C)\cdot P(C) \\ &= (0.4)\cdot(0.3)+(0.2)\cdot(0.6)+(0.7)\cdot(0.1)=0.31 \end{aligned}$$

**9.**
$$\begin{aligned} P(A|E) &= \frac{P(A)\cdot P(E|A)}{P(E)}=\frac{P(A)\cdot P(E\mid A)}{P(A)\cdot P(E\mid A)+P(B)\cdot P(E\mid B)+P(C)\cdot P(E\mid C)} \\ &= \frac{(0.3)(0.4)}{(0.3)\cdot(0.4)+(0.6)\cdot(0.2)+(0.1)\cdot(0.7)}=\frac{0.12}{0.31}=\frac{12}{31}\approx 0.387 \end{aligned}$$

**11.**
$$\begin{aligned} P(C|E) &= \frac{P(C)\cdot P(E|C)}{P(E)}=\frac{P(C)\cdot P(E|C)}{P(A)\cdot P(E\mid A)+P(B)\cdot P(E\mid B)+P(C)\cdot P(E\mid C)} \\ &= \frac{(0.1)(0.7)}{(0.3)(0.4)+(0.6)(0.2)+(0.1)(0.7)}=\frac{.07}{.31}=\frac{7}{31}\approx 0.226 \end{aligned}$$

**13.**
$$\begin{aligned} P(B|E) &= \frac{P(B)\cdot P(E|B)}{P(E)}=\frac{P(B)\cdot P(E|B)}{P(A)\cdot P(E\mid A)+P(B)\cdot P(E\mid B)+P(C)\cdot P(E\mid C)} \\ &= \frac{(0.6)(0.2)}{(0.3)(0.4)+(0.6)(0.2)+(0.1)(0.7)}=\frac{(0.6)(0.2)}{0.31}=\frac{0.12}{0.31}=\frac{12}{31}\approx 0.387 \end{aligned}$$

**15.** $P(E) = P(A_1)\cdot P(E|A_1) + P(A_2)\cdot P(E|A_2) = (0.4)(0.03) + (0.6)(0.02) = 0.024$

**17.** $P(E) = P(A_1)\cdot P(E|A_1) + P(A_2)\cdot P(E|A_2) + P(A_3)\cdot P(E|A_3)$
$= (0.6)(0.01) + (0.2)(0.03) + (0.2)(0.02) = 0.016$

**19.** $P(A_1|E) = \frac{P(A_1)\cdot P(E|A_1)}{P(E)} = \frac{(0.4)(0.03)}{0.024} = \frac{12}{24} = \frac{1}{2} = 0.5$

$P(A_2|E) = \frac{P(A_2)\cdot P(E|A_2)}{P(E)} = \frac{(0.6)(0.02)}{0.024} = \frac{12}{24} = \frac{1}{2} = 0.5$

**21.** $P(A_1|E) = \frac{P(A_1)\cdot P(E|A_1)}{P(E)} = \frac{(0.6)(0.01)}{0.016} = \frac{6}{16} = \frac{3}{8} = 0.375$

$P(A_2|E) = \frac{P(A_2)\cdot P(E|A_2)}{P(E)} = \frac{(0.2)(0.03)}{0.016} = \frac{6}{16} = \frac{3}{8} = 0.375$

$P(A_3|E) = \frac{P(A_3)\cdot P(E|A_3)}{P(E)} = \frac{(0.2)(0.02)}{0.016} = \frac{4}{16} = \frac{1}{4} = 0.25$

**23.** From Example 3, $A_1$ = "Item produced by machine I," $A_2$ = "Item produced by machine II," $A_3$ = "Item produced by machine III," and $E$ = "Item is defective."

$P(A_1|E) = \frac{P(A_1)\cdot P(E|A_1)}{P(E)} = \frac{(0.4)(0.02)}{0.029} = \frac{8}{29} \approx 0.276.$

$P(A_2|E) = \frac{P(A_2)\cdot P(E|A_2)}{P(E)} = \frac{(0.5)(0.04)}{0.029} = \frac{20}{29} \approx 0.690.$

$P(A_3|E) = \frac{P(A_3)\cdot P(E|A_3)}{P(E)} = \frac{(0.1)(0.01)}{0.029} = \frac{1}{29} \approx 0.034.$

**25.** $P(A_2|E) = \frac{P(A_2)\cdot P(E|A_2)}{P(E)} = \frac{(0.2)(0)}{0.31} = 0,$

$P(A_3|E) = \frac{P(A_3)\cdot P(E|A_3)}{P(E)} = \frac{(0.1)(0.2)}{0.31} = \frac{2}{31} \approx 0.065,$

$P(A_4|E) = \frac{P(A_4)\cdot P(E|A_4)}{P(E)} = \frac{(0.3)(0)}{0.31} = 0,$

$P(A_5|E) = \frac{P(A_5)\cdot P(E|A_5)}{P(E)} = \frac{(0.1)(0.2)}{0.31} = \frac{2}{31} \approx 0.065.$

**27.** Note that each of jars I, II and III contains 16 balls.

$P(\text{"The ball is red"}) = P(E) = P(U_I)\cdot P(E\,|\,U_I) + P(U_{II})\cdot P(E\,|\,U_{II}) + P(U_{III})\cdot P(E\,|\,U_{III})$
$= \left(\frac{1}{3}\right)\left(\frac{5}{16}\right) + \left(\frac{1}{3}\right)\left(\frac{3}{16}\right) + \left(\frac{1}{3}\right)\left(\frac{7}{16}\right) = \frac{5}{16}$

$$P(\text{"The ball is from jar I given it is red"}) = P(U_{\text{I}}|E) = \frac{P(U_{\text{I}}) \cdot P(E|U_{\text{I}})}{P(E)} = \frac{(1/3)(5/16)}{5/16}$$
$$= \frac{1}{3} \approx 0.333.$$

$$P(\text{"The ball is from jar II given it is red"}) = P(U_{\text{II}}|E) = \frac{P(U_{\text{II}}) \cdot P(E|U_{\text{II}})}{P(E)} = \frac{(1/3)(3/16)}{5/16}$$
$$= \frac{1}{5} = 0.2.$$

$$P(\text{"The ball is from jar III given it is red"}) = P(U_{\text{III}}|E) = \frac{P(U_{\text{III}}) \cdot P(E|U_{\text{III}})}{P(E)} = \frac{(1/3)(7/16)}{5/16}$$
$$= \frac{7}{15} \approx 0.467.$$

**29.** Define the events: $M$:male; $F$: female; $C$: person is colorblind. Then:

$$P(M|C) = \frac{P(M) \cdot P(C|M)}{P(C)}$$
$$= \frac{P(M) \cdot P(C|M)}{P(M) \cdot P(C|M) + P(F) \cdot P(C|F)}$$
$$= \frac{(0.51)(0.05)}{(0.51)(0.05) + (0.49)(0.003)} = 0.945$$

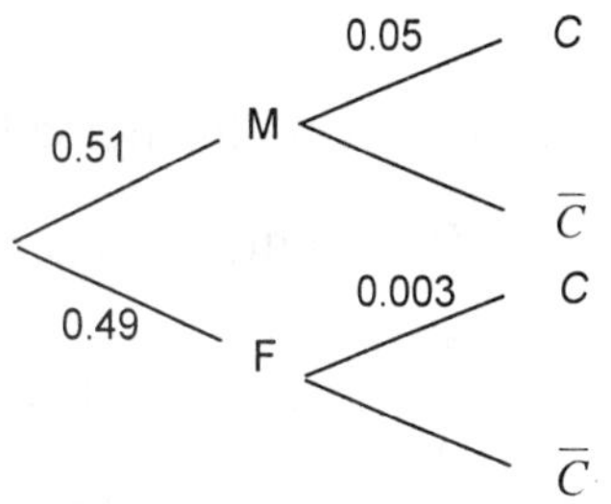


**31.** Define the events: $D$: democrat; $R$: republican; $I$: independent; $V$: voted. Then:

$$P(D|V) = \frac{P(D) \cdot P(V|D)}{P(D) \cdot P(V|D) + P(R) \cdot P(V|R) + P(I) \cdot P(V|I)}$$
$$= \frac{(0.55)(0.35)}{(0.55)(0.35) + (0.30)(0.65) + (0.15)(0.75)} = 0.385$$

$$P(R|V) = \frac{P(R) \cdot P(V|R)}{P(D) \cdot P(V|D) + P(R) \cdot P(V|R) + P(I) \cdot P(V|I)}$$
$$= \frac{(0.30)(0.65)}{(0.55)(0.35) + (0.30)(0.65) + (0.15)(0.75)} = 0.39$$

$$P(I|V) = \frac{P(I) \cdot P(V|I)}{P(D) \cdot P(V|D) + P(R) \cdot P(V|R) + P(I) \cdot P(V|I)}$$
$$= \frac{(0.15)(0.75)}{(0.55)(0.35) + (0.30)(0.65) + (0.15)(0.75)} = 0.225$$

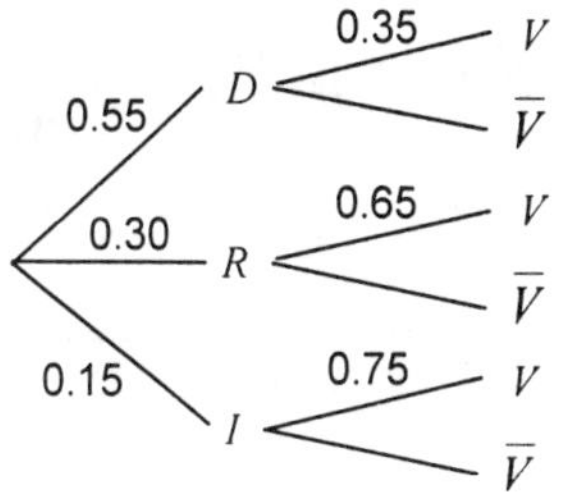


**33.** Define events $E$ = "the test gives a positive result," $R$ = "the soil is rock," $C$ = "the soil is clay," $S$ = "the soil is sand." $P(E) = P(R) \cdot P(E|R) + P(C) \cdot P(E|C) + P(S) \cdot P(E|S)$ $= (0.53)(0.35) + (0.21)(0.48) + (0.26)(0.75) = 0.4813.$

$$P(R|E) = \frac{P(R)\cdot P(E\mid R)}{P(E)} = \frac{(0.53)(0.35)}{0.4813} \approx 0.385.$$
$$P(C|E) = \frac{P(C)\cdot P(E\mid C)}{P(E)} = \frac{(0.21)(0.48)}{0.4813} \approx 0.209.$$
$$P(S|E) = \frac{P(S)\cdot P(E\mid S)}{P(E)} = \frac{(0.26)(0.75)}{0.4813} \approx 0.405.$$

**35.** Define events $R$ = "the person votes Republican," $N$ = "the person is from the Northeast," $S$ = "the person is from the South," $M$ = "the person is from the Midwest," and $W$ = "the person is from the West." The probability that a person will vote republican is:
$$P(R) = P(N)\cdot P(R|N) + P(S)\cdot P(R|S) + P(M)\cdot P(R|M) + P(W)\cdot P(R|W)$$
$$= (0.4)(0.4) + (0.1)(0.56) + (0.25)(0.48) + (0.25)(0.52) = 0.466.$$
The probability that a person voting republican is from the Northeast is:
$$P(N|R) = \frac{P(N)\cdot P(R|N)}{P(R)} = \frac{(0.4)(0.4)}{0.466} \approx 0.343.$$

**37.** Let $A_1$ be the event "the nurse forgot to give Mr. Brown the pill", $A_2 = \overline{A_1}$, and $E$ be the event "Mr. Brown died." Then $P(A_1) = \frac{2}{3}$, $P(A_2) = 1 - \frac{2}{3} = \frac{1}{3}$, $P(E|A_1) = \frac{3}{4}$ and $P(E|A_2) = \frac{1}{3}$. The probability that the nurse forgot to give Mr. Brown the pill is:
$$P(A_1|E) = \frac{P(A_1)\cdot P(E|A_1)}{P(A_1)\cdot P(E|A_1) + P(A_2)\cdot P(E|A_2)} = \frac{(2/3)(3/4)}{(2/3)(3/4) + (1/3)(1/3)} = \frac{9}{11} \approx 0.818.$$

**39.**
$$P(En|F) = \frac{P(En)\cdot P(F|En)}{\begin{gathered}P(En)\cdot P(F|En) + P(B)\cdot P(F|B) + P(Ed)\cdot P(F|Ed) + P(S)\cdot P(F|S)\\ + P(N)\cdot P(F|N) + P(H)\cdot P(F|H) + P(O)\cdot P(F|O)\end{gathered}}$$
$$= \frac{(0.26)(0.40)}{\begin{gathered}(0.26)(0.40) + (0.30)(0.35) + (0.09)(0.80) + (0.12)(0.52) + (0.12)(0.56)\\ + (0.09)(0.65) + (0.02)(0.51)\end{gathered}}$$
$$\approx 0.217$$
The probability that a female student selected at random from the freshment class is majoring in engineering is approximately 0.217.

**41.** Define events $D =$ "has disease" and $+ =$ "tests positive".

(a) By Bayes' Formula, $P(D|+) = \dfrac{P(D)\cdot P(+|D)}{P(+)}$. Here,

$P(+) = P(D)\cdot P(+|D) + P(\bar{D})\cdot P(+|\bar{D}) = (0.2)(0.97) + (0.8)(0.04) = 0.226$, so

$P(D|+) = \dfrac{(0.2)(0.97)}{0.226} \approx 0.858$.

(b) By Bayes' Formula, $P(D|+) = \dfrac{P(D)\cdot P(+|D)}{P(+)}$. Here,

$P(+) = P(D)\cdot P(+|D) + P(\bar{D})\cdot P(+|\bar{D}) = (0.04)(0.97) + (0.96)(0.04) = 0.0772$, so

$P(D|+) = \dfrac{(0.04)(0.97)}{0.0772} \approx 0.503$.

(c) Define event $++ =$ "tested positive twice in two trials". Under the assumption that the test results are independent,

$P(++) = P(D)\cdot P(++|D) + P(\bar{D})\cdot P(++|\bar{D}) = (0.04)(0.97^2) + (0.96)(0.04^2)$
$= 0.039172$,so

$$P(D|++) = \frac{P(D)\cdot P(++|D)}{P(++)} = \frac{(0.04)(0.97^2)}{0.39172} \approx 0.961.$$

**43.** Since $F \subseteq E$, then $P(E \cap F) = P(F)$, thus $P(E|F) = \dfrac{P(E \cap F)}{P(F)} = \dfrac{P(F)}{P(F)} = 1$.

## 8.2 The Binomial Probability Model

**1.** $b(7,4;0.20) = \binom{7}{4}(0.2)^4(0.8)^3 = \dfrac{7!}{4!\cdot 3!}(0.2)^4(0.8)^3 = 35\cdot(0.2)^4(0.8)^3 = 0.0287$

**3.** $b(15,8;0.80) = \binom{15}{8}(0.8)^8(0.2)^7 = \dfrac{15!}{8!\cdot 7!}(0.8)^8(0.2)^7 = 6435(0.8)^8(0.2)^7 = 0.0138$

**5.** $b\left(15,10;\frac{1}{2}\right) = \binom{15}{10}\left(\frac{1}{2}\right)^{10}\left(\frac{1}{2}\right)^5 = \dfrac{15!}{10!\cdot 5!}\left(\frac{1}{2}\right)^{15} = \dfrac{3003}{2^{15}} = \dfrac{3003}{32768} = .0916$

**7.** $b(15,3;0.3) + b(15,2;0.3) + b(15,1;0.3) + b(15,0;0.3)$

$= \binom{15}{3}(0.3)^3(0.7)^{12} + \binom{15}{2}(0.3)^2(0.7)^{13} + \binom{15}{1}(0.3)(0.7)^{14} + \binom{15}{0}(0.3)^0(0.7)^{15}$

$= 0.1700 + 0.0916 + 0.0305 + 0.0047 = 0.2968$

**9.** $b\left(3,2;\frac{1}{3}\right) = \binom{3}{2}\left(\frac{1}{3}\right)^2\left(\frac{2}{3}\right) = 3\cdot\frac{1}{9}\cdot\frac{2}{3} = \frac{2}{9} = 0.2222$

**11.** $b\left(3,0;\frac{1}{6}\right) = \binom{3}{0}\left(\frac{1}{6}\right)^0\left(\frac{5}{6}\right)^3 = \left(\frac{5}{6}\right)^3 = \frac{125}{216} = 0.5787$

**13.** $b\left(5,3;\frac{2}{3}\right)=\binom{5}{3}\left(\frac{2}{3}\right)^3\left(\frac{1}{3}\right)^2=10\cdot\frac{8}{27}\cdot\frac{1}{9}=\frac{80}{243}=0.3292$

**15.** $b(10,6;0.3)=\binom{10}{6}(0.3)^6(0.7)^4=0.0368$ **17.** $b(12,9;0.8)=\binom{12}{9}(0.8)^9(0.2)^3=0.2362$

**19.** $b(8,5;0.3)+b(8,6;0.3)+b(8,7;0.3)+b(8,8;0.3)$

$=\binom{8}{5}(0.3)^5(0.7)^3+\binom{8}{6}(0.3)^6(0.7)^2+\binom{8}{7}(0.3)^7(0.7)^1+\binom{8}{8}(0.3)^8(0.7)^0$

$=0.0467+0.0100+0.0012+0.0001=0.0580$

**21.** $b\left(8,1;\frac{1}{2}\right)=\binom{8}{1}\left(\frac{1}{2}\right)^1\left(\frac{1}{2}\right)^7=\frac{1}{32}=.03125$

**23.** $b\left(8,5;\frac{1}{2}\right)+b\left(8,6;\frac{1}{2}\right)+b\left(8,7;\frac{1}{2}\right)+b\left(8,8;\frac{1}{2}\right)=\frac{7}{32}+\frac{7}{64}+\frac{1}{32}+\frac{1}{256}=\frac{93}{256}=0.3633$

**25.** $P(\text{exactly two heads})=b\left(8,2;\frac{1}{2}\right)=\frac{7}{64}$. $P(\text{at least one head})=1-P(\text{no heads})$

$=1-b\left(8,0;\frac{1}{2}\right)=1-\frac{1}{256}=\frac{255}{256}$. Note that "at least one head" intersect "exactly two heads" = "exactly two heads". So

$P(\text{exactly two heads}\mid\text{at least one head})=\frac{P(\text{exactly two heads})}{P(\text{at least one head})}=\frac{7/64}{255/256}=\frac{28}{255}=0.1098$

**27.** The probability of a "sum of 7" in a single roll is $\frac{6}{36}=\frac{1}{6}$. The probability of obtaining exactly two 7's in 5 rolls is: $b\left(5,2;\frac{1}{6}\right)=\frac{5!}{2!3!}\left(\frac{1}{6}\right)^2\left(\frac{5}{6}\right)^3=\frac{625}{3888}=0.1608$

**29.** (a) $b(8,1;0.05)=\frac{8!}{1!7!}(.05)^1(.95)^7\approx 0.2793$

(b) $b(8,2;0.05)=\frac{8!}{2!6!}(.05)^2(.95)^6\approx 0.0515$

(c) $P(\text{at least one is defective})=1-P(\text{none are defective})$

$=1-b(8,0;0.05)=1-\frac{8!}{0!8!}(.05)^0(.95)^8\approx 1-0.6634=0.3366$

(d) $b(8,0;0.05)+b(8,1;0.05)+b(8,2;0.05)$

$=\frac{8!}{0!8!}(.05)^0(.95)^8+\frac{8!}{1!7!}(.05)^1(.95)^7+\frac{8!}{2!6!}(.05)^2(.95)^6\approx 0.9942$

**31.** Assuming boys and girls are equally probable:

$$b\left(6,3;\frac{1}{2}\right)=\frac{6!}{3!3!}\left(\frac{1}{2}\right)^3\left(\frac{1}{2}\right)^3=\frac{5}{16}=0.3125$$

**33.** (a) See tree diagram at the right.

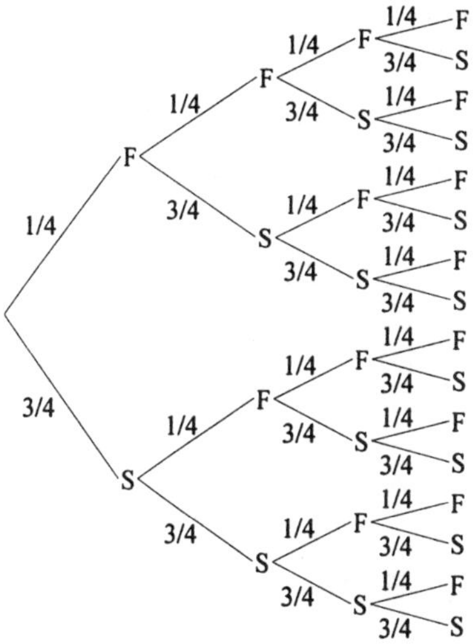


(b) $P(FFSS$ or $FSFS$ or $FSSF$ or $SFFS$ or $SFSF$ or $SSFF)=\left(\frac{1}{4}\right)\left(\frac{1}{4}\right)\left(\frac{3}{4}\right)\left(\frac{3}{4}\right)+\left(\frac{1}{4}\right)\left(\frac{3}{4}\right)\left(\frac{1}{4}\right)\left(\frac{3}{4}\right)$

$+\left(\frac{1}{4}\right)\left(\frac{3}{4}\right)\left(\frac{3}{4}\right)\left(\frac{1}{4}\right)+\left(\frac{3}{4}\right)\left(\frac{1}{4}\right)\left(\frac{1}{4}\right)\left(\frac{3}{4}\right)$

$+\left(\frac{3}{4}\right)\left(\frac{1}{4}\right)\left(\frac{3}{4}\right)\left(\frac{1}{4}\right)+\left(\frac{3}{4}\right)\left(\frac{3}{4}\right)\left(\frac{1}{4}\right)\left(\frac{1}{4}\right)=6\cdot\left(\frac{9}{256}\right)$

$=\frac{27}{128}\approx 0.2109$

(c) $P(\text{exactly 2 failures})=b\left(4,2;\frac{1}{4}\right)=\frac{4!}{2!2!}\left(\frac{1}{4}\right)^2\left(\frac{3}{4}\right)^2$

$=6\cdot\left(\frac{9}{256}\right)=\frac{27}{128}\approx 0.2109$

**35.** $1-P(\text{"the target is not hit or it is hit just once"})$

$$=1-\left[b\left(10,0;\frac{2}{3}\right)+b\left(10,1;\frac{2}{3}\right)\right]=1-\frac{10!}{0!10!}\left(\frac{2}{3}\right)^0\left(\frac{1}{3}\right)^{10}-\frac{10!}{1!9!}\left(\frac{2}{3}\right)^1\left(\frac{1}{3}\right)^9=0.9996$$

**37.** The student's probability of guessing a correct answer on a question is $\frac{1}{2}$. Thus, the probability of at least 10 correct guesses on the 15 item examination is

$$b\left(15,10;\frac{1}{2}\right)+b\left(15,11;\frac{1}{2}\right)+b\left(15,12;\frac{1}{2}\right)+b\left(15,13;\frac{1}{2}\right)+b\left(15,14;\frac{1}{2}\right)+b\left(15,15;\frac{1}{2}\right)$$

$$=\binom{15}{10}\left(\frac{1}{2}\right)^{10}\left(\frac{1}{2}\right)^5+\binom{15}{11}\left(\frac{1}{2}\right)^{11}\left(\frac{1}{2}\right)^4+\binom{15}{12}\left(\frac{1}{2}\right)^{12}\left(\frac{1}{2}\right)^3+\binom{15}{13}\left(\frac{1}{2}\right)^{13}\left(\frac{1}{2}\right)^2$$

$$+\binom{15}{14}\left(\frac{1}{2}\right)^{14}\left(\frac{1}{2}\right)^1+\binom{15}{15}\left(\frac{1}{2}\right)^{15}\left(\frac{1}{2}\right)^0$$

$$=0.09616+0.0417+0.0139+0.0032+0.0005+0.00003=0.1509$$

The probability that the student with a probability of 0.8 for correctly answering each question will correctly answer at least 12 questions is

$$b(15,12;0.8)+b(15,13;0.8)+b(15,14;0.8)+b(15,15;0.8)$$

$$=\binom{15}{12}(0.8)^{12}(0.2)^3+\binom{15}{13}(0.8)^{13}(0.2)^2+\binom{15}{14}(0.8)^{14}(0.2)^1+\binom{15}{15}(0.8)^{15}(0.2)^0.$$

$$=0.2501+0.2309+0.1319+0.0352=0.6481$$

**39.** $b(8,8;0.4)=\binom{8}{8}(0.4)^8(0.6)^0=(0.4)^8=.0007$

**41.** $p=0.23;\ b(10,4;0.23)=\binom{10}{4}(0.23)^4(0.77)^6=0.1225$

**43.** (a) $b\left(6,5;\frac{1}{2}\right)+b\left(6,6;\frac{1}{2}\right)=\binom{6}{5}\left(\frac{1}{2}\right)^5\left(\frac{1}{2}\right)^1+\binom{6}{6}\left(\frac{1}{2}\right)^6\left(\frac{1}{2}\right)^0=\frac{7}{64}=.1094$

(b) $1-[b(6,5;0.8)+b(6,6;0.8)]=1-\binom{6}{5}(0.8)^5(0.2)-\binom{6}{6}(0.8)^6$

$1-0.3932-0.2621=0.3447$

## Technology Exercises

**1.** Actual values for $P(k)$ are computed by $P(k)=b(4,k;0.5)$

| $k$: | 0 | 1 | 2 | 3 | 4 |
|---|---|---|---|---|---|
| $P(k)$: | 0.0625 | 0.25 | 0.375 | 0.25 | 0.0625 |

**3.** Actual value for $P(3)$ is $b(8,3;0.5)=0.21875$.

## 8.3 Expected Value

**1.** Expected value $E=m_1\cdot p_1+m_2\cdot p_2+m_3\cdot p_3+m_4\cdot p_4$
$=(2)(0.4)+(3)(0.2)+(-2)(0.1)+(0)(0.3)=1.2$

**3.** The expected number of fans $E=m_1\cdot p_1+m_2\cdot p_2+m_3\cdot p_3+m_4\cdot p_4$
$=(30{,}000)(0.08)+(40{,}000)(0.42)+(60{,}000)(0.42)+(80{,}000)(0.08)=50{,}800$.

**5.** Her expected winnings are $\left(\frac{1}{10}\right)(\$8)=\$0.80$, so she should pay 80¢ for one draw if it is a fair game.

**7.** His expected winnings are $\left(\frac{1}{6}\right)(\$10)\approx\$1.67$, so he should pay \$1.67 for a throw if it is a fair game.

**9.** A ticket's expected value is $(\$100)(0.001)+(\$50)(0.003)+(\$0)(0.996)=\$0.25$. Thus, the price of a ticket exceeds its expected value by $\$1.00-\$0.25=\$0.75$, i.e., by 75¢.

| Outcome | 1st place | 2nd place | 3rd place |
|---|---|---|---|
| Probability | 0.001 | 0.003 | 0.996 |
| Payoff | \$100 | \$50 | \$0 |

**11.** (a) The expected value of the game is $(\$3)\left(\frac{1}{8}\right)+(\$2)\left(\frac{3}{8}\right)+(\$0)\left(\frac{3}{8}\right)+(-\$3)\left(\frac{1}{8}\right)=\$0.75$.

| Outcome | 3 tails | 2 tails | 1 tail | 0 tails |
|---|---|---|---|---|
| Probability | 1/8 | 3/8 | 3/8 | 1/8 |
| Payoff | \$3 | \$2 | \$0 | –\$3 |

(b) The game is not fair since its expected value is different from zero.

(c) Solve $(3)\left(\frac{1}{8}\right)+(2)\left(\frac{3}{8}\right)+(x)\left(\frac{3}{8}\right)+(-3)\left(\frac{1}{8}\right)=0$; $\left(\frac{3}{8}\right)(x)=\frac{-6}{8}$; $x=-2$. The game will be "fair" if a player loses \$2 for a toss of exactly 1 tail.

**13.** If team A wins, you lose \$4; if team B wins, you win \$6. The expected value of the game for you is $(-\$4)\left(\frac{9}{14}\right)+(\$6)\left(\frac{5}{14}\right)=-\$\frac{3}{7}$, so your expected loss is $\$\frac{3}{7}\approx 0.43$. The bet is not fair to you.

| Outcome | A wins | B wins |
|---|---|---|
| Probability | 9/14 | 5/14 |
| Payoff | –\$4 | \$6 |

**15.** The expected value of a card drawn in this game is $(40¢)\left(\frac{12}{52}\right)+(50¢)\left(\frac{3}{52}\right)+(90¢)\left(\frac{1}{52}\right)+(0¢)\left(\frac{36}{52}\right)=\left(\frac{180}{13}\right)¢=\left(13\frac{11}{13}\right)¢$. Since Sarah pays 15¢ to draw a card, she should expect to lose $\left(1\frac{2}{13}\right)¢$, so Sarah should not play the game.

| Outcome | heart | ace | ace of ♥ | other |
|---|---|---|---|---|
| Probability | 12/52 | 3/52 | 1/52 | 36/52 |
| Payoff | 40¢ | 50¢ | 90¢ | 0¢ |

**17.** Set $x=$ the amount bet. The two outcomes of this game are "the horse wins" and "the horse loses," with probabilities $\frac{7}{12}$ and $\frac{5}{12}$, respectively. The expected value of the game is $(\$5)\left(\frac{7}{12}\right)-\$x\left(\frac{5}{12}\right)=\$(7-x)\left(\frac{5}{12}\right)$ dollars. The game is fair when its expected value is zero, i.e., when $x=7$. The bettor should bet \$7 to make the game fair.

**19.**

| 1st site outcome | succeeds | fails |
|---|---|---|
| Probability | 1/2 | 1/2 |
| Payoff | \$15,000 | –\$3000 |

| 2nd site outcome | succeeds | fails |
|---|---|---|
| Probability | 1/2 | 1/2 |
| Payoff | \$20,000 | –\$6000 |

The expected profit at the first site is $(\$15{,}000)\left(\frac{1}{2}\right)-(\$3000)\left(\frac{1}{2}\right)=\$6000$. The expected profit at the second site is $(\$20{,}000)\left(\frac{1}{2}\right)-(\$6000)\left(\frac{1}{2}\right)=\$7000$. The expected profit is \$1000 greater if the company opens the supermarket at the second location.

**21.** The probability of a five in one throw is $\frac{1}{6}$. The expected number of five's in 2000 throws is $\left(\frac{1}{6}\right)(2000)=\frac{1000}{3}=333\frac{1}{3}$.

**23.** Expect $(0.02)(500)=10$ defective light bulbs.

**25.** Expect $(0.002)(500)=1$ unfavorable reaction.

**27.** The sample space is $S=\{H, TH, TTH, TTTH, TTTT\}$.

| Outcome | $H$ | $TH$ | $TTH$ | $TTTH$ | $TTTT$ |
|---|---|---|---|---|---|
| Probability | $\frac{1}{4}$ | $\left(\frac{3}{4}\right)\left(\frac{1}{4}\right)$ | $\left(\frac{3}{4}\right)^2\left(\frac{1}{4}\right)$ | $\left(\frac{3}{4}\right)^3\left(\frac{1}{4}\right)$ | $\left(\frac{3}{4}\right)^4$ |
| # tosses | 1 | 2 | 3 | 4 | 4 |

Expected number of tosses $=\left(\frac{1}{4}\right)(1)+\left(\frac{3}{4}\right)\left(\frac{1}{4}\right)(2)+\left(\frac{3}{4}\right)^2\left(\frac{1}{4}\right)(3)+\left(\frac{3}{4}\right)^3\left(\frac{1}{4}\right)(4)+\left(\frac{3}{4}\right)^4(4)$

$=\frac{175}{64}\approx 2.734$.

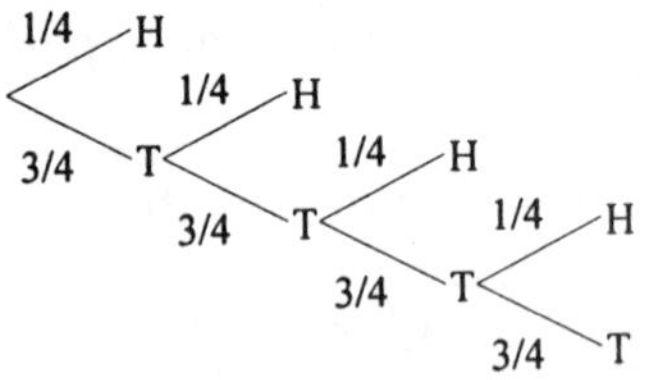


**29.** The expected number of passengers using aircraft A is $150(0.2)+180(0.3)+200(0.5)=184$. The profit using aircraft A is $500\cdot 184-(16{,}000+200\cdot 184)=\$39{,}200$. The expected number of passengers using aircraft B is $150(0.2)+180(0.3)+200(0.2)+250(0.2)+300(0.1)=204$. The profit using aircraft B is $500\cdot 204-(18{,}000+230\cdot 204)=\$37{,}080$. The airline will have greater profit if it uses aircraft A for this flight.

## 8.4 Applications

**1.** The expected number of customers is $(7)(.10)+(8)(.20)+(9)(.40)+(10)(.20)+(11)(.10)=9$.

| Number of Available Cars | Cost at \$10 per Car | Expected Profit |
|---|---|---|
| 7 | \$70 | $(\$30\cdot 7)1-\$70=\$140$ |
| 8 | \$80 | $(\$30\cdot 7)(0.1)+(\$30\cdot 8)(0.9)-\$80=\$157$ |
| 9 | \$90 | $(\$30\cdot 7)(0.1)+(\$30\cdot 8)(0.2)+(\$30\cdot 9)(0.7)-\$90=\$168$ |
| 10 | \$100 | $(\$30\cdot 7)(0.1)+(\$30\cdot 8)(0.2)+(\$30\cdot 9)(0.4)+(\$30\cdot 10)(0.3)-\$100=\$167$ |
| 11 | \$110 | $(\$30\cdot 7)(0.1)+(\$30\cdot 8)(0.2)+(\$30\cdot 9)(0.4)+(\$30\cdot 10)(0.2)+(\$30\cdot 11)(0.1)-\$110=\$160$ |

The optimal number of cars is 9. The expected daily profit is \$168.

**3.**

| Group size | $p^n-\frac{1}{n}$ |
|---|---|
| 2 | $(.95)^2-0.5=.4025$ |

| | |
|---|---|
| 3 | $(.95)^3 - 0.333 = .524$ |
| 4 | $(.95)^4 - 0.25 = .565$ |
| 5 | $(.95)^5 - 0.2 = .574$ |
| 6 | $(.95)^6 - 0.167 = .568$ |

The optimal group size is 5 since this is the group size having highest expected tests saved per component.

**5.** (a) $E(x) = \$75{,}000 - \$75{,}000(0.05)^x - \$500x$.

(b) $E(1) = \$75{,}000 - \$75{,}000(0.05)^1 - \$500(1) = \$70{,}750$

$E(2) = \$75{,}000 - \$75{,}000(0.05)^2 - \$500(2) = \$73{,}812.50$

$E(3) = \$75{,}000 - \$75{,}000(0.05)^3 - \$500(3) = \$73{,}490.63$

$E(4) = \$75{,}000 - \$75{,}000(0.05)^4 - \$500(4) = \$72{,}999.53$

$E(5) = \$75{,}000 - \$75{,}000(0.05)^5 - \$500(5) = \$72{,}499.98$

The expected net gain is optimal when 2 scuba divers are hired.

**7.** $$b(15,15;0.98) + b(15,14;0.98) = \binom{15}{15}(0.98)^{15} + \binom{15}{14}(0.98)^{14}(0.02)$$
$$= (0.98)^{15} + 15(0.98)^{14}(0.02) = 0.9647$$

## 8.5 Random Variables

**1.** $$P(X=0) = b(2,0;0.5) = \binom{2}{0}\left(\frac{1}{2}\right)^0\left(\frac{1}{2}\right)^2 = \frac{1}{4}$$
$$P(X=1) = b(2,1;0.5) = \binom{2}{1}\left(\frac{1}{2}\right)^1\left(\frac{1}{2}\right)^1 = \frac{1}{2}$$
$$P(X=2) = b(2,2;0.5) = \binom{2}{2}\left(\frac{1}{2}\right)^2\left(\frac{1}{2}\right)^0 = \frac{1}{4}$$

**3.** $$P(X=0) = b(3,0;0.5) = \binom{3}{0}\left(\frac{1}{2}\right)^0\left(\frac{1}{2}\right)^3 = \frac{1}{8}$$
$$P(X=1) = b(3,2;0.5) = \binom{3}{1}\left(\frac{1}{2}\right)^1\left(\frac{1}{2}\right)^2 = \frac{3}{8}$$
$$P(X=2) = b(3,2;0.5) = \binom{3}{2}\left(\frac{1}{2}\right)^2\left(\frac{1}{2}\right)^1 = \frac{3}{8}$$
$$P(X=3) = b(3,3;0.5) = \binom{3}{3}\left(\frac{1}{2}\right)^3\left(\frac{1}{2}\right)^0 = \frac{1}{8}$$

5. $P(X=0)=b(3,0;0.4)=\binom{3}{0}(0.4)^0(0.6)^3=0.216$

$P(X=1)=b(3,1;0.4)=\binom{3}{1}(0.4)^1(0.6)^2=0.432$

$P(X=2)=b(3,2;0.4)=\binom{3}{2}(0.4)^2(0.6)^1=0.288$

$P(X=3)=b(3,3;0.4)=\binom{3}{3}(0.4)^3(0.6)^0=0.064$

7. Expected value $=(2)(0.4)+(3)(0.2)+(-2)(0.1)+(0)(0.3)=1.2$

## Technology Exercises

1. Actual values for $P(X=k)$ are computed by $P(X=k)=\frac{1}{6}\approx 0.167$

| $k$: | 0 | 1 | 2 | 3 | 4 | 5 | 6 |
|---|---|---|---|---|---|---|---|
| $P(X=k)$: | 0.167 | 0.167 | 0.167 | 0.167 | 0.167 | 0.167 | 0.167 |

3. $P(0.1\le X\le 0.3)=0.3-0.1=0.2$

5. $P(X=2)=\frac{1}{12}\approx 0.083333$

## Chapter 8 Review

### True or False

1. True

3. False

5. False

### Fill in the Blank

1. Bayes' Formula

3. expected value

5. expected value

### Review Exercises

1. $P(E|A)=0.5$

3. $P(E|B)=0.4$

5. $P(E|C)=0.3$

7. $$P(E)=P(A)P(E\mid A)+P(B)P(E\mid B)+P(C)P(E\mid C)$$
$$=(0.4)(0.5)+(0.5)(0.4)+(0.1)(0.3)=0.43$$

9. $P(A|E)=\frac{P(A)P(E|A)}{P(E)}=\frac{(0.4)(0.5)}{0.43}\approx 0.4651$

**11.** $P(B|E) = \dfrac{P(B)P(E|B)}{P(E)} = \dfrac{(0.5)(0.4)}{0.43} \approx 0.4651$

**13.** $P(C|E) = \dfrac{P(C)P(E|C)}{P(E)} = \dfrac{(0.1)(0.3)}{0.43} \approx 0.0698$

**15.** (a) $P(E|G) = \dfrac{180}{180+60+20} = \dfrac{9}{13} \approx 0.6923$

(b) $P(G|E) = \dfrac{180}{180+110+55} = \dfrac{12}{23} \approx 0.5217$

(c) $P(H|E) = \dfrac{110}{180+110+55} = \dfrac{22}{69} \approx 0.3188$

(d) $P(K|E) = \dfrac{55}{180+110+55} = \dfrac{11}{69} \approx 0.1594$

(e) $P(F|G) = \dfrac{60}{180+60+20} = \dfrac{3}{13} \approx 0.2308$

(f) $P(G|F) = \dfrac{60}{60+85+65} = \dfrac{2}{7} \approx 0.2857$

(g) $P(H|F) = \dfrac{85}{60+85+65} = \dfrac{17}{42} \approx 0.4048$

(h) $P(K|F) = \dfrac{65}{60+85+65} = \dfrac{13}{42} \approx 0.3095$

**17.** Define events $C$: the individual has cancer and $D$: the test detects cancer. Then $P(C) = 0.018$, $P(\overline{C}) = 0.982$, $P(D|C) = 0.85$ and $P(D|\overline{C}) = 0.08$. See the tree diagram.

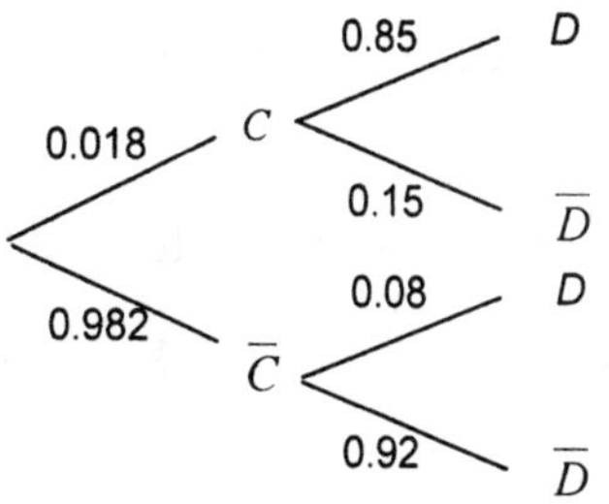


The probability that the person has cancer given that the test detected cancer is:

$$P(C|D) = \frac{P(C)\cdot P(D\mid C)}{P(D)}$$

$$P(C|D) = \frac{P(C)\cdot P(D\mid C)}{P(C)\cdot P(D\mid C) + P(\overline{C})\cdot P(D|\overline{C})}$$

$$= \frac{(0.018)(0.85)}{(0.018)(0.85)+(0.982)(0.08)} = 0.1630$$

**19.** The probability that none will purchase the product is

$$b\left(5,0;\frac{1}{5}\right) = \frac{5!}{0!5!}\left(\frac{1}{5}\right)^0\left(\frac{4}{5}\right)^5 = \left(\frac{4}{5}\right)^5 = 0.32768$$

The probability that exactly three will purchase the product is

$$b\left(5,3;\frac{1}{5}\right) = \frac{5!}{3!2!}\left(\frac{1}{5}\right)^3\left(\frac{4}{5}\right)^2 = 0.0512$$

**21.** (a) $\left(\frac{1}{2}\right)^{12} = \frac{1}{4096} = .0002$

(b) $b\left(12,7;\frac{1}{2}\right)+b\left(12,8;\frac{1}{2}\right)+b\left(12,9;\frac{1}{2}\right)+b\left(12,10;\frac{1}{2}\right)+b\left(12,11;\frac{1}{2}\right)+b\left(12,12;\frac{1}{2}\right)$

$= \frac{99}{512}+\frac{495}{4096}+\frac{55}{1024}+\frac{33}{2048}+\frac{3}{1024}+\frac{1}{4096}=\frac{793}{2048}\approx 0.3872$

(c) The odds in favor of passing are 793 to $(2048-793)$, or 793 to 1255.

**23.** The probability of throwing a sum of 11 in one throw is $\frac{2}{36}=\frac{1}{18}$. The probability of throwing at least 3 11's in 5 throws is

$b\left(5,3;\frac{1}{18}\right)+b\left(5,4;\frac{1}{18}\right)+b\left(5,5;\frac{1}{18}\right)=\frac{2890+85+1}{1{,}889{,}568}=\frac{31}{19{,}683}\approx 0.001575$.

**25.**

| # red balls drawn | 0 | 1 | 2 |
|---|---|---|---|
| probability | $\frac{\binom{4}{2}}{\binom{6}{2}}=\frac{6}{15}$ | $\frac{\binom{2}{1}\binom{4}{1}}{\binom{6}{2}}=\frac{8}{15}$ | $\frac{\binom{2}{2}}{\binom{6}{2}}=\frac{1}{15}$ |
| net gain in dollars (winnings – 0.70) | $-0.7$ | $1-0.7=0.3$ | $2-0.7=1.3$ |

The expected value of this game is $(-\$0.7)\left(\frac{6}{15}\right)+(\$0.3)\left(\frac{8}{15}\right)+(\$1.3)\left(\frac{1}{15}\right)=-\frac{1}{30}$ dollars.

Frank paid $3\frac{1}{3}$¢ too much.

**27.** The expected value of the game is (the expected payoff) – (cost to play)

$=\left[(\$0.80)\left(\frac{1}{6}\right)+(\$0.30)\left(\frac{1}{3}\right)+(\$0.10)\left(\frac{1}{2}\right)\right]-\$0.30=-\$0.0167$ or $-1\frac{2}{3}$¢. The game is not fair.

**29.** The expected number of heads is $\left(\frac{1}{4}\right)\cdot 200 = 50$.

**31.** Let $q = 1-p =$ the probability that an instance of the test is negative.

(a) The probability that a test for a pooled sample of 30 people will be positive, i.e., that at least one of the 30 individuals in the pool tests positive, is 1–probability all test negative or $1-q^{30}$.

(b) Under plan 2, the expected number of tests for a pool of 30 people is $(1)q^{30}+31(1-q^{30})=31-30q^{30}$.

## Professional Exam Questions

**1.** d: $1-2\cdot\left(\frac{1}{2}\right)^{10}=1-\left(\frac{1}{2}\right)^{9}$

**3.** a: The expected selling price is \$10,200.

**5.** b: $10,200=(.2)(6000)+(.2)(8000)+(.2)(10,000)+(.2)(12,000)+(.1)(14,000)$
$+(.1)(16,000)$

**7.** b *or* c (the two answers are equal)

# Chapter 9

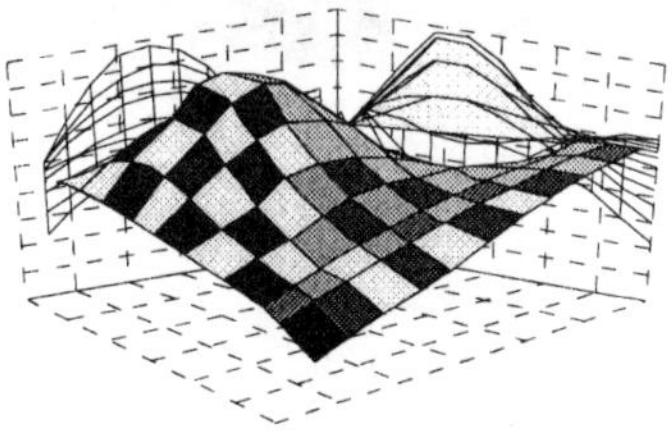

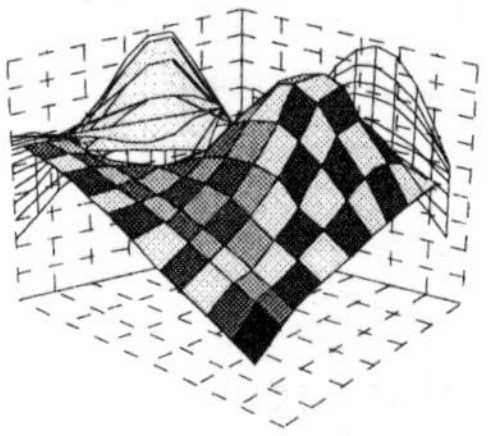

# Statistics

## 9.1 Bar Graphs; Pie Charts

**1.** A study of the opinions of people about a certain television program. A poll should be taken either door-to-door or by means of the telephone.

**3.** A study of the opinions of people toward Medicare: A poll should be taken door-to-door in which people are asked to fill out a questionnaire.

**5.** A study of the number of savings accounts per family in the United States. The data should be gathered from all different kinds of banks.

**7.** (a) Asking a group of children if they like candy to determine what percentage of people like candy.
(b) Asking a group of people over 65 their opinion toward Medicare to determine the opinion of people in general about Medicare.

**9.** By taking a poll downtown, you would question mostly people who are either shopping or working downtown. For instance, you would question few students.

**11.** (a) Northwest (b) TWA (c) About 78%

**13.** (a) Housing, fuel and utilities
(b) Misc. goods and services

(c) Senior citizens would feel like the CPI weight of 7% for healthcare is too low because they spend twice as much (14%) of their income on health care.

**15.** (a)

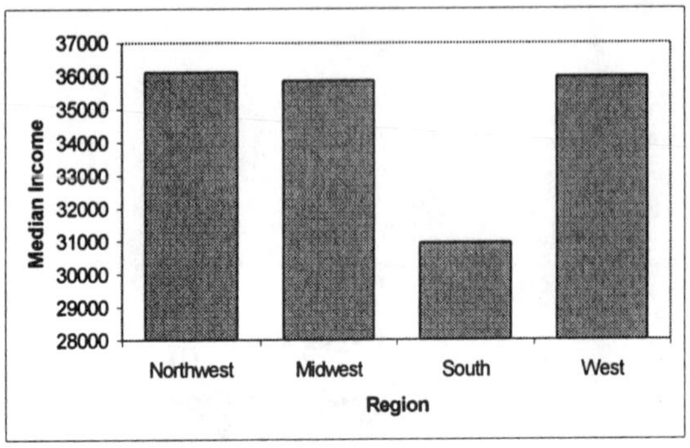


(b)

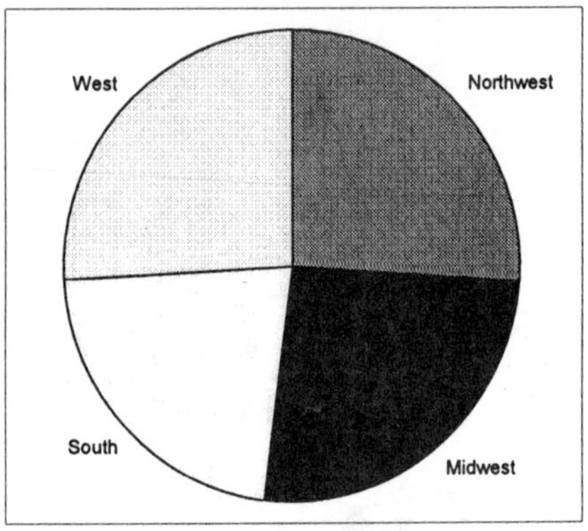


(c) The bar graph seems to summarize the data better

(d) Northwest

(e) South

**17.** (a)

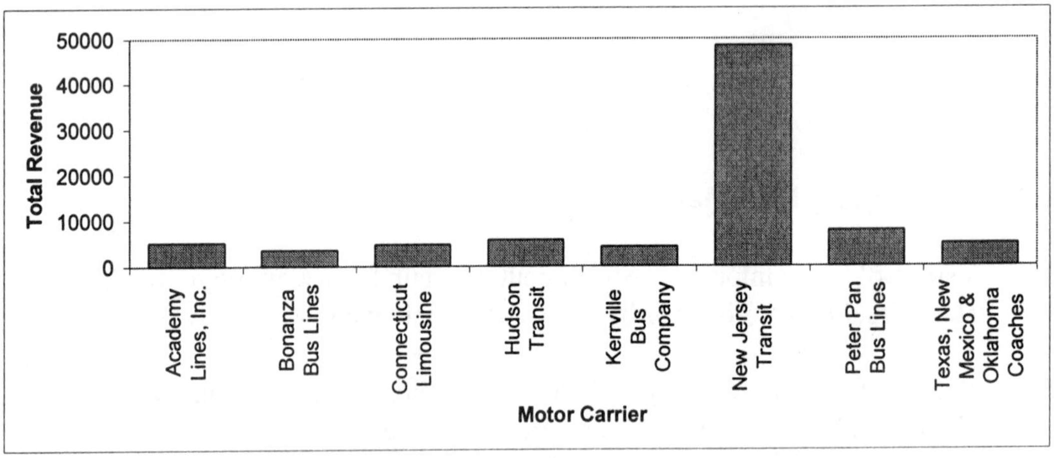


(b)

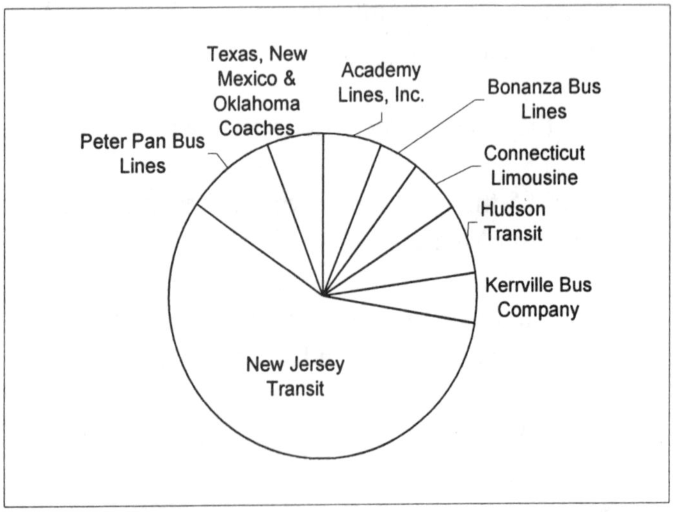


(c) Answers will vary.

(d) New Jersey Transit

(e) Bonanza Bus Lines

**19.** (a)

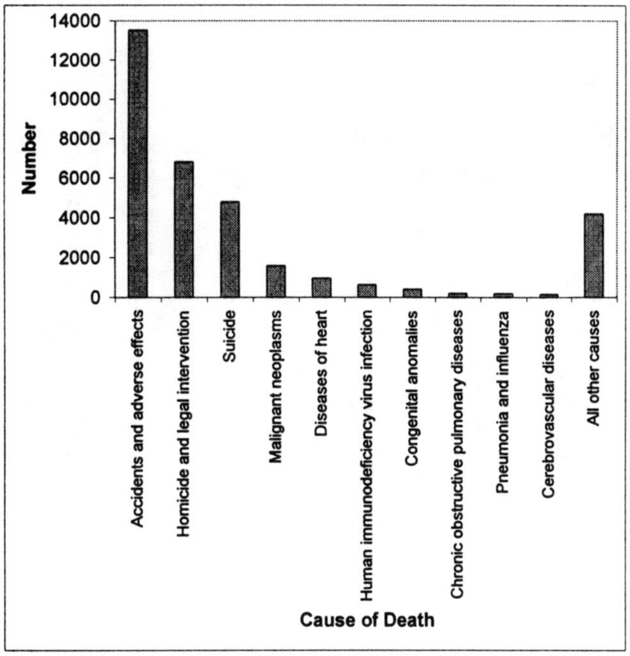


(b)

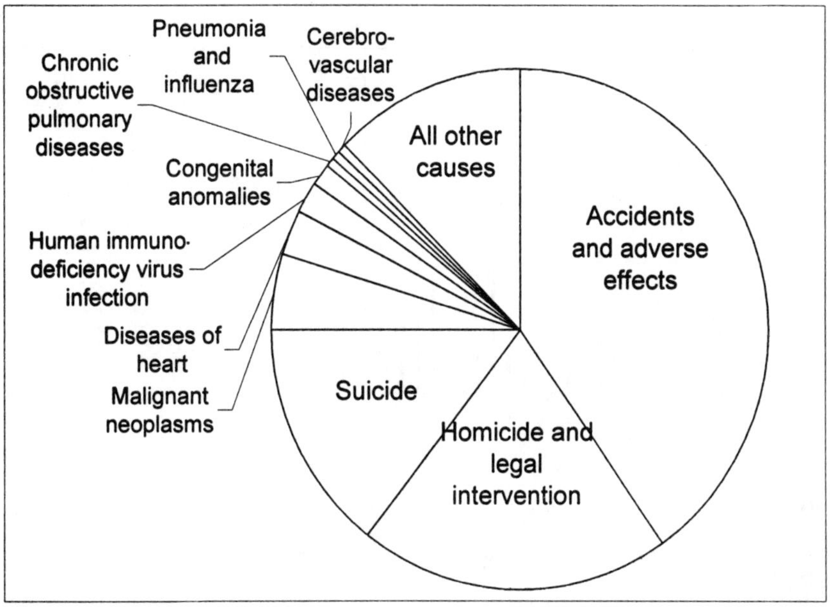


(c) Answers will vary

(d) Accidents and adverse effects

## 9.2 Organization of Data

**1.** (a) Frequency Table

| Score | Tally | Frequency f | Score | Tally | Frequency f |
|---|---|---|---|---|---|
| 25 | \| | 1 | 41 | ~~\|\|\|\|~~ | 5 |
| 26 | \| | 1 | 42 | \|\|\| | 3 |
| 28 | \| | 1 | 43 | \| | 1 |
| 29 | \| | 1 | 44 | \|\| | 2 |
| 30 | \|\|\| | 3 | 45 | \| | 1 |
| 31 | \|\| | 2 | 46 | \|\| | 2 |
| 32 | \| | 1 | 47 | \| | 1 |
| 33 | \|\| | 2 | 48 | \|\|\| | 3 |
| 34 | \|\| | 2 | 49 | \| | 1 |
| 35 | \| | 1 | 50 | \| | 1 |
| 36 | \|\| | 2 | 51 | \| | 1 |
| 37 | \|\|\|\| | 4 | 52 | \|\|\| | 3 |
| 38 | \| | 1 | 53 | \|\| | 2 |
| 39 | \| | 1 | 54 | \|\| | 2 |
| 40 | \| | 1 | 55 | \| | 1 |

range = 55 – 25 = 30

(b) Line Chart

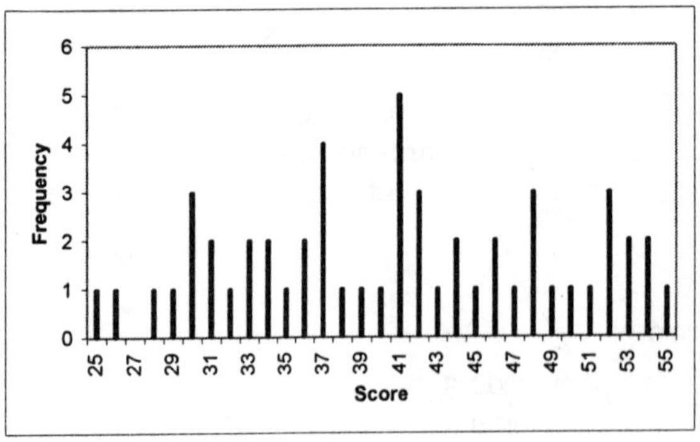


(c) Histogram

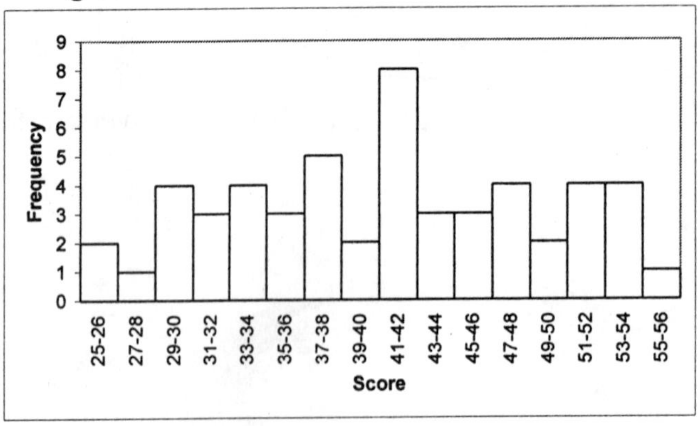

(d) Frequency Polygon

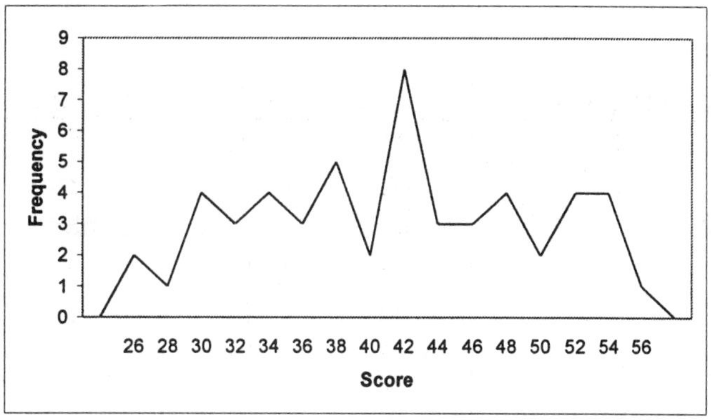


(e) Cumulative (less than) frequencies

| Class Interval | f | cf |
|---|---|---|
| 25-26 | 2 | 2 |
| 27-28 | 1 | 3 |
| 29-30 | 4 | 7 |
| 31-32 | 3 | 10 |
| 33-34 | 4 | 14 |
| 35-36 | 3 | 17 |
| 37-38 | 5 | 22 |
| 39-40 | 2 | 24 |
| 41-42 | 8 | 32 |
| 43-44 | 3 | 35 |
| 45-46 | 3 | 38 |
| 47-48 | 4 | 42 |
| 49-50 | 2 | 44 |
| 51-52 | 4 | 48 |
| 53-54 | 4 | 52 |
| 55-56 | 1 | 53 |

(f) Cumulative (less than) Frequency Distribution

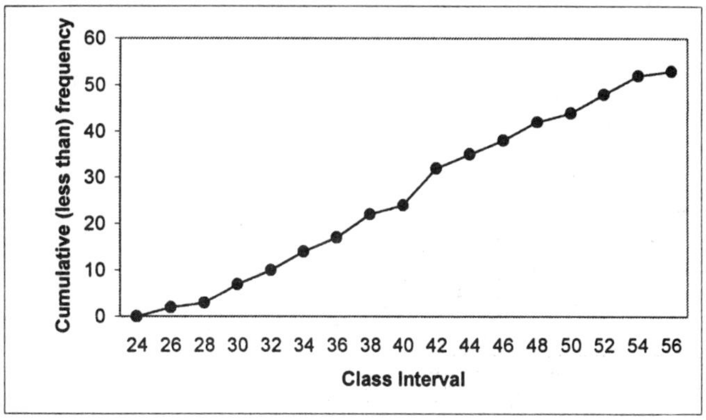

**3.** (a) 13
(b) 20; 24
(c) Class Width = 25 – 20 = 5
(d) About 65,000 + 50,000 + 30,000 = 145,000
(e) 30 – 34
(f) 80 – 84
(g) Place a dot at the top of each rectangle at the midpoint of the class interval, then connect the dots.

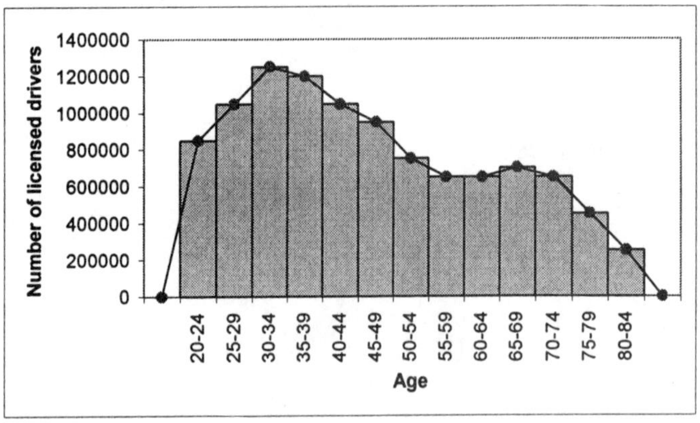


**5.** (a) 13 (b) 20; 24 (c) Class width = 25 – 20 = 5
(d)

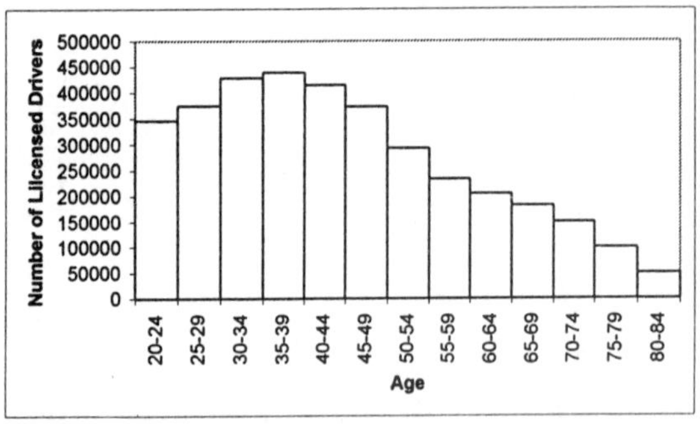


(e)

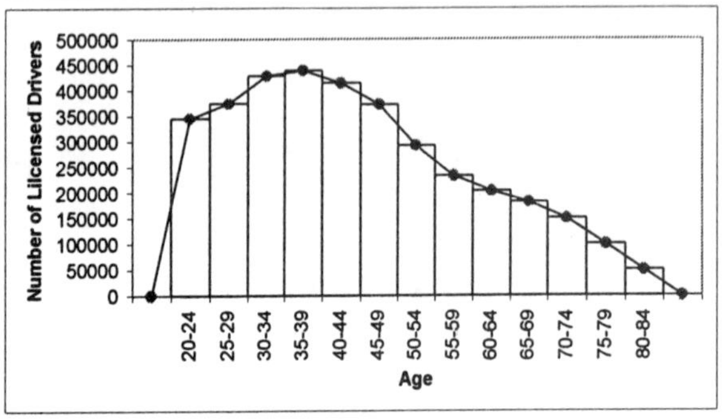


(f) 35-39 (g) 80-84

**7.** (a) 15 (b) 0; 999 (c) Class width = 1000 – 0 = 1000

(d)

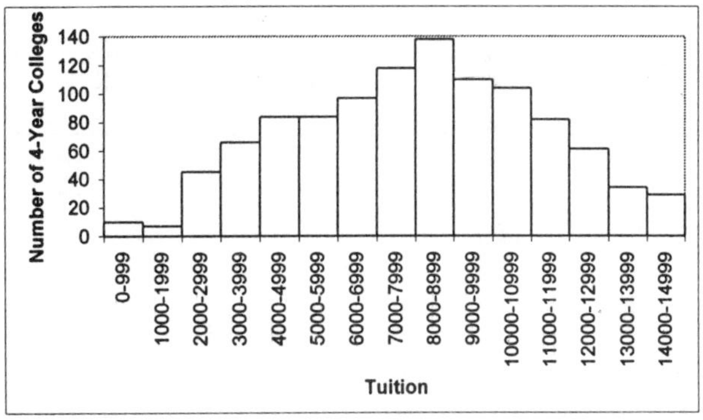


(e)

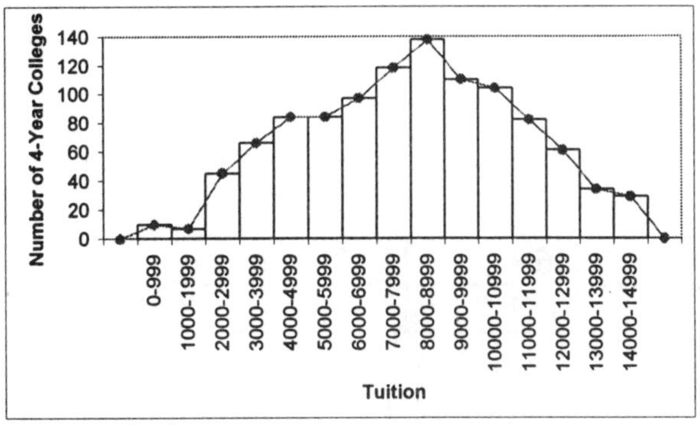


(f) 8000-8999

## Technology Exercises

**1.** Frequency distribution

| Class Interval: | 13 – 13.9 | 14 – 14.9 | 15 – 15.9 | 16 – 16.9 |
|---|---|---|---|---|
| Frequency: | 1 | 8 | 5 | 2 |

| Class Interval: | 17 – 17.9 | 18 – 18.9 | 19 – 19.9 | 20 – 20.9 |
|---|---|---|---|---|
| Frequency: | 1 | 0 | 1 | 2 |

Frequency Histogram:

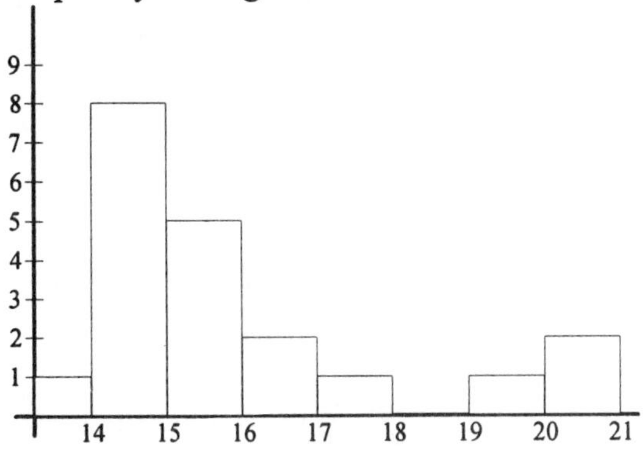

**3.** Frequency distribution

| Class Interval: | 85 – 119 | 120 – 154 | 155 – 189 | 190 – 224 | 225 – 259 |
|---|---|---|---|---|---|
| Frequency: | 1 | 1 | 1 | 10 | 7 |

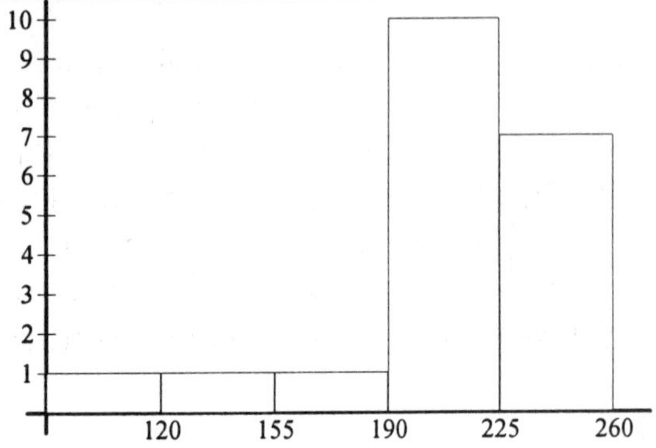


**5.** Frequency distribution

| Class Interval: | 9.36 – 10.35 | 10.36 – 11.35 | 11.36 – 12.35 |
|---|---|---|---|
| Frequency: | 3 | 2 | 8 |

| Class Interval: | 12.36 – 13.35 | 13.36 – 14.35 | 14.36 – 15.35 |
|---|---|---|---|
| Frequency: | 5 | 1 | 1 |

Frequency Histogram:

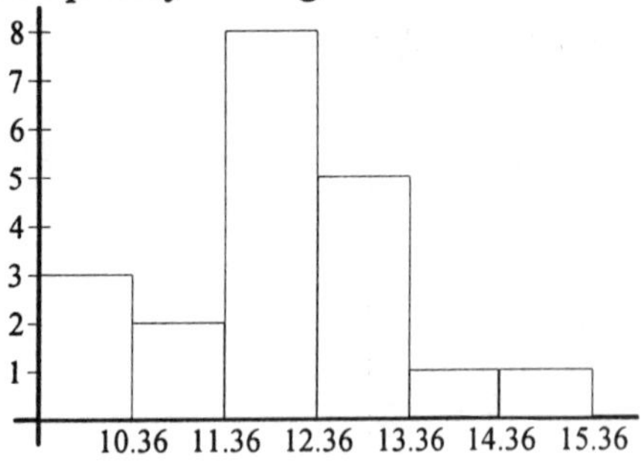


## 9.3 Measures of Central Tendency

**1.** Mean $= \dfrac{21+25+43+36}{4} = 31.25$; median $= \dfrac{25+36}{2} = 30.5$; no mode

**3.** Mean $= \dfrac{55+55+80+92+70}{5} = 70.4$; median $= 70$; mode $= 55$

**5.** Mean $= \dfrac{65+82+82+95+70}{5} = 78.8$; median $= 82$; mode $= 82$

**7.** Mean $= \dfrac{48+65+80+92+80}{5} = 73$; median $= 80$; mode $= 80$

**9.** Average cost per share $= \dfrac{50 \cdot \$85 + 90 \cdot \$105 + 120 \cdot \$110 + 75 \cdot \$130}{50 + 90 + 120 + 75} = \dfrac{\$36{,}650}{335} \approx \$109.40$.

**11.** Mean $= \dfrac{\$34{,}000 + \$35{,}000 + \$36{,}000 + \$36{,}500 + \$65{,}000}{5} = \$41{,}300$; median $= \$36{,}000$.

The median describes the "typical" faculty member situation more realistically since it more nearly reflects the salary of the majority of the faculty in this sample. One of the four lower paid faculty members might use the mean salary as an argument that she is underpaid compared to her peers. Collectively, the five faculty members might use the median to argue that as a group their salary is low compared to other institutions. The highest paid faculty member might compare her salary to the mean or to the median, depending on whether or not she wishes to emphasize her relative status with respect to her colleagues.

**13.** Mean

$$= \frac{25 \cdot \$1075 + 55 \cdot \$1325 + 325 \cdot \$1575 + 410 \cdot \$1825 + 215 \cdot \$2075 + 75 \cdot \$2325 + 50 \cdot \$2575}{25 + 55 + 325 + 410 + 215 + 75 + 50}$$

$\approx \$1826.08$

The total number of items is 1155. The median is the number at 50% of 1155 = 577.5. Through the interval \$1450 - \$1699.99, there are 405 data items. The next interval, \$1700 - \$1949.99, contains 410 data items, so the median will lie in this interval. There are 577.5 – 405 = 172.5 data items left in this interval, so $p = 172.5$. The frequency $q$ of this interval is 410; the size, $i$, is 1950 – 1700 = 250. Thus the

$$\text{median} = 1700 + \frac{p}{q} i = 1700 + \frac{172.5}{410} \cdot 250 = 1700 + 105.18 = \$1805.18.$$

**15.**

| Class Interval Age | $f_i$ Number of Licensed Drivers | $m_i$ | $f_i m_i$ |
|---|---|---|---|
| 20-24 | 345,941 | 22 | 7,610,702 |
| 25-29 | 374,629 | 27 | 10,114,983 |
| 30-34 | 418,748 | 32 | 13,719,936 |
| 35-39 | 439,137 | 37 | 16,248,069 |
| 40-44 | 414,344 | 42 | 17,402,448 |
| 45-49 | 372,814 | 47 | 17,522,258 |
| 50-54 | 292,460 | 52 | 15,207,920 |
| 55-59 | 233,615 | 57 | 13,316,055 |
| 60-64 | 204,235 | 62 | 12,662,570 |
| 65-69 | 181,977 | 67 | 12,192,459 |
| 70-74 | 150,347 | 72 | 10,824,984 |
| 75-79 | 100,068 | 77 | 7,705,236 |
| 80-84 | 50,190 | 82 | 4,115,580 |
| Source: FHWA | $n = 3{,}588{,}505$ | | 158,643,200 |

= Sum of $f_i m_i$

The average age of a licensed driver in Tennessee is $\dfrac{158{,}643{,}200}{3{,}588{,}505} = 44.2$ years.

**17.**

| Class Interval Tuition (Dollars) | $f_i$ Number of 4-Year Colleges | $m_i$ | $f_i m_i$ |
|---|---|---|---|
| 0-999 | 10 | 499.5 | 4995 |
| 1000-1999 | 7 | 1,499.5 | 10,496.5 |
| 2000-2999 | 45 | 2,499.5 | 112,477.5 |
| 3000-3999 | 66 | 3,499.5 | 230,967 |
| 4000-4999 | 84 | 4,499.5 | 377,958 |
| 5000-5999 | 84 | 5,499.5 | 461,958 |
| 6000-6999 | 97 | 6,499.5 | 630,451.5 |
| 7000-7999 | 118 | 7,499.5 | 884,941 |
| 8000-8999 | 138 | 8,499.5 | 1,172,931 |
| 9000-9999 | 110 | 9,499.5 | 1,044,945 |
| 10,000-10,999 | 104 | 10,499.5 | 1,091,948 |
| 11,000-11,999 | 82 | 11,499.5 | 942,959 |
| 12,000-12,999 | 61 | 12,499.5 | 762,469.5 |
| 13,000-13,999 | 34 | 13,499.5 | 458,983 |
| 14,000-14,999 | 29 | 14,499.5 | 420,485.5 |
| | $n = 1069$ | | 8,608,965.5 |

= Sum of $f_i m_i$

**Source: The College Board, New York, NY, Annual Survey of Colleges 1992 and 1993.**

The average tuition at a 4-year college in 1992-1993 was $\frac{8{,}608{,}965.5}{1069} = \$8053.29$.

**19.** The data in increasing order is: 3.25, 3.28, 3.3, 3.4, 3.45, 3.45, 3.45, 3.5, 3.55, 3.6, 3.79, 3.8, 3.9

$$\text{Mean} = \frac{3.25 + 3.28 + 3.3 + 3.4 + 3.45 + 3.45 + 3.45 + 3.5 + 3.55 + 3.6 + 3.79 + 3.8 + 3.9}{13}$$

$\approx 3.52$ million units;

median $= 3.45$ million units

**21.** $\text{Mean} = \frac{2.5 + 5.5 + 7.2 + 7.8 + 8.3 + 8.7 + 10.2 + 12}{8} = 7.8\%$; $\text{Median} = \frac{7.8 + 8.3}{2} = 8.05\%$

## 9.4 Measures of Dispersion

**1.** (b)

**3.** $\overline{X} = \frac{4 + 5 + 9 + 9 + 10 + 14 + 25}{7} \approx 10.86$

| $x$ | $x-\overline{X}$ | $(x-\overline{X})^2$ |
|---|---|---|
| 4 | −6.86 | 47.0596 |
| 5 | −5.86 | 34.3396 |
| 9 | −1.86 | 3.4596 |
| 9 | −1.86 | 3.4596 |
| 10 | −0.86 | .7396 |
| 14 | 3.14 | 9.8596 |
| 25 | 14.14 | 199.9396 |
| | | 298.8572 |

$$\sigma = \sqrt{\frac{298.8572}{7}} = \sqrt{42.694} \approx 6.53$$

**5.** $\overline{X} = \frac{58+62+70+70}{4} = \frac{260}{4} = 65$

| $x$ | $x-\overline{X}$ | $(x-\overline{X})^2$ |
|---|---|---|
| 58 | −7 | 49 |
| 62 | −3 | 9 |
| 70 | 5 | 25 |
| 70 | 5 | 25 |
| | | 108 |

$$\sigma = \sqrt{\frac{108}{4}} = \sqrt{27} = 5.2$$

**7.** $\overline{X} = \frac{62+75+78+85+100}{5} = 80$

| $x$ | $x-\overline{X}$ | $(x-\overline{X})^2$ |
|---|---|---|
| 62 | −18 | 324 |
| 75 | −5 | 25 |
| 78 | −2 | 4 |
| 85 | 5 | 25 |
| 100 | 20 | 400 |
| | | 778 |

$$\sigma = \sqrt{\frac{778}{5}} = \sqrt{155.6} = 12.47$$

**9.** $\text{mean} = \frac{(13)(1)+(20)(3)+(27)(10)+(34)(12)+(41)(5)+(48)(2)}{1+3+10+12+5+2} = \frac{1052}{33} \approx 31.9$

$$\sigma^2 = \frac{\begin{pmatrix}(13-\text{mean})^2(1)+(20-\text{mean})^2(3)+(27-\text{mean})^2(10)\\ +(34-\text{mean})^2(12)+(41-\text{mean})^2(5)+(48-\text{mean})^2(2)\end{pmatrix}}{33} = \frac{2007.53}{33} = 60.834$$

$$\sigma = \sqrt{60.834} = 7.8$$

**11.** $\overline{X} = \frac{968+893+769+845+922+815}{6} = \frac{5212}{6} = 868.67$.

The mean lifetime is 868.67 hours.

| $x$ | $x-\overline{X}$ | $(x-\overline{X})^2$ |
|---|---|---|
| 769 | −99.67 | 9934.11 |
| 815 | −53.67 | 2880.47 |
| 845 | −23.67 | 560.27 |
| 893 | 24.33 | 591.95 |
| 922 | 53.33 | 2844.09 |
| 968 | 99.33 | 9866.45 |
| $n=6$ | | 26,677.34 |

$\sigma = \sqrt{\frac{26{,}677.34}{6}} = \sqrt{4446.22} = 66.68$. The standard deviation is 66.68 hours.

**13.** River I:

mean $= \frac{(1000)(4)+(2000)(8)+(3000)(2)+(4000)(1)}{4+8+2+1} = 2000$, $\sigma \approx 816.5$

River II:

mean $= \frac{(1000)(2)+(1575)(3)+(2025)(4)+(2475)(4)+(2925)(2)}{15} \approx 2038.3$, $\sigma \approx 585.2$

Average number of salmon caught on the River II is greater than that on River I, and the standard deviation is lower for River II, indicating a more predictable experience at River II. The fisherperson looking for fish should opt for River II; the one looking for a more solitary experience might go to River I.

**15.** From problem 15, Exercise 9.2, $\overline{X} = 44.2$ years and $n = 3{,}588{,}505$.

| Class Interval Age | $f_i$ Number of Licensed Drivers | $m_i$ | $m_i - \overline{X}$ | $(m_i - \overline{X})^2$ | $(m_i - \overline{X})^2 \cdot f_i$ |
|---|---|---|---|---|---|
| 20-24 | 345,941 | 22 | −22.2 | 492.84 | 170,493,562.40 |
| 25-29 | 374,629 | 27 | −17.2 | 295.84 | 110,830,243.40 |
| 30-34 | 418,748 | 32 | −12.2 | 148.84 | 63,814,852.32 |
| 35-39 | 439,137 | 37 | −7.2 | 51.84 | 22,764,862.08 |
| 40-44 | 414,344 | 42 | −2.2 | 4.84 | 2,005,424.96 |
| 45-49 | 372,814 | 47 | 2.8 | 7.84 | 2,922,861.76 |
| 50-54 | 292,460 | 52 | 7.8 | 60.84 | 17,793,266.40 |
| 55-59 | 233,615 | 57 | 12.8 | 163.84 | 38,275,481.60 |
| 60-64 | 204,235 | 62 | 17.8 | 316.84 | 64,709,817.40 |
| 65-69 | 181,977 | 67 | 22.8 | 519.84 | 94,598,923.68 |
| 70-74 | 150,347 | 72 | 27.8 | 772.84 | 116,194,175.50 |
| 75-79 | 100,068 | 77 | 32.8 | 1075.84 | 107,657,157.10 |
| 80-84 | 50,190 | 82 | 37.8 | 1428.84 | 71,713,479.60 |
| Sum: | $n = 3{,}588{,}505$ | | | | 883,774,108.20 |

Source: FHWA

$$\sigma = \sqrt{\frac{883{,}774{,}108.20}{3{,}588{,}505}} = \sqrt{246.28} = 15.7 \text{ years.}$$

**17.** From problem 17, Exercise 9.2, $\overline{X} = \$8053.29$ years and $n = 1069$.

| Class Interval Tuition (Dollars) | $f_i$ Number of 4-Year Colleges | $m_i$ | $m_i - \overline{X}$ | $(m_i - \overline{X})^2$ | $(m_i - \overline{X})^2 \cdot f_i$ |
|---|---|---|---|---|---|
| 0-999 | 10 | 499.5 | −7553.79 | 57,059,743.36 | 570,597,433.64 |
| 1000-1999 | 7 | 1,499.5 | −6553.79 | 42,952,163.36 | 300,665,143.55 |
| 2000-2999 | 45 | 2,499.5 | −5553.79 | 30,844,583.36 | 1,388,006,251.38 |
| 3000-3999 | 66 | 3,499.5 | −4553.79 | 20,737,003.36 | 1,368,642,222.03 |
| 4000-4999 | 84 | 4,499.5 | −3553.79 | 12,629,423.36 | 1,060,871,562.58 |
| 5000-5999 | 84 | 5,499.5 | −2553.79 | 6,521,843.36 | 547,834,842.58 |
| 6000-6999 | 97 | 6,499.5 | −1553.79 | 2,414,263.36 | 234,183,546.32 |
| 7000-7999 | 118 | 7,499.5 | −553.79 | 306,683.36 | 36,188,636.96 |
| 8000-8999 | 138 | 8,499.5 | 446.21 | 199,103.36 | 27,476,264.25 |
| 9000-9999 | 110 | 9,499.5 | 1446.21 | 2,091,523.36 | 230,067,570.05 |
| 10,000-10,999 | 104 | 10,499.5 | 2446.21 | 5,983,943.36 | 622,330,109.87 |
| 11,000-11,999 | 82 | 11,499.5 | 3446.21 | 11,876,363.36 | 973,861,795.86 |
| 12,000-12,999 | 61 | 12,499.5 | 4446.21 | 19,768,783.36 | 1,205,895,785.21 |
| 13,000-13,999 | 34 | 13,499.5 | 5446.21 | 29,661,203.36 | 1,008,480,914.38 |
| 14,000-14,999 | 29 | 14,499.5 | 6446.21 | 41,553,623.36 | 1,205,055,077.56 |
| Sum: | $n = 1069$ | | | | 10,780,157,156.22 |

**Source: The College Board, New York, NY, Annual Survey of Colleges 1992 and 1993.**

$$\sigma = \sqrt{\frac{10{,}780{,}157{,}156.22}{1069}} = \sqrt{10{,}084{,}337.84} = \$3{,}175.58.$$

**19.** $\overline{X} = 25$, $\sigma = 3$

(a) $k = \overline{X} - 19 = 25 - 19 = 6$. The probability that an outcome lies between 19 and 31 $(\overline{X} - 6, \overline{X} + 6)$ is at least $1 - \frac{\sigma^2}{k^2} = 1 - \frac{3^2}{6^2} = 1 - \frac{9}{36} = .75$. Thus, at least 75% of the outcomes lie between 19 and 31.

(b) $k = \overline{X} - 20 = 25 - 20 = 5$. The probability that an outcome lies between 20 and 30 $(\overline{X} - 5, \overline{X} + 5)$ is at least $1 - \frac{\sigma^2}{k^2} = 1 - \frac{3^2}{5^2} = 1 - \frac{9}{25} = .64$. Thus, at least 64% of the outcomes lie between 20 and 30.

(c) $k = \overline{X} - 16 = 25 - 16 = 9$. The probability that an outcome lies between 16 and 34 $(\overline{X} - 9, \overline{X} + 9)$ is at least $1 - \frac{\sigma^2}{k^2} = 1 - \frac{3^2}{9^2} = 1 - \frac{9}{81} = .88\frac{8}{9}$. Thus, at least $88\frac{8}{9}$% of the outcomes lie between 16 and 34.

(d) Since approximately 75% of the outcomes lie between 19 and 31, approximately 25% will be less than 19 or more than 31.

(e) Since approximately $88\frac{8}{9}\%$ of the outcomes lie between 16 and 34, approximately $11\frac{1}{9}\%$ will be less than 16 or more than 34.

**21.** $\overline{X} = 51.25$, $\sigma = 8.50$, $1 - \frac{\sigma^2}{k^2} = 0.90 \Rightarrow 0.1 = \frac{\sigma^2}{k^2} \Rightarrow k^2 = \frac{\sigma^2}{0.1} \Rightarrow k = \frac{\sigma}{\sqrt{0.1}} = \frac{8.50}{\sqrt{0.1}} = 26.88$, $(\overline{X} - k, \overline{X} + k) = (\$24.37, \$78.13)$

## 9.5 The Normal Distribution

**1.** $\mu = 8,\ \sigma = 9 - 8 = 1$

**3.** $\mu = 18,\ \sigma = 19 - 18 = 1$

**5.**

| $x$ | 7 | 9 | 13 | 15 |
|---|---|---|---|---|
| Z-score | $\frac{7-13.1}{9.3} \approx -0.6559$ | $\frac{9-13.1}{9.3} \approx -0.4409$ | $\frac{13-13.1}{9.3} \approx -0.0108$ | $\frac{15-13.1}{9.3} \approx 0.2043$ |

| $x$ | 29 | 37 | 41 |
|---|---|---|---|
| Z-score | $\frac{29-13.1}{9.3} \approx 1.7097$ | $\frac{37-13.1}{9.3} \approx 2.5699$ | $\frac{41-13.1}{9.3} = 3$ |

**7.** (a) 0.3133 (b) 0.3642 (c) 0.4938
(d) 0.4987 (e) 0.2734 (f) 0.4896
(g) 0.2881 (h) 0.4988

**9.** $0.5 - 0.1915 = 0.3085$

**11.** $0.5 - 0.4332 = 0.0668$

**13.** (a) The number within $\pm 1\sigma$ is $0.6827 \cdot 2000 \approx 1365$ women
(b) The number within $\pm 2\sigma$ is $0.9545 \cdot 2000 \approx 1909$ women
(c) The number within $\pm 3\sigma$ is $0.9973 \cdot 2000 \approx 1995$ women

**15.** (a) The number above $\frac{142-130}{5.2}\sigma \approx 2.31\sigma$ is $(0.5 - 0.4896) \cdot 100 \approx 1$ student.
(b) For the middle 70% of students, 35% are above the mean. From Table III in the back of the book, the nearest area to 0.35 is 0.3508 which corresponds to a Z-score of 1.04. The range of weights within $\pm 1.04\sigma$ of the mean is from $130 - 1.04 \cdot 5.2 \approx 124.6$ pounds to $130 + 1.04 \cdot 5.2 \approx 135.4$ pounds.

**17.** $z = \frac{1-2.2}{1.7} = -0.71$. The number of pairs of shoes lasting less than one year is the number that are $0.71\sigma$ below the mean or $(0.5 - 0.2611) \cdot 1000 \approx 239$ pairs of shoes.

**19.** Colleen's Z-score is $\frac{76-82}{7}=-\frac{6}{7}$, so her score is $\frac{6}{7}\sigma$ below the mean score for her group; Mary's Z-score is $\frac{89-93}{2}=-2$, so her score is $2\sigma$ below the mean score for her group; Kathleen's Z-score is $\frac{21-24}{9}=-\frac{1}{3}$, so her score is $\frac{1}{3}\sigma$ below the mean score for her group. Kathleen (whose score is the least below the corresponding group mean) has the highest relative standing.

**21.**

| $k$ | 0 | 1 | 2 | 3 | 4 | 5 | 6 | 7 |
|---|---|---|---|---|---|---|---|---|
| $b(15, k; 0.30)$ | 0.0047 | 0.0305 | 0.0916 | 0.1700 | 0.2186 | 0.2061 | 0.1472 | 0.0811 |
| $k$ | 8 | 9 | 10 | 11 | 12 | 13 | 14 | 15 |
| $b(15, k; 0.30)$ | 0.0348 | 0.0116 | 0.0030 | 0.0006 | 0.0001 | 0.0000 | 0.0000 | 0.0000 |

Line Chart:

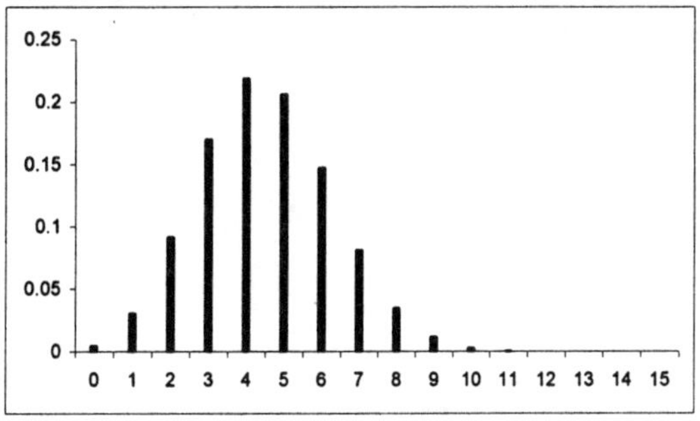


Frequency curve:

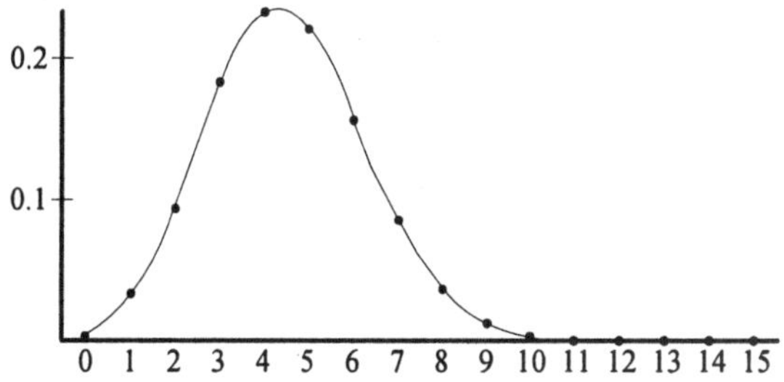


**23.** The probability of obtaining a number of successes within $\pm\frac{15}{13}\sigma \approx \pm 1.15\sigma$ of the mean is $2 \cdot 0.3749 = 0.7498$.

**25.** The probability of obtaining a number of successes which is at or above the mean is 0.5.

**27.** The probability of obtaining a number of successes at least $\frac{25}{13}\sigma \approx 1.92\sigma$ above the mean is $0.5-0.4726=0.0274$.

**29.** Since $n=300$ is large, we use a normal curve approximation for the binomial distribution. Then with $n=300$ and $p=0.250$, $\mu=np=300(0.250)=75$, $\sigma=\sqrt{npq}=\sqrt{(300)(.25)(.75)}=7.5$. To find the probability that at least 80 and no more than 90 hits occur, we find the area under a normal curve from $x=80$ to $x=90$.

For $x=80$, $z_1=\frac{80-75}{7.5}=\frac{2}{3}=.67$ $\quad A_1=.2486$

For $x=90$, $z_2=\frac{90-75}{7.5}=2$ $\quad A_2=.4772$

The area is $A_2-A_1=.4772-.2486=.2286$, thus the approximate probability is .2286.

To find the probability that 85 or more hits occur, we find the area under the curve above

$x = 85$. $z = \frac{85-75}{7.5} = \frac{4}{3} = 1.33$, area = 0.5 – 0.4082 = 0.0918, thus the approximate probability is 0.0918.

**31.** Since $n = 500$ is large, we use a normal curve approximation for the binomial distribution. Then with $n = 500$ and $p = .01$, $\mu = np = 500(.01) = 5$, $\sigma = \sqrt{npq} = \sqrt{(500)(0.01)(0.99)}$ $= 2.22$. The probability that at least 10 are not properly sealed is the area under a normal curve above $x = 10$: $z = \frac{10-5}{2.22} = 2.25$, area = .5 – .4878 = .0122.

## Technology Exercises

**1.** The function assumes its maximum at $x = 0$ $\left(\text{where } y = \frac{1}{\sqrt{2\pi}} \approx 0.398942\right)$. Graph of

$y = \frac{1}{\sqrt{2\pi}} e^{-(1/2)x^2}$:

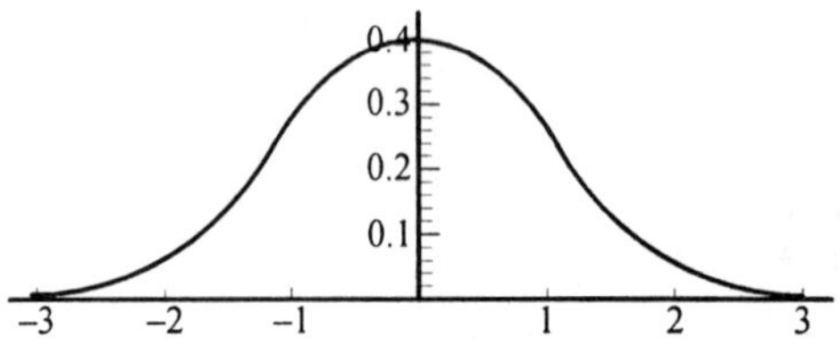


**2.** Graph of $y = \frac{1}{2\sqrt{2\pi}} e^{-(1/8)(x-10)^2}$ :

The function assumes its maximum at $x = 10$ (where $y = \frac{1}{2\sqrt{2\pi}} = 0.19947$).

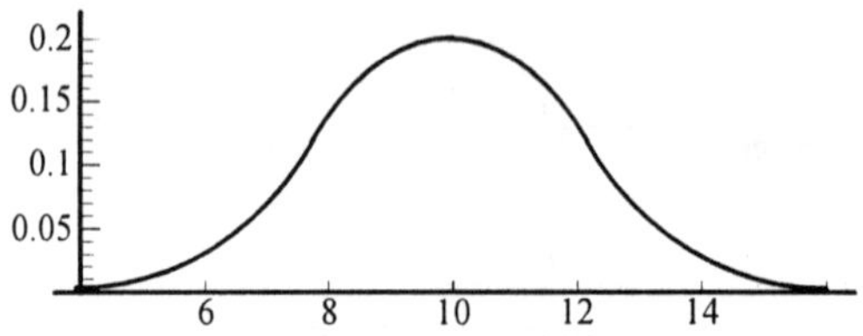


## Chapter 9 Review

### True or False

**1.** False

**3.** False

**5.** False

### Fill in the Blank

**1.** mean, median, mode

**3.** bell

**5.** $\overline{X} - k, \overline{X} + k$

## Review Exercises

**1.** Bar graph:

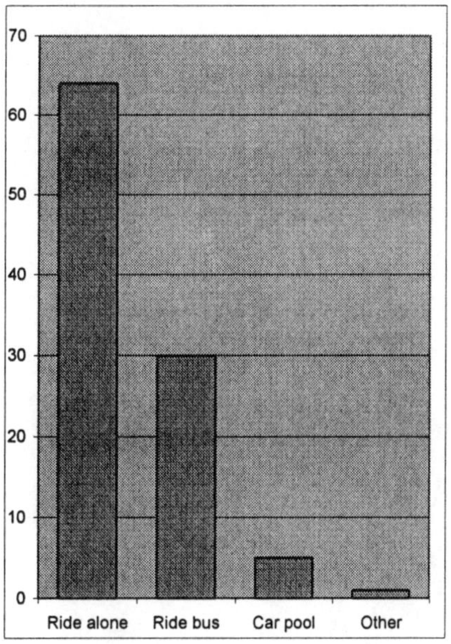


Pie Chart:

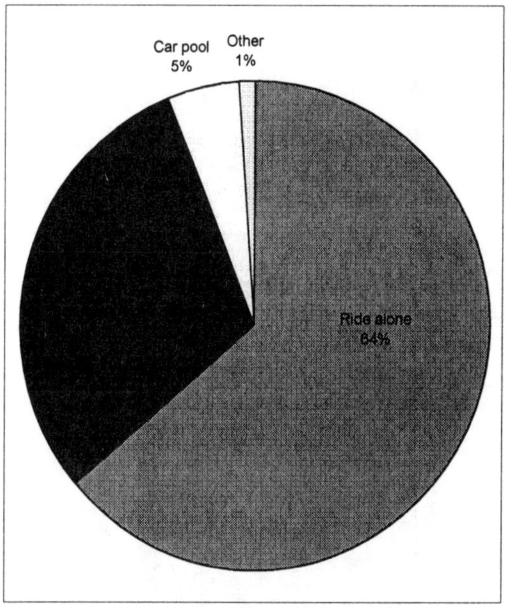


**3.** (a) Frequency table:

| Score | freq | Score | freq | Score | freq | Score | freq |
|---|---|---|---|---|---|---|---|
| 8 | 1 | 33 | 1 | 70 | 1 | 85 | 2 |
| 10 | 1 | 42 | 1 | 72 | 2 | 87 | 2 |
| 12 | 1 | 44 | 1 | 73 | 2 | 89 | 1 |
| 14 | 2 | 48 | 1 | 74 | 1 | 90 | 1 |
| 17 | 1 | 52 | 2 | 75 | 1 | 92 | 1 |
| 19 | 1 | 55 | 1 | 77 | 1 | 95 | 1 |
| 20 | 1 | 60 | 1 | 78 | 2 | 99 | 1 |
| 21 | 1 | 63 | 2 | 80 | 3 | 100 | 2 |
| 26 | 1 | 66 | 2 | 82 | 1 | | |
| 30 | 1 | 69 | 1 | 83 | 1 | | |

Range = 100 − 8 = 92

(b) Line chart:

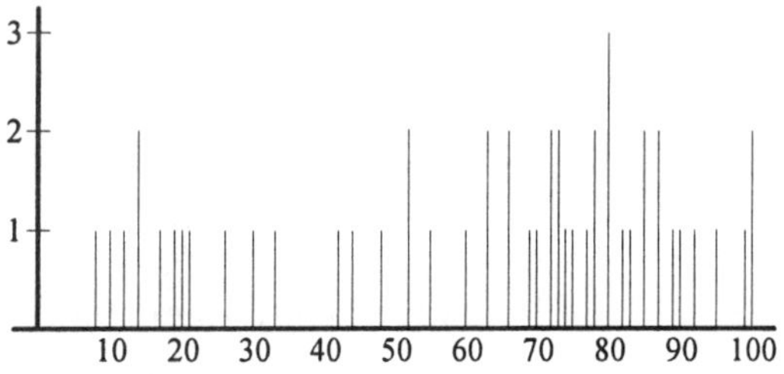

(c) Histogram

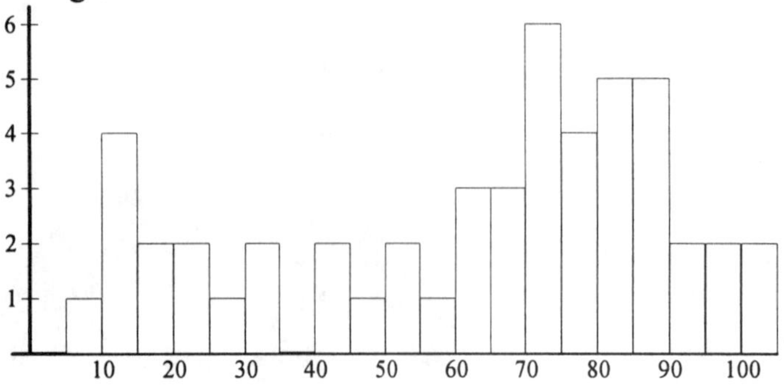


(d) Frequency polygon

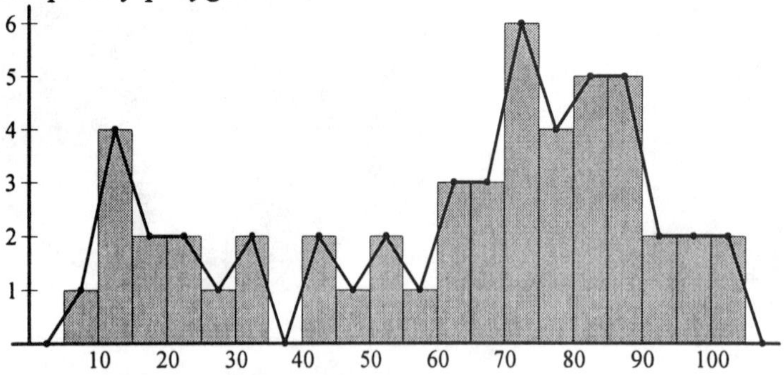


(e) Cumulative (less than) frequencies:

| Class Interval | < cum freq | Class Interval | < cum freq | Class Interval | < cum freq | Class Interval | <cum freq |
|---|---|---|---|---|---|---|---|
| 4.5 - 9.5 | 1 | 29.5 - 34.5 | 12 | 54.5 - 59.5 | 18 | 79.5 - 84.5 | 39 |
| 9.5 - 14.5 | 5 | 34.5 - 39.5 | 12 | 59.5 - 64.5 | 21 | 84.5 - 89.5 | 44 |
| 14.5 - 19.5 | 7 | 39.5 - 44.5 | 14 | 64.5 - 69.5 | 24 | 89.5 - 94.5 | 46 |
| 19.5 - 24.5 | 9 | 44.5 - 49.5 | 15 | 69.5 - 74.5 | 30 | 94.5 - 99.5 | 48 |
| 24.5 - 29.5 | 10 | 49.5 - 54.5 | 17 | 74.5 - 79.5 | 34 | 99.5 - 104.5 | 50 |

Drawing of the cumulative (less than) frequencies:

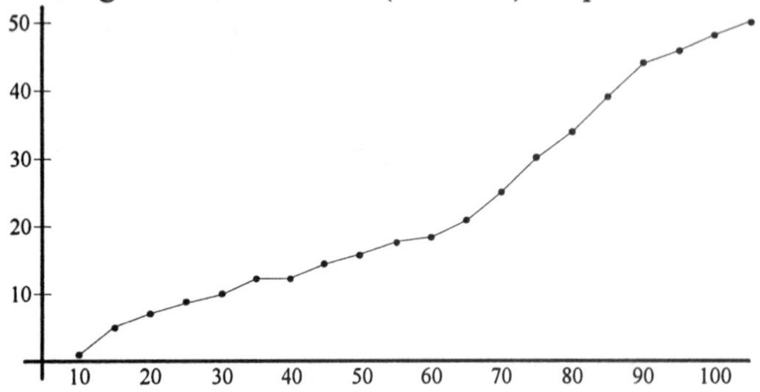


**5.** (a) Mean $= \frac{67}{12} = 5.58$; median $= \frac{4+5}{2} = 4.5$; mode = 4 (with a frequency of 3).

(b) Mean $= \frac{209}{8} = 26.125$; median $= 2$; mode = 2 (with a frequency of 4).

(c) Mean $= \frac{62}{9} = 6.89$; median $= 7$; mode = 7 (with a frequency of 3).

**7.** Consider data sets $A = \{-1, 1\}$ and $B = \{-10, 10\}$. Both sets have mean 0, but the standard deviation for $A$ is 1, while the standard deviation for set $B$ is 10.

**9.** The mean is $\overline{X} = \frac{529}{7} = 75\frac{5}{7}$

$$\sigma = \sqrt{\frac{(74-\overline{X})^2 + (72-\overline{X})^2 + (76-\overline{X})^2 + (81-\overline{X})^2 + (77-\overline{X})^2 + (76-\overline{X})^2 + (73-\overline{X})^2}{7}}$$

$\approx 2.7701$ strokes.

**11.** (a) $0.6827 \cdot 600 = 410$ scores (b) $(0.4987 - 0.3413) \cdot 600 \approx 94$ scores
(c) $2 \cdot 0.2486 \cdot 600 \approx 298$ scores

**13.** (a) $0.4970 - 0.4115 = 0.0855$ (b) $0.4599 - 0.3849 = 0.075$

**15.** The variance, $\sigma^2 = 25$. Using Chebychev's theorem, the probability that the week's production is within $\pm 10$ items of the mean is at least $1 - \frac{25}{10^2} = 0.75$.

## Professional Exam Questions

**1.** e: $K = 2\sigma$, where $\sigma = \sqrt{180 \cdot \frac{1}{6} \cdot \left(1 - \frac{1}{6}\right)} = 5$ from the binomial distribution.

**3.** c: This is the probability that the weight is within $\pm 1$ standard deviation of the mean weight.

# Chapter 10

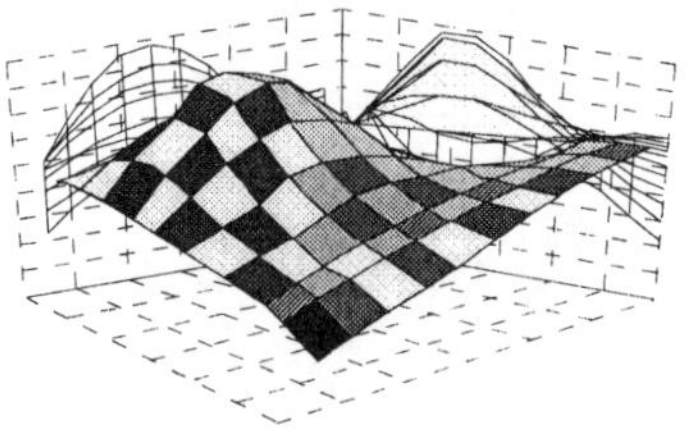

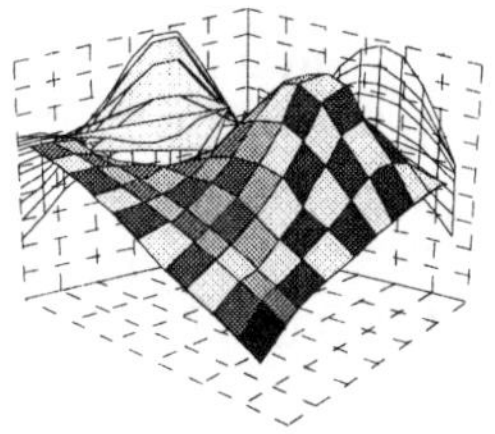

# Markov Chains; Games

## 10.1 Markov Chains and Transition Matrices

**1.** The entry in the 3rd row, 2nd column is negative; entries in a transition matrix must be nonnegative. The sum of the entries in a row must equal 1, but the third row sums to 0.

**3.** (a) $\frac{1}{4}$ = the probability of a transition from state 2 to state 1.

(b) If the system is initially in state 1, the probability distribution one observation later is $\begin{bmatrix} 1 & 0 \end{bmatrix} P = \begin{bmatrix} \frac{1}{3} & \frac{2}{3} \end{bmatrix}$. The probability distribution two observations later is $\begin{bmatrix} \frac{1}{3} & \frac{2}{3} \end{bmatrix} P = \begin{bmatrix} \frac{5}{18} & \frac{13}{18} \end{bmatrix}$.

(c) If the system is initially in state 2, the probability distribution one observation later is $\begin{bmatrix} 0 & 1 \end{bmatrix} P = \begin{bmatrix} \frac{1}{4} & \frac{3}{4} \end{bmatrix}$. The probability distribution two observations later is $\begin{bmatrix} \frac{1}{4} & \frac{3}{4} \end{bmatrix} P = \begin{bmatrix} \frac{13}{48} & \frac{35}{48} \end{bmatrix}$.

**5.** $\begin{bmatrix} 0.25 & 0.75 \end{bmatrix} \begin{bmatrix} 0.3 & 0.7 \\ 0.4 & 0.6 \end{bmatrix}^2 = \begin{bmatrix} 0.3625 & 0.6375 \end{bmatrix}$

**7.** $0.2 + a + 0.4 = 1$, so $a = 0.4$; $b + 0.6 + 0.3 = 1$, so $b = 0.1$; $0 + c + 0 = 1$, so $c = 1$

**9.** $\begin{bmatrix}0.7 & 0.3\end{bmatrix}\begin{bmatrix}0.93 & 0.07\\0.01 & 0.99\end{bmatrix}^5 \approx \begin{bmatrix}0.50397 & 0.49603\end{bmatrix}$

**11.** (a) The experiment measures the proportion of mayors in these cities who are Democrats and the proportion who are Republicans during a given term of office. These proportions depend only on the proportions during the preceding mayoral terms, so the sequence of experiments can be represented as a Markov chain.

(b) $P = \begin{matrix} & D & R \\ D & 0.6 & 0.4 \\ R & 0.3 & 0.7\end{matrix}$

(c) $P^2 = \begin{bmatrix}0.48 & 0.52\\0.39 & 0.61\end{bmatrix}$; $P^3 = \begin{bmatrix}0.444 & 0.556\\0.417 & 0.583\end{bmatrix}$

**13.** With the states being "drink brand $X$" and "drink another brand $O$," respectively, the transition matrix is $P = \begin{matrix} & X & O \\ X & 0.75 & 0.25 \\ O & 0.35 & 0.65\end{matrix}$. The initial probability distribution is [0.5 0.5].

$\begin{bmatrix}0.5 & 0.5\end{bmatrix}\begin{bmatrix}0.75 & 0.25\\0.35 & 0.65\end{bmatrix}^2 = \begin{bmatrix}0.57 & 0.43\end{bmatrix}$. 57% of wine drinkers drink brand X after 2 months.

**15.** With the states being "insured by Travelers, T" "insured by General American, GA" and "insured by other companies, O," respectively, the transition matrix is

$\begin{matrix} & T & GA & O \\ T & 0.92 & 0.08 & 0.00 \\ GA & 0.04 & 0.90 & 0.06 \\ O & 0.10 & 0.08 & 0.82\end{matrix}$. The initial (1998) probability distribution is [0.45 0.3 0.25].

(a) $\begin{bmatrix}0.45 & 0.3 & 0.25\end{bmatrix}\cdot\begin{bmatrix}0.92 & 0.08 & 0.00\\0.04 & 0.90 & 0.06\\0.10 & 0.08 & 0.82\end{bmatrix} = \begin{bmatrix}0.451 & 0.326 & 0.223\end{bmatrix}$. 45.1% were insured by Travelers and 32.6% were insured by General American in 1999.

(b) $\begin{bmatrix}0.45 & 0.3 & 0.25\end{bmatrix}\cdot\begin{bmatrix}0.92 & 0.08 & 0.00\\0.04 & 0.90 & 0.06\\0.10 & 0.08 & 0.82\end{bmatrix}^2 = \begin{bmatrix}0.45026 & 0.34732 & 0.20242\end{bmatrix}$. 45.026% were insured by Travelers and 34.732% were insured by General American in 2000.

**17.** $uA = [u_1a_{11} + u_2a_{21} \quad u_1a_{12} + u_2a_{22}]$. Note that since the entries of $u$ and of $A$ are nonnegative, the entries of $uA$ (being sums of products of nonnegative numbers) are nonnegative. Also,
$$(u_1a_{11} + u_2a_{21}) + (u_1a_{12} + u_2a_{22}) = u_1(a_{11} + a_{12}) + u_2(a_{21} + a_{22}) = u_1(1) + u_2(1)$$
$$= u_1 + u_2 = 1.$$
Since $uA$ is a row-vector whose entries are nonnegative and sum to 1, it is a probability vector.

## Technology Exercises

**1.** $v \cdot P^{10} \approx [0.32000 \quad 0.28923 \quad 0.16308 \quad 0.22769]$

**3.** (a) The entry in row i, column j, is the probability that a mouse in room i moves to room j in one transition.
$$P = \begin{bmatrix} 0 & 1/2 & 0 & 1/2 & 0 & 0 \\ 1/3 & 0 & 1/3 & 0 & 1/3 & 0 \\ 0 & 1/2 & 0 & 0 & 0 & 1/2 \\ 1/2 & 0 & 0 & 0 & 1/2 & 0 \\ 0 & 0 & 0 & 0 & 0 & 1 \\ 0 & 0 & 0 & 0 & 0 & 1 \end{bmatrix}$$
(b) $v^{(0)} = [0 \quad 1 \quad 0 \quad 0 \quad 0 \quad 0]$

(c) $v^{(0)} \cdot P^{10} \approx [0 \quad 0.01875 \quad 0 \quad 0.01250 \quad 0 \quad 0.96875]$ (rounded to five decimal places)

(d) The mouse is most likely to be in room 6 (with probability 0.96875.)

## 10.2 Regular Markov Chains

**1.** $P = \begin{bmatrix} 1/2 & 1/2 \\ 1 & 0 \end{bmatrix}$ is regular since $P^2 = \begin{bmatrix} 3/4 & 1/4 \\ 1/2 & 1/2 \end{bmatrix}$ has only positive entries. To find the fixed probability vector, solve $\begin{cases} t_1 + t_2 = 1 \\ [t_1 \quad t_2]P = [t_1 \quad t_2] \end{cases}$, or $\begin{cases} t_1 + t_2 = 1 \\ \frac{1}{2}t_1 + t_2 = t_1 \\ \frac{1}{2}t_1 = t_2 \end{cases}$, or $\begin{cases} t_1 + t_2 = 1 \\ -1/2\,t_1 + t_2 = 0 \\ 1/2\,t_1 - t_2 = 0 \end{cases}$, to get $[t_1 \quad t_2] = [2/3 \quad 1/3]$.

**3.** $P=\begin{bmatrix}0 & 1\\ 1/4 & 3/4\end{bmatrix}$ is regular since $P^2=\begin{bmatrix}1/4 & 3/4\\ 3/16 & 13/16\end{bmatrix}$ has only positive entries. To find the fixed probability vector, solve $\begin{cases}t_1+t_2 & =1\\ [t_1\ \ t_2]P=[t_1\ \ t_2]\end{cases}$, or $\begin{cases}t_1+t_2=1\\ \frac{1}{4}t_2=t_1\\ t_1+\frac{3}{4}t_2=t_2\end{cases}$, or $\begin{cases}t_1+\quad t_2=1\\ -t_1+1/4\,t_2=0\\ t_1-1/4\,t_2=0\end{cases}$, to get $[t_1\ \ t_2]=[1/5\ \ 4/5]$.

**5.** $P=\begin{bmatrix}1 & 0 & 0\\ 1/4 & 1/2 & 1/4\\ 0 & 1 & 0\end{bmatrix}$ is not regular since every power of $P$ has as its first row $[1\ \ 0\ \ 0]$, so every power of $P$ contains a zero entry.

**7.** First, note that $[1/2\ \ 1/2]$ is a probability vector since its entries are nonnegative and sum to 1. Also, $[1/2\ \ 1/2]P=[(1/2)(1-p)+(1/2)p\ \ \ (1/2)p+(1/2)(1-p)]=[1/2\ \ 1/2]$, so $[1/2\ \ 1/2]$ is the fixed probability vector for $P$.

**9.** With the states being $A=$ "the grocer restocks with brand A," $B=$ "she restocks with brand B" and $C=$ "she restocks with brand C," respectively, the transition matrix is

$$P=\begin{matrix} & A & B & C\\ A & 0.7 & 0.15 & 0.15\\ B & 0.1 & 0.8 & 0.1\\ C & 0.2 & 0.2 & 0.6\end{matrix}$$

. Note that $P$ has only positive entries, so it is regular. To find the fixed probability vector, solve $\begin{cases}t_1+t_2+t_3 & =1\\ [t_1\ \ t_2\ \ t_3]P=[t_1\ \ t_2\ \ t_3]\end{cases}$, or $\begin{cases}t_1+\quad t_2+\quad t_3=1\\ -0.3t_1+0.1t_2+0.2t_3=0\\ 0.15t_1-0.2t_2+0.2t_3=0\\ 0.15t_1+0.1t_2-0.4t_3=0\end{cases}$, to get $[t_1\ \ t_2\ \ t_3]=[4/13\ \ 6/13\ \ 3/13]$. In the long run, $\frac{4}{13}(\approx 30.8\%)$ of the detergent stock will be brand A, $\frac{6}{13}(\approx 46.2\%)$ of the stock will be brand B and $\frac{3}{13}(\approx 23.1\%)$ of the stock will be brand C.

**11.** With states being "votes Conservative," "votes Labour" and "votes Socialist," respectively, the transition matrix (from English father to English sons) is

$$P=\begin{matrix} & C & L & S\\ C & 0.7 & 0.3 & 0\\ L & 0.4 & 0.5 & 0.1\\ S & 0.2 & 0.4 & 0.4\end{matrix}.$$

(Note that, since $P^2$ has only positive entries, $P$ is a regular matrix.) The probability that the grandson of a Labourite will vote Socialist is the 2, 3-entry of $P^2$, or 0.09. To find the fixed probability vector, solve $\begin{cases} t_1+t_2+t_3 &=1 \\ [t_1 \quad t_2 \quad t_3]P=[t_1 \quad t_2 \quad t_3] \end{cases}$, or $\begin{cases} t_1+ \quad t_2+ \quad t_3=1 \\ -0.3t_1+0.4t_2+0.2t_3=0 \\ 0.3t_1-0.5t_2+0.4t_3=0 \\ 0.1t_2-0.6t_3=0 \end{cases}$, to get $[t_1 \quad t_2 \quad t_3]=[26/47 \quad 18/47 \quad 3/47]$. The membership distribution in the long run is $\frac{26}{47}(\approx 55.3\%)$ Conservative, $\frac{18}{47}(\approx 38.3\%)$ Labourite and $\frac{3}{47}(\approx 6.4\%)$ Socialist.

**13.** With the states being "blonde *BL*," "brunette *BR*" and "redheaded *R*," respectively, the transition matrix (from mother to daughter) is $P=\begin{array}{c} \\ BL \\ BR \\ R \end{array}\!\!\begin{array}{c} \begin{array}{ccc} BL & BR & R \end{array} \\ \begin{bmatrix} 0.6 & 0.2 & 0.2 \\ 0.1 & 0.7 & 0.2 \\ 0.4 & 0.2 & 0.4 \end{bmatrix} \end{array}$. The probability that a blond woman is the grandmother of a brunette is the 1, 2-entry of $P^2$, or 0.3. For parts (a) and (b), assume the probability vector for the current distribution of women's hair color is $[0.3 \quad 0.5 \quad 0.2]$.

(a) The distribution of women's hair color after two generations will be $[0.3 \quad 0.5 \quad 0.2]P^2=[0.327 \quad 0.425 \quad 0.248]$. Thus, after two generations 32.7% of the population of women will be blonds, 42.5% will be brunettes and 24.8% will be redheads.

(b) To find the fixed probability vector, solve $\begin{cases} t_1+t_2+t_3 &=1 \\ [t_1 \quad t_2 \quad t_3]P=[t_1 \quad t_2 \quad t_3] \end{cases}$, or $\begin{cases} t_1+ \quad t_2+ \quad t_3=1 \\ -0.4t_1+0.1t_2+0.4t_3=0 \\ 0.2t_1-0.3t_2+0.2t_3=0 \\ 0.2t_1+0.2t_2-0.6t_3=0 \end{cases}$, to get $[t_1 \quad t_2 \quad t_3]=[0.35 \quad 0.4 \quad 0.25]$. In the long run, 35% of the population will be blonds, 40% will be brunettes and 25% will be redheads.

## Technology Exercises

**1.** By successive squaring, $P=\begin{bmatrix} .93 & .07 \\ .01 & .99 \end{bmatrix}$; $P^2=\begin{bmatrix} .8656 & .1344 \\ .0192 & .9808 \end{bmatrix}$; $P^4\approx\begin{bmatrix} .751844 & .248156 \\ .035451 & .964549 \end{bmatrix}$; $P^8\approx\begin{bmatrix} .574067 & .425933 \\ .060848 & .939152 \end{bmatrix}$; ...; $P^{128}\approx\begin{bmatrix} .12502 & .87498 \\ .124997 & .875003 \end{bmatrix}$; $P^{512}\approx\begin{bmatrix} .125 & .875 \\ .125 & .875 \end{bmatrix}$; The fixed probability vector $\mathbf{t}=[0.125 \quad 0.875]=[1/8 \quad 7/8]$.

**3.** $P^8 \approx \begin{bmatrix} .444443 & .222183 & .333374 \\ .444466 & .22226 & .333274 \\ .444433 & .22225 & .333318 \end{bmatrix}$; $P^{32} \approx \begin{bmatrix} .444444 & .222222 & .333333 \\ .444444 & .222222 & .333333 \\ .444444 & .222222 & .333333 \end{bmatrix}$; The fixed probability vector $\mathbf{t} = [0.444444 \quad 0.222222 \quad 0.333333] = [4/9 \quad 2/9 \quad 1/3]$.

**5.** $P^8 \approx \begin{bmatrix} .267647 & .208824 & .205882 & .317647 \\ .267647 & .208824 & .205882 & .317647 \\ .267647 & .208823 & .205882 & .317647 \\ .267647 & .208824 & .205882 & .317647 \end{bmatrix}$;

$P^{16} \approx \begin{bmatrix} .267647 & .208824 & .205882 & .317647 \\ .267647 & .208824 & .205882 & .317647 \\ .267647 & .208824 & .205882 & .317647 \\ .267647 & .208824 & .205882 & .317647 \end{bmatrix}$; The fixed probability vector

$\mathbf{t} = [.267647 \quad .208824 \quad .205882 \quad .317647] = [91/340 \quad 71/340 \quad 7/34 \quad 27/85]$.

## 10.3 Absorbing Markov Chains

**1.** There are no absorbing states, so this Markov chain is *not* absorbing.

**3.** States 1 and 3 are absorbing and it is possible to pass from state 2 to state 1 and state 3, so this Markov chain *is* absorbing.

**5.** State 3 is absorbing but it is not possible to pass from state 1 or state 2 to state 3, so this Markov chain is *not* absorbing.

**7.** Write the states in the order 1, 3, 2, so that the absorbing states (1 & 3) are first, to obtain the transition matrix

| | 1 | 3 | 2 |
|---|---|---|---|
| 1 | 1 | 0 | 0 |
| 3 | 0 | 1 | 0 |
| 2 | 1/8 | 2/8 | 5/8 |

. Then $T = \left[1 - \frac{5}{8}\right]^{-1} = \left[\frac{8}{3}\right]$, $S = [1/8 \quad 2/8]$ and

$T \cdot S = [1/3 \quad 2/3]$.

**9.** (a) The expected number of times a person will have \$3 given that she started with \$1 is given by the 1, 3-entry of $T$, 0.5. If she started with \$2, the expected number of times she will have \$3 is the (2, 3)-entry of $T$, 1.

(b) If she starts with \$3, the expected number of times she will play before absorption is given by the sum of the entries in the third row of $T$, $0.5 + 1 + 1.5 = 3$.

**11.** With the states being 0, 1, 2, 3, for "gambler is broke," "gambler has \$1," "gambler has \$2" and "gambler has \$3," respectively, the transition matrix is

$$P = \begin{array}{c} \\ 0 \\ 1 \\ 2 \\ 3 \end{array} \begin{array}{c} \begin{array}{cccc} 0 & 1 & 2 & 3 \end{array} \\ \begin{bmatrix} 1 & 0 & 0 & 0 \\ 0.6 & 0 & 0.4 & 0 \\ 0 & 0.6 & 0 & 0.4 \\ 0 & 0 & 0 & 1 \end{bmatrix} \end{array}.$$

Reorganizing the states in order 0, 3, 1, 2 (so that the absorbing states are first), the transition matrix becomes

$$\begin{array}{c|cc|cc} & 0 & 3 & 1 & 2 \\ \hline 0: & 1 & 0 & 0 & 0 \\ 3: & 0 & 1 & 0 & 0 \\ \hline 1: & 0.6 & 0 & 0 & 0.4 \\ 2: & 0 & 0.4 & 0.6 & 0 \end{array},$$

so

$$T = \left(\begin{bmatrix} 1 & 0 \\ 0 & 1 \end{bmatrix} - \begin{bmatrix} 0 & 0.4 \\ 0.6 & 0 \end{bmatrix}\right)^{-1} = \begin{bmatrix} 1 & -0.4 \\ -0.6 & 1 \end{bmatrix}^{-1} = \begin{bmatrix} 25/19 & 10/19 \\ 15/19 & 25/19 \end{bmatrix}, \ S = \begin{bmatrix} 0.6 & 0 \\ 0 & 0.4 \end{bmatrix} \text{ and}$$

$$T \cdot S = \begin{array}{c} \\ 1: \\ 2: \end{array} \begin{array}{c} \begin{array}{cc} 0 & 3 \end{array} \\ \begin{bmatrix} 15/19 & 4/19 \\ 9/19 & 10/19 \end{bmatrix} \end{array}.$$

The probability of eventually accumulating \$3 if the gambler starts with \$1 is $\frac{4}{19} \approx 0.2105$. The probability of eventually accumulating \$3 if the gambler starts with \$2 is $\frac{10}{19} \approx 0.5263$.

**13.** With the states being 0, 1, 2, 3, 4, for "gambler is broke," "gambler has \$1000," "gambler has \$2000," "gambler has \$3000" and "gambler has \$4000," respectively, the transition matrix is

$$P = \begin{array}{c} \\ 0 \\ 1 \\ 2 \\ 3 \\ 4 \end{array} \begin{array}{c} \begin{array}{ccccc} 0 & 1 & 2 & 3 & 4 \end{array} \\ \begin{bmatrix} 1 & 0 & 0 & 0 & 0 \\ 0.6 & 0 & 0.4 & 0 & 0 \\ 0.6 & 0 & 0 & 0 & 0.4 \\ 0 & 0 & 0.6 & 0 & 0.4 \\ 0 & 0 & 0 & 0 & 1 \end{bmatrix} \end{array}.$$

Reorganizing the states in the order 0, 4, 1, 2, 3 (so that the absorbing states are first), the transition matrix becomes

$$\begin{array}{c} \\ 0: \\ 4: \\ 1: \\ 2: \\ 3: \end{array}\begin{array}{cc|ccc} 0 & 4 & 1 & 2 & 3 \\ \hline 1 & 0 & 0 & 0 & 0 \\ 0 & 1 & 0 & 0 & 0 \\ \hline 0.6 & 0 & 0 & 0.4 & 0 \\ 0.6 & 0.4 & 0 & 0 & 0 \\ 0 & 0.4 & 0 & 0.6 & 0 \end{array}, \text{ so } T = \begin{bmatrix} 1 & -0.4 & 0 \\ 0 & 1 & 0 \\ 0 & -0.6 & 1 \end{bmatrix}^{-1} = \begin{bmatrix} 1 & 0.4 & 0 \\ 0 & 1 & 0 \\ 0 & 0.6 & 1 \end{bmatrix}, \; S = \begin{bmatrix} 0.6 & 0 \\ 0.6 & 0.4 \\ 0 & 0.4 \end{bmatrix} \text{ and}$$

$$T \cdot S = \begin{array}{c} \\ 1: \\ 2: \\ 3: \end{array}\begin{array}{cc} 0 & 4 \\ \end{array}\begin{bmatrix} 0.84 & 0.16 \\ 0.6 & 0.4 \\ 0.36 & 0.64 \end{bmatrix}.$$

(a) The expected number of wagers placed before the game ends is the sum of the entries in the first row of $T$, or 1.4.
(b) The probability that Colleen is wiped out is 0.84 (the (1, 1)-entry of $T \cdot S$).
(c) The probability that Colleen wins \$4000 is 0.16 (the (1, 2)-entry of $T \cdot S$).

**15.** With the states being 0, 1, 2, 3, 4, for "gambler is broke," "gambler has \$1000," "gambler has \$2000," "gambler has \$3000" and "gambler has \$4000," respectively, the transition matrix is

$$P = \begin{array}{c} \\ 0 \\ 1 \\ 2 \\ 3 \\ 4 \end{array}\begin{array}{ccccc} 0 & 1 & 2 & 3 & 4 \\ 1 & 0 & 0 & 0 & 0 \\ 0.4 & 0 & 0.6 & 0 & 0 \\ 0.4 & 0 & 0 & 0 & 0.6 \\ 0 & 0 & 0.4 & 0 & 0.6 \\ 0 & 0 & 0 & 0 & 1 \end{array}.$$

Reorganizing the states in the order 0, 4, 1, 2, 3 (so that the absorbing states are first), the transition matrix becomes

$$\begin{array}{c} \\ 0: \\ 4: \\ 1: \\ 2: \\ 3: \end{array}\begin{array}{cc|ccc} 0 & 4 & 1 & 2 & 3 \\ \hline 1 & 0 & 0 & 0 & 0 \\ 0 & 1 & 0 & 0 & 0 \\ \hline 0.4 & 0 & 0 & 0.6 & 0 \\ 0.4 & 0.6 & 0 & 0 & 0 \\ 0 & 0.6 & 0 & 0.4 & 0 \end{array}, \text{ so } T = \begin{bmatrix} 1 & -0.6 & 0 \\ 0 & 1 & 0 \\ 0 & -0.4 & 1 \end{bmatrix}^{-1} = \begin{bmatrix} 1 & 0.6 & 0 \\ 0 & 1 & 0 \\ 0 & 0.4 & 1 \end{bmatrix}, \; S = \begin{bmatrix} 0.4 & 0 \\ 0.4 & 0.6 \\ 0 & 0.6 \end{bmatrix} \text{ and}$$

$$T \cdot S = \begin{array}{c} \\ 1: \\ 2: \\ 3: \end{array}\begin{array}{cc} 0 & 4 \\ \end{array}\begin{bmatrix} 0.64 & 0.36 \\ 0.4 & 0.6 \\ 0.16 & 0.84 \end{bmatrix}.$$

(a) The expected number of wagers placed before the game ends is the sum of the entries in the first row of $T$, or 1.6.
(b) The probability that Colleen is wiped out is 0.64 (the (1, 1)-entry of $T \cdot S$).
(c) The probability that Colleen wins \$4000 is 0.36 (the (1, 2)-entry of $T \cdot S$).

**17.** Modify the model so that states $I$ and $D$ are absorbing states to obtain the transition matrix

$$P'' = \begin{array}{c} \\ I: \\ D: \\ N: \end{array} \begin{array}{|ccc|} I & D & N \\ \hline 1 & 0 & 0 \\ 0 & 1 & 0 \\ \hline 0.079 & 0.064 & 0.857 \\ \hline \end{array}$$

. The fundamental matrix of this absorbing Markov chain is $T = ([1]-[0.857])^{-1} \approx [6.993]$. The average time spent in state $N$ having begun in state $N$ is 6.993 days.

## Technology Exercises

**1.** $I_2 = \begin{bmatrix} 1 & 0 \\ 0 & 1 \end{bmatrix}$, $S = \begin{bmatrix} 0 & 0 \\ 0 & 0.2 \\ 0.75 & 0.25 \end{bmatrix}$, $Q = \begin{bmatrix} 0.05 & 0.95 & 0 \\ 0 & 0.5 & 0.3 \\ 0 & 0 & 0 \end{bmatrix}$,

$T = \begin{bmatrix} 20/19 & 2 & 3/5 \\ 0 & 2 & 3/5 \\ 0 & 0 & 1 \end{bmatrix} \approx \begin{bmatrix} 1.05263 & 2 & 0.6 \\ 0 & 2 & 0.6 \\ 0 & 0 & 1 \end{bmatrix}$, $T \cdot S = \begin{bmatrix} 0.45 & 0.55 \\ 0.45 & 0.55 \\ 0.75 & 0.25 \end{bmatrix}$

## 10.4 Two-Person Games

**1.** Katy's payoff matrix: $\begin{array}{r} \\ 1\text{ finger}: \\ 2\text{ fingers}: \end{array} \begin{array}{c} \begin{array}{cc} 1 & 2 \end{array} \\ \begin{bmatrix} -1 & 1 \\ 1 & -1 \end{bmatrix} \end{array}$, where the entries are in dimes.

**3.** Katy's payoff matrix: $\begin{array}{r} \\ 1: \\ 4: \\ 7: \end{array} \begin{array}{c} \begin{array}{ccc} 1 & 4 & 7 \end{array} \\ \begin{bmatrix} -2 & 5 & -8 \\ 5 & -8 & 11 \\ -8 & 11 & -14 \end{bmatrix} \end{array}$, where the entries are in dimes.

**5.** The game is strictly determined: –2 is the smallest entry in its row and it is the largest entry in its column. The value of the game is –2.

**7.** The game is strictly determined: the (1, 2)-entry, 3 is the smallest entry in its row and it is the largest entry in its column. The value of the game is 3.

**9.** The game is not strictly determined.

**11.** The game is strictly determined: each of the 2's in the third row is the smallest entry in its row and is the largest entry in its column. The value of the game is 2.

**13.** The game is not strictly determined.

**15.** The smallest entry in the first row is either $a$ or 3. If it is $a$, then $a \le 3$ and the game will be strictly determined with value $a$ if $0 \le a$. If the smallest entry in the first row is 3, then $3 \le a$, and the game will be strictly determined with value 3 only if $a \le 3$, so $a = 3$. In either case, the game will be strictly determined with value $a$ in the first row only if $0 \le a \le 3$.

The smallest entry of the second row is either $a$ or –9. If it is $a$, then $a \le -9$, but then $a$ is not largest in the second column. Also, –9 is not the largest entry in the third column. Thus, if the game is strictly determined, its value can't come from the second row.
The smallest entry in the third row is either –5 or $a$. If it is $a$, then $a \le -5$, but then $a$ is not largest in the third column. Also, –5 is not the largest entry in the first column. Thus, if the game is strictly determined, its value can't come from the third row.
Thus, if the game is strictly determined then the value, $a$, comes from the first row, and this happens only for $0 \le a \le 3$.

**17.** Suppose the game is strictly determined and the value for the game comes from the first row. If $a < 0$ then $a$ is smallest in its row but not largest in its column. If $a = 0$ then the value of the game is 0 (from the (1, 1)-entry.) If $a > 0$ then 0 is smallest in the first row and largest in the second column only if $b \le 0$. Thus, the game is strictly determined and the value for the game comes from the first row when ($a = 0$) or ($a > 0$ and $b \le 0$).
Suppose the game is strictly determined and the value for the game comes from the second row. If $b < 0$, then $b$ is smallest in its row but not largest in its column. If $b = 0$, then the value of the game is 0 (from the (2, 2)-entry). If $b > 0$ then 0 is smallest in the second row and largest in the first column only if $a \le 0$. Thus, the game is strictly determined and the value for the game comes from the second row when ($b = 0$) or ($b > 0$ and $a \le 0$.)
The game is strictly determined if ($a = 0$) or ($a > 0$ and $b \le 0$) or ($b = 0$) or ($b > 0$ and $a \le 0$.) More compactly, the game is strictly determined if $ab \le 0$.

## 10.5 Mixed Strategies

**1.** $E = PAQ = \begin{bmatrix} 0.3 & 0.7 \end{bmatrix} \begin{bmatrix} 6 & 0 \\ -2 & 3 \end{bmatrix} \begin{bmatrix} 0.4 \\ 0.6 \end{bmatrix} = 1.42$

**3.** $E = \begin{bmatrix} 1/2 & 1/2 \end{bmatrix} \begin{bmatrix} 4 & 0 \\ 2 & 3 \end{bmatrix} \begin{bmatrix} 1/2 \\ 1/2 \end{bmatrix} = \frac{9}{4} = 2.25$

**5.** $E = \begin{bmatrix} 1/4 & 3/4 \end{bmatrix} \begin{bmatrix} 4 & 0 \\ 2 & 3 \end{bmatrix} \begin{bmatrix} 1/2 \\ 1/2 \end{bmatrix} = \frac{19}{8} = 2.375$

**7.** $E = \begin{bmatrix} 2/3 & 1/3 \end{bmatrix} \begin{bmatrix} 4 & 0 \\ -3 & 6 \end{bmatrix} \begin{bmatrix} 1/3 \\ 2/3 \end{bmatrix} = \frac{17}{9} \approx 1.889$

**9.** $E = \begin{bmatrix} 1/3 & 1/3 & 1/3 \end{bmatrix} \begin{bmatrix} 1 & 0 & 0 \\ 0 & 1 & 0 \\ 0 & 0 & 1 \end{bmatrix} \begin{bmatrix} 1/3 \\ 1/3 \\ 1/3 \end{bmatrix} = \frac{1}{3} \approx 0.333$

**11.** If the game is not strictly determined then none of the entries $a_{11}$, $a_{12}$, $a_{21}$, $a_{22}$ can be a saddle point. Comparing $a_{11}$ and $a_{12}$, either $a_{11} < a_{12}$, $a_{11} = a_{12}$, or $a_{11} > a_{12}$. If $a_{11} < a_{12}$ then $a_{11} < a_{21}$, for otherwise $a_{11}$ would be a saddle point. Then $a_{21} > a_{22}$, else $a_{21}$ would be a saddle point. Then $a_{12} > a_{22}$ so that $a_{22}$ won't be a saddle point. Thus, if $a_{11} < a_{12}$ and the game is not strictly determined, then the inequalities listed in (b) follow. Note that $a_{11} \neq a_{12}$ for otherwise there would be a saddle point in the matrix. Finally, if $a_{11} > a_{12}$, then $a_{12} < a_{22}$, as $a_{12}$ isn't a saddle point. Then $a_{21} < a_{22}$ and $a_{11} > a_{21}$ to prevent $a_{22}$ and $a_{21}$ from being saddle points. Thus, if $a_{11} > a_{12}$ and the game is not strictly determined, then the inequalities in (a) follow.

## 10.6 Optimal Strategy in Two-Person Zero-Sum Games with 2 × 2 Matrices

**1.** Suppose Player I chooses row 1 with probability $p$ (and row 2 with probability $1-p$). If Player II chooses column 1, then Player I expects to earn $E_I = (1)p + 4(1-p) = 4 - 3p$. If Player II chooses column 2, then Player I expects to earn $E_I = (2)p + (1-p) = p + 1$. The graphs of $E_I = 4 - 3p$ and $E_I = p + 1$ intersect at $p = \frac{3}{4}$, so $P = \begin{bmatrix} \frac{3}{4} & \frac{1}{4} \end{bmatrix}$.

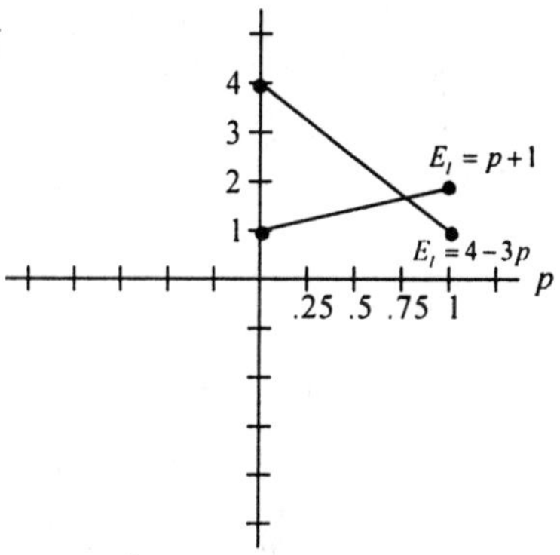


The optimal strategy for Player I is to play row 1 with probability $\frac{3}{4}$ and to play row 2 with probability $\frac{1}{4}$.

$\left(\text{Check: from formula (3), } p_1 = \frac{1-4}{1+1-2-4} = \frac{-3}{-4} = \frac{3}{4} \text{ and } p_2 = \frac{1-2}{1+1-2-4} = \frac{-1}{-4} = \frac{1}{4}.\right)$

Suppose Player II chooses column 1 with probability $q$ (and column 2 with probability $1-q$.) If Player I chooses row 1, then Player II expects to earn $E_{II} = (-1)q + (-2)(1-q)$ $= q - 2$. If Player I chooses row 2, then Player II expects to earn $E_{II} = (-4)q + (-1)(1-q) = -3q - 1$. The graphs of $E_{II} = q - 2$ and $E_{II} = -3q - 1$ intersect at $q = \frac{1}{4}$, so $Q = \begin{bmatrix} 1/4 \\ 3/4 \end{bmatrix}$. The optimal strategy for Player II is to play

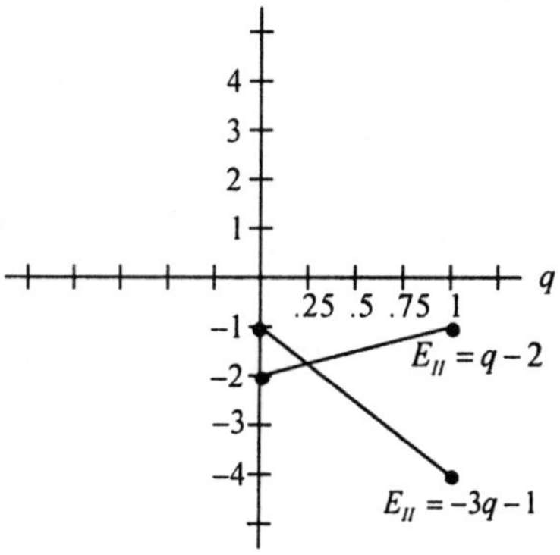

column 1 with probability $\frac{1}{4}$ and column 2 with probability $\frac{3}{4}$.

$\left(\text{Check: from formula (4), } q_1 = \frac{1-2}{1+1-2-4} = \frac{-1}{-4} = \frac{1}{4} \text{ and } q_2 = \frac{1-4}{1+1-2-4} = \frac{-3}{-4} = \frac{3}{4}.\right)$

The expected payoff of the game is $E = [3/4 \quad 1/4]\begin{bmatrix} 1 & 2 \\ 4 & 1 \end{bmatrix}\begin{bmatrix} 1/4 \\ 3/4 \end{bmatrix} = \frac{7}{4}$.

**3.** Suppose Player I chooses row 1 with probability $p$ (and row 2 with probability $1-p$). If Player II chooses column 1, then Player I expects to earn $E_I = (-3)p + (1)(1-p) = 1-4p$. If Player II chooses column 2, then Player I expects to earn $E_I = (2)p + (0)(1-p) = 2p$. The graphs of $E_I = 1-4p$ and $E_I = 2p$ intersect at $p = \frac{1}{6}$, so $P = [1/6 \quad 5/6]$.

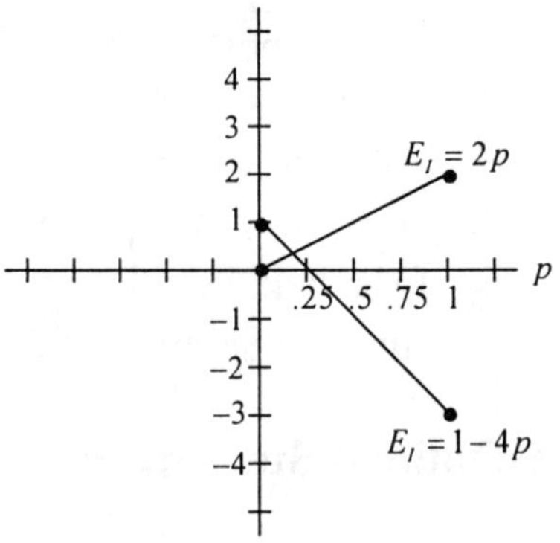


The optimal strategy for Player I is to play row 1 with probability $\frac{1}{6}$ and to play row 2 with probability $\frac{5}{6}$.

$\left(\text{Check: from formula (3), } p_1 = \frac{0-1}{-3+0-2-1} = \frac{-1}{-6} = \frac{1}{6} \text{ and } p_2 = \frac{-3-2}{-3+0-2-1} = \frac{-5}{-6} = \frac{5}{6}.\right)$

Suppose Player II chooses column 1 with probability $q$ (and column 2 with probability $1-q$.) If Player I chooses row 1, then Player II expects to earn $E_{II} = (3)q + (-2)(1-q) = 5q-2$. If Player I chooses row 2, then Player II expects to earn $E_{II} = (-1)q + (0)(1-q) = -q$. The graphs of $E_{II} = 5q-2$ and $E_{II} = -q$ intersect at $q = \frac{1}{3}$, so $Q = \begin{bmatrix} 1/3 \\ 2/3 \end{bmatrix}$.

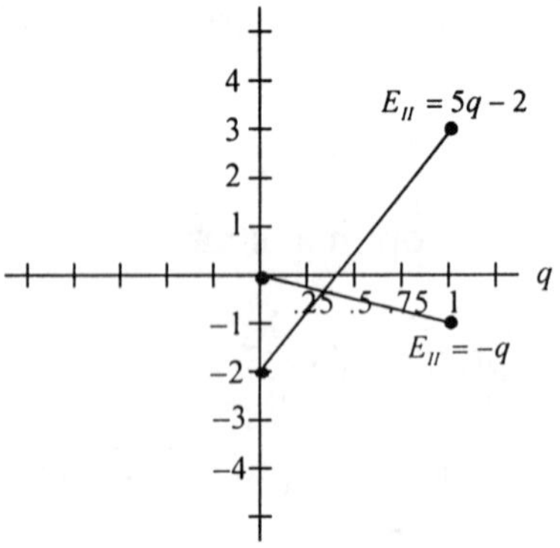


The optimal strategy for Player II is to play column 1 with probability $\frac{1}{3}$ and column 2 with probability $\frac{2}{3}$.

$\left(\text{Check: from formula (4), } q_1 = \frac{0-2}{-3+0-2-1} = \frac{-2}{-6} = \frac{1}{3} \text{ and } q_2 = \frac{-3-1}{-3+0-2-1} = \frac{-4}{-6} = \frac{2}{3}.\right)$

The expected payoff of the game is $E = [1/6 \quad 5/6]\begin{bmatrix} -3 & 2 \\ 1 & 0 \end{bmatrix}\begin{bmatrix} 1/3 \\ 2/3 \end{bmatrix} = \frac{1}{3}$.

**5.** Suppose Player I chooses row 1 with probability $p$ (and row 2 with probability $1-p$). If Player II chooses column 1, then Player I expects to earn $E_I = (2)p + (-1)(1-p) = 3p - 1$. If Player II chooses column 2, then Player I expects to earn $E_I = (-1)p + (4)(1-p) = 4 - 5p$. The graphs of $E_I = 3p - 1$ and $E_I = 4 - 5p$ intersect at $p = \frac{5}{8}$, so $P = [5/8 \quad 3/8]$.

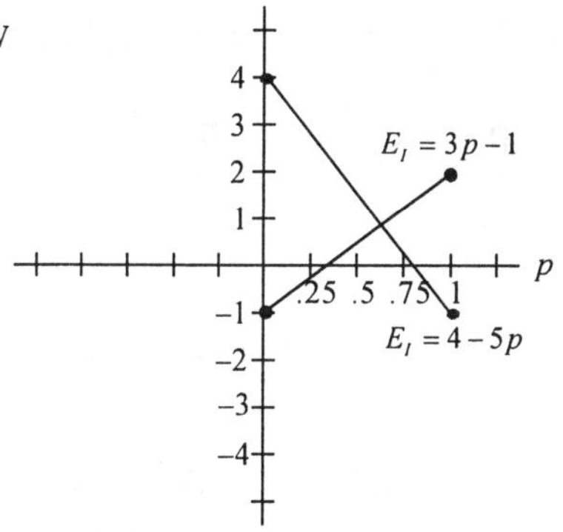


The optimal strategy for Player I is to play row 1 with probability $\frac{5}{8}$ and to play row 2 with probability $\frac{3}{8}$.

$$\left(\text{Check : from formula (3), } p_1 = \frac{4-(-1)}{2+4-(-1)-(-1)} = \frac{5}{8} \text{ and } p_2 = \frac{2-(-1)}{2+4-(-1)-(-1)} = \frac{3}{8}.\right)$$

Suppose Player II chooses column 1 with probability $q$ (and column 2 with probability $1-q$). If Player I chooses row 1, then Player II expects to earn $E_{II} = (-2)q + (1)(1-q) = 1 - 3q$. If Player I chooses row 2, then Player II expects to earn $E_{II} = (1)q + (-4)(1-q) = 5q - 4$. The graphs of $E_{II} = 1 - 3q$ and $E_{II} = 5q - 4$ intersect at $q = \frac{5}{8}$, so $Q = \begin{bmatrix} 5/8 \\ 3/8 \end{bmatrix}$.

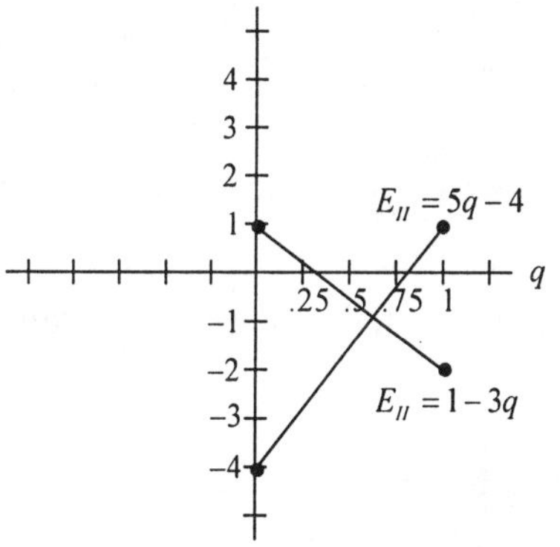


The optimal strategy for Player II is to play column 1 with probability $\frac{5}{8}$ and column 2 with probability $\frac{3}{8}$.

$$\left(\text{Check : from formula (4), } q_1 = \frac{4-(-1)}{2+4-(-1)-(-1)} = \frac{5}{8} \text{ and } q_2 = \frac{2-(-1)}{2+4-(-1)-(-1)} = \frac{3}{8}.\right)$$

The expected payoff of the game is $E = [5/8 \quad 3/8]\begin{bmatrix} 2 & -1 \\ -1 & 4 \end{bmatrix}\begin{bmatrix} 5/8 \\ 3/8 \end{bmatrix} = \frac{7}{8}$.

**7.** From formula (3), $p_1 = \frac{3-0}{4+3-(-1)-0} = \frac{3}{8}$ and $p_2 = \frac{4-(-1)}{4+3-(-1)-0} = \frac{5}{8}$. The Democrat should spend $\frac{3}{8}(= 37.5\%)$ of her/his time on domestic issues and $\frac{5}{8}(= 62.5\%)$ on foreign issues. From formula (4), $q_1 = \frac{3-(-1)}{4+3-(-1)-0} = \frac{4}{8} = \frac{1}{2}$ and $q_2 = \frac{4-0}{4+3-(-1)-0} = \frac{4}{8} = \frac{1}{2}$.

The Republican should divide his/her time evenly between the two issues. The expected payoff of the game is $\begin{bmatrix}3/8 & 5/8\end{bmatrix}\begin{bmatrix}4 & -1\\0 & 3\end{bmatrix}\begin{bmatrix}1/2\\1/2\end{bmatrix}=\frac{3}{2}$, favoring the Democratic candidate.

**9.** The matrix for this game is $\begin{matrix} & D & HU \\ \text{deserted exit } (D): & -100 & 30 \\ \text{heavily used exit } (HU): & 10 & -2\end{matrix}$, where the spy chooses row strategies and his opponent chooses column strategies. From formula (3), $p_1=\frac{(-2)-10}{-100+(-2)-30-10}=\frac{-12}{-142}=\frac{6}{71}$ and $p_2=\frac{-100-30}{-100+(-2)-30-10}=\frac{-130}{-142}=\frac{65}{71}$.

The spy should choose the deserted exit $\frac{6}{71}(\approx 8.5\%)$ of the time and the heavily used exit $\frac{65}{71}(\approx 91.5\%)$ of the time. From formula (4), $q_1=\frac{(-2)-30}{-100+(-2)-30-10}=\frac{-32}{-142}=\frac{16}{71}$ and $q_2=\frac{-100-10}{-100+(-2)-30-10}=\frac{-110}{-142}=\frac{55}{71}$. The spy's opponent should choose the deserted exit $\frac{16}{71}(\approx 22.5\%)$ of the time and the heavily used exit $\frac{55}{71}(\approx 77.5\%)$ of the time. The expected payoff of the game is $\begin{bmatrix}6/71 & 65/71\end{bmatrix}\begin{bmatrix}-100 & 30\\10 & -2\end{bmatrix}\begin{bmatrix}16/71\\55/71\end{bmatrix}=\frac{50}{71}$, favoring the spy.

## Chapter 10 Review

### True or False

**1.** False **3.** False **5.** True **7.** False

### Fill in the Blank

**1.** $m$ **3.** $v^{(k)}=v^{(0)}P^k$ **5.** payoff

### Review Exercises

**1.** (a) Solve $\begin{cases} t_1+t_2=1\\ \begin{bmatrix}t_1 & t_2\end{bmatrix}P=\begin{bmatrix}t_1 & t_2\end{bmatrix}\end{cases}$, or $\begin{cases} t_1+t_2=1\\ -3/4t_1+1/2t_2=0\\ 3/4t_1-1/2t_2=0\end{cases}$, to get $[t_1\ \ t_2]=[2/5\ \ 3/5]$.

(b) Solve $\begin{cases} t_1+t_2=1\\ \begin{bmatrix}t_1 & t_2\end{bmatrix}P=\begin{bmatrix}t_1 & t_2\end{bmatrix}\end{cases}$, or $\begin{cases} t_1+t_2=1\\ -2/3t_1+2/3t_2=0\\ 2/3t_1-2/3t_2=0\end{cases}$, to get $[t_1\ \ t_2]=[1/2\ \ 1/2]$.

(c) Solve $[t_1 \quad t_2 \quad t_3] \cdot P = [t_1 \quad t_2 \quad t_3]$, where $t_1 + t_2 + t_3 = 1$. I.e., solve the system

$$\begin{cases} t_1 + t_2 + t_3 = 1 \\ 3t_1 - 6t_2 - 4t_3 = 0 \\ -t_1 + 9t_2 - 2t_3 = 0 \\ -2t_1 - 3t_2 + 6t_3 = 0 \end{cases}$$

to get $[t_1 \quad t_2 \quad t_3] = [48/79 \quad 10/79 \quad 21/79] \approx [0.607595 \quad 0.126582 \quad 0.265823]$

**3.** With the states corresponding to purchases at distributors A, B and C, respectively, the transition matrix $P$ is

$$P = \begin{matrix} & \begin{matrix} A & B & C \end{matrix} \\ \begin{matrix} A \\ B \\ C \end{matrix} & \begin{bmatrix} 0.5 & 0.2 & 0.3 \\ 0.4 & 0.4 & 0.2 \\ 0.5 & 0.25 & 0.25 \end{bmatrix} \end{matrix}.$$

The market share after 2 years is given by $[1/3 \quad 1/3 \quad 1/3]P^2 = [283/600 \quad 323/1200 \quad 311/1200]$. After 2 years, A holds $283/600 \approx 47.17\%$, B holds $323/1200 \approx 26.92\%$ and C holds $311/1200 \approx 25.92\%$ of the beer market. To find the fixed probability vector, solve $\begin{cases} t_1 + t_2 + t_3 = 1 \\ [t_1 \quad t_2 \quad t_3]P = [t_1 \quad t_2 \quad t_3] \end{cases}$, or

$$\begin{cases} t_1 + t_2 + t_3 = 1 \\ -0.5t_1 + 0.4t_2 + 0.5t_3 = 0 \\ 0.2t_1 - 0.6t_2 + 0.25t_3 = 0 \\ 0.3t_1 + 0.2t_2 - 0.75t_3 = 0 \end{cases},$$

to get $[t_1 \quad t_2 \quad t_3] = [80/169 \quad 45/169 \quad 44/169]$. The distribution of the beer market in the long run is a $\frac{80}{169} \approx 47.34\%$ market share for A, a $\frac{45}{169} \approx 26.63\%$ market share for B and a $\frac{44}{169} \approx 26.04\%$ market share for C.

**5.** With the states being the university, $U_1$, $U_2$, or $U_3$, respectively, at which the book representative sells in a month, the transition matrix $P$ is

$$P = \begin{matrix} & \begin{matrix} U_1 & U_2 & U_3 \end{matrix} \\ \begin{matrix} U_1 \\ U_2 \\ U_3 \end{matrix} & \begin{bmatrix} 0 & 1 & 0 \\ 0.75 & 0 & 0.25 \\ 0.75 & 0.25 & 0 \end{bmatrix} \end{matrix}.$$

(Note that $P^4$ has only positive entries, so $P$ is a regular matrix.) To find the fixed probability vector, solve $\begin{cases} t_1 + t_2 + t_3 = 1 \\ [t_1 \quad t_2 \quad t_3]P = [t_1 \quad t_2 \quad t_3] \end{cases}$, or

$$\begin{cases} t_1 + t_2 + t_3 = 1 \\ -t_1 + 0.75t_2 + 0.75t_3 = 0 \\ t_1 - t_2 + 0.25t_3 = 0 \\ 0.25t_2 - t_3 = 0 \end{cases},$$

to get

$[t_1 \quad t_2 \quad t_3] = [3/7 \quad 16/35 \quad 4/35]$. In the long run, she will sell at $U_1 \frac{3}{7} \approx 42.86\%$ of the time, at $U_2 \frac{16}{35} \approx 45.71\%$ of the time, and at $U_3 \frac{4}{35} \approx 11.43\%$ of the time.

**7.** Labeling the states by the gambler's assets in the order \$0, \$5, \$1, \$2, \$3, \$4 (so that the absorbing states are first), the transition matrix is

| | 0 | 5 | 1 | 2 | 3 | 4 |
|---|---|---|---|---|---|---|
| \$0 : | 1 | 0 | 0 | 0 | 0 | 0 |
| \$5 : | 0 | 1 | 0 | 0 | 0 | 0 |
| \$1 : | 0.55 | 0 | 0 | 0.45 | 0 | 0 |
| \$2 : | 0 | 0 | 0.55 | 0 | 0.45 | 0 |
| \$3 : | 0 | 0 | 0 | 0.55 | 0 | 0.45 |
| \$4 : | 0 | 0.45 | 0 | 0 | 0.55 | 0 |

(a) Since the gambler starts with \$2, the expended number of times he is in each of the nonabsorbing states is given by the second row of the fundamental matrix $T$.

| Nonabsorbing state \$$x$ | \$1 | \$2 | \$3 | \$4 |
|---|---|---|---|---|
| Expected # times in state \$$x$ | 1.298406 | 2.360738 | 1.411737 | 0.635282 |

(b) The expected number of bets before absorption is the sum of the entries in the second row of $T$: 5.706163.

(c) Here, $S = \begin{bmatrix} 0.55 & 0 \\ 0 & 0 \\ 0 & 0 \\ 0 & 0.45 \end{bmatrix}$. The probability that the gambler loses his money is the (2, 1)-entry of the matrix product $T \cdot S$, which is 0.714123. The probability that the gambler wins \$5 is $1 - 0.74123 = 0.285877$. (This can also be found by inspecting the (2, 2)-entry of $T \cdot S$.)

**9.** (a) $E = [1/3 \quad 2/3]\begin{bmatrix} -1 & 1 \\ 1 & -1 \end{bmatrix}\begin{bmatrix} 1 \\ 0 \end{bmatrix} = -\frac{1}{3}$ (b) $E = [0 \quad 1]\begin{bmatrix} -1 & 1 \\ 1 & -1 \end{bmatrix}\begin{bmatrix} 1/2 \\ 1/2 \end{bmatrix} = 0$

(c) $E = [1/2 \quad 1/2]\begin{bmatrix} -1 & 1 \\ 1 & -1 \end{bmatrix}\begin{bmatrix} 1/2 \\ 1/2 \end{bmatrix} = 0$

**11.** (a) Using formula (3) of section 10.6, the row player's optimal strategy is $P = \frac{1}{10 + 20 - 5 - (-5)}[20 - (-5) \quad 10 - 5] = [5/6 \quad 1/6]$. The investor should allocate $\frac{5}{6}$, or approximately 83.3%, of his investment in A and $\frac{1}{6}$, or approximately 16.7%, of his investment in B.

(b) The expected payoff of this game is $E = \dfrac{(10)(20)-(5)(-5)}{30} = \dfrac{15}{2} = 7.5$.

(c) The investor expects to gain 7.5% on his investment if he uses the optimal investment strategy.

**13.** Using formula (3) of Section10.6,

$$P = \frac{1}{1.5+1-(-1)-(-1)}[1-(-1) \quad 1.5-(-1)] = [4/9 \quad 5/9].$$

The expected payoff of the game is $E = \dfrac{(1.5)(1)-(-1)(-1)}{4.5} = \dfrac{1}{9}$.

# Chapter 11

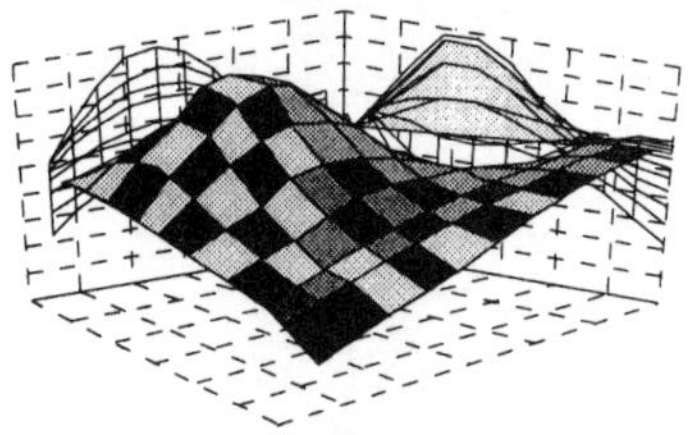

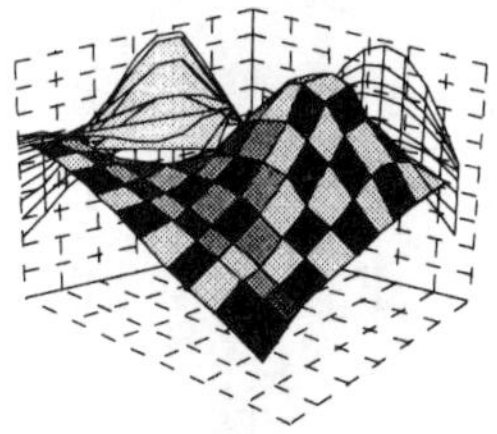

# Logic and Logic Circuits

## 11.1 Propositions

**1.** Proposition

**3.** Not a proposition

**5.** Proposition

**7.** Proposition

**9.** A fox is not an animal.

**11.** I am not buying stocks and bonds.

**13.** Someone wants to buy my house.

**15.** Every person has a car.

**17.** John is an economics major or a sociology major.

**19.** John is an economics major and a sociology major.

**21.** John is not an economics major or he is not a sociology major.

**23.** John is not an economics major or he is a sociology major.

## 11.2 Truth Tables

**1.**

| $p$ | $q$ | $\sim q$ | $p \vee \sim q$ |
|---|---|---|---|
| T | T | F | T |
| T | F | T | T |
| F | T | F | F |
| F | F | T | T |

**3.**

| $p$ | $q$ | $\sim p$ | $\sim q$ | $\sim p \wedge \sim q$ |
|---|---|---|---|---|
| T | T | F | F | F |
| T | F | F | T | F |
| F | T | T | F | F |
| F | F | T | T | T |

**5.**

| $p$ | $q$ | $\sim p$ | $\sim p \wedge q$ | $\sim(\sim p \wedge q)$ |
|---|---|---|---|---|
| T | T | F | F | T |
| T | F | F | F | T |
| F | T | T | T | F |
| F | F | T | F | T |

**7.**

| $p$ | $q$ | $\sim p$ | $\sim q$ | $\sim p \vee \sim q$ | $\sim(\sim p \vee \sim q)$ |
|---|---|---|---|---|---|
| T | T | F | F | F | T |
| T | F | F | T | T | F |
| F | T | T | F | T | F |
| F | F | T | T | T | F |

**9.**

| $p$ | $q$ | $\sim q$ | $p \vee \sim q$ | $(p \vee \sim q) \wedge p$ |
|---|---|---|---|---|
| T | T | F | T | T |
| T | F | T | T | T |
| F | T | F | F | F |
| F | F | T | T | F |

**11.**

| $p$ | $q$ | $\sim q$ | $p \underline{\vee} q$ | $p \wedge \sim q$ | $(p \underline{\vee} q) \wedge (p \wedge \sim q)$ |
|---|---|---|---|---|---|
| T | T | F | F | F | F |
| T | F | T | T | T | T |
| T | T | F | T | F | F |
| F | F | T | F | F | F |

**13.**

| $p$ | $q$ | $\sim p$ | $\sim q$ | $p \wedge q$ | $\sim p \wedge \sim q$ | $(p \wedge q) \vee (\sim p \wedge \sim q)$ |
|---|---|---|---|---|---|---|
| T | T | F | F | T | F | T |
| T | F | F | T | F | F | F |
| F | T | T | F | F | F | F |
| F | F | T | T | F | T | T |

**15.**

| $p$ | $q$ | $r$ | $\sim q$ | $p \wedge \sim q$ | $(p \wedge \sim q) \underline{\vee} r$ |
|---|---|---|---|---|---|
| T | T | T | F | F | T |
| T | T | F | F | F | F |
| T | F | T | T | T | F |
| T | F | F | T | T | T |
| F | T | T | F | F | T |
| F | T | F | F | F | F |
| F | F | T | T | F | T |
| F | F | F | T | F | F |

**17.**

| $p$ | $p \wedge p$ | $p \vee p$ |
|---|---|---|
| T | T | T |
| F | F | F |

Since each column is the same, $p \equiv p \wedge p \equiv p \vee p$.

**19.**

| $p$ | $q$ | $r$ | $p \wedge q$ | $q \wedge r$ | $(p \wedge q) \wedge r$ | $p \wedge (q \wedge r)$ |
|---|---|---|---|---|---|---|
| T | T | T | T | T | T | T |
| T | T | F | T | F | F | F |
| T | F | T | F | F | F | F |
| T | F | F | F | F | F | F |
| F | T | T | F | T | F | F |
| F | T | F | F | F | F | F |
| F | F | T | F | F | F | F |
| F | F | F | F | F | F | F |

The last two columns are the same, so $(p \wedge q) \wedge r \equiv p \wedge (q \wedge r)$.

| $p$ | $q$ | $r$ | $p \vee q$ | $q \vee r$ | $(p \vee q) \vee r$ | $p \vee (q \vee r)$ |
|---|---|---|---|---|---|---|
| T | T | T | T | T | T | T |
| T | T | F | T | T | T | T |
| T | F | T | T | T | T | T |
| T | F | F | T | F | T | T |
| F | T | T | T | T | T | T |
| F | T | F | T | T | T | T |
| F | F | T | F | T | T | T |
| F | F | F | F | F | F | F |

The last two columns are the same, so $(p \vee q) \vee r \equiv p \vee (q \vee r)$.

**21.**

| 1 | 2 | 3 | 4 | 5 | 6 |
|---|---|---|---|---|---|
| $p$ | $q$ | $p \vee q$ | $p \wedge q$ | $p \wedge (p \vee q)$ | $p \vee (p \wedge q)$ |
| T | T | T | T | T | T |
| T | F | T | F | T | T |
| F | T | T | F | F | F |
| F | F | F | F | F | F |

Since columns 1 and 5 are the same, $p \equiv p \wedge (p \vee q)$. Since columns 1 and 6 are the same, $p \equiv p \vee (p \wedge q)$.

**23.**

| 1 | 2 | 3 | 4 | 5 |
|---|---|---|---|---|
| $p$ | $q$ | $\sim q$ | $\sim q \vee q$ | $p \wedge (\sim q \vee q)$ |
| T | T | F | T | T |
| T | F | T | T | T |
| F | T | F | T | F |
| F | F | T | T | F |

Since columns 1 and 5 are the same, $p \equiv p \wedge (\sim q \vee q)$.

**25.**

| 1 | 2 | 3 |
|---|---|---|
| $p$ | $\sim p$ | $\sim (\sim p)$ |
| T | F | T |
| F | T | F |

Since columns 1 and 3 are the same, $p \equiv \sim (\sim p)$.

**27.**

| $p$ | $q$ | $\sim p$ | $q \wedge (\sim p)$ | $p \wedge (q \wedge \sim p)$ |
|---|---|---|---|---|
| T | T | F | F | F |
| T | F | F | F | F |
| F | T | T | T | F |
| F | F | T | F | F |

**29.**

| $p$ | $q$ | $\sim p$ | $\sim q$ | $p \wedge q$ | $\sim p \wedge \sim q$ | $(p \wedge q) \vee (\sim p \wedge \sim q)$ | $[(p \wedge q) \vee (\sim p \wedge \sim q)] \wedge p$ |
|---|---|---|---|---|---|---|---|
| T | T | F | F | T | F | T | T |
| T | F | F | T | F | F | F | F |
| F | T | T | F | F | F | F | F |
| F | F | T | T | F | T | T | F |

**31.** Smith is an ex-convict and he is an ex-convict ≡ Smith is an ex-convict or he is an ex-convict ≡ Smith is an ex-convict.

**33.** "It is not true that Smith is an ex-convict or rehabilitated" means the same as the statement "Smith is not an ex-convict and he is not rehabilitated."
"It is not true that Smith is an ex-convict and he is rehabilitated" means the same as the statement "Smith is not an ex-convict or he is not rehabilitated."

**35.** $(p \wedge q) \vee r \equiv r \vee (p \wedge q) \equiv (r \vee p) \wedge (r \vee q) \equiv (p \vee r) \wedge (r \vee q) \equiv (p \vee r) \wedge (q \vee r)$

**37.** Let $p \equiv$ Michael will sell his car, $q \equiv$ Michael will buy a bicycle and $r \equiv$ Michael will rent a truck. Then $a \equiv p \wedge (q \vee r) \equiv (p \wedge q) \vee (p \wedge r)$, $b \equiv (p \wedge q) \vee r$. Use a truth table to show that $b$ is true and $a$ is false if Michael rents a truck and does not sell his car.

**39.** Katy is not a good volleyball player or she is conceited.

## 11.3 Implications; The Biconditional Connective; Tautologies

**1.** $\sim p \Rightarrow q$; Converse: $q \Rightarrow \sim p$; Contrapositive: $\sim q \Rightarrow p$; Inverse: $p \Rightarrow \sim q$

**3.** $\sim q \Rightarrow \sim p$; Converse: $\sim p \Rightarrow \sim q$; Contrapositive: $p \Rightarrow q$; Inverse: $q \Rightarrow p$

**5.** If it is raining, the grass is wet. Converse: If the grass is wet, it is raining. Contrapositive: If the grass is not wet, it is not raining. Inverse: If it is not raining, the grass is not wet.

**7.** "If it is not raining, it is not cloudy," Converse: If it is not cloudy, it is not raining. Contrapositive: If it is cloudy, it is raining. Inverse: If it is raining, it is cloudy.

**9.** If it is raining, it is cloudy. Converse: If it is cloudy, it is raining. Contrapositive: If it is not cloudy, it is not raining. Inverse: If it is not raining, it is not cloudy.

**11.** (a) If Jack studies psychology, then Mary studies sociology.
(b) If Mary studies sociology, then Jack studies psychology.
(c) If Jack does not study psychology, then Mary studies sociology.

**13.** (a)

| $p$ | $q$ | $r$ | $q \vee r$ | $p \Rightarrow (q \vee r)$ | $p \wedge \sim q$ | $(p \wedge \sim q) \Rightarrow r$ |
|---|---|---|---|---|---|---|
| T | T | T | T | T | F | T |
| T | T | F | T | T | F | T |
| T | F | T | T | T | T | T |
| T | F | F | F | F | T | F |
| F | T | T | T | T | F | T |
| F | T | F | T | T | F | T |
| F | F | T | T | T | F | T |
| F | F | F | F | T | F | T |
| | | | | ↑ | $\equiv$ | ↑ |

(b) $p \Rightarrow (q \vee r) \equiv \sim p \vee (q \vee r)$

$$(p \wedge \sim q) \Rightarrow r \equiv \sim (p \wedge \sim q) \vee r$$
$$\equiv (\sim p \vee \sim (\sim q)) \vee r$$
$$\equiv (\sim p \vee q) \vee r$$

**15.**

| $p$ | $q$ | $\sim p$ | $p\wedge q$ | $\sim p\vee(p\wedge q)$ |
|---|---|---|---|---|
| T | T | F | T | T |
| T | F | F | F | F |
| F | T | T | F | T |
| F | F | T | F | T |

**17.**

| $p$ | $q$ | $\sim p$ | $\sim p\wedge q$ | $p\vee(\sim p\wedge q)$ |
|---|---|---|---|---|
| T | T | F | F | T |
| T | F | F | F | T |
| F | T | T | T | T |
| F | F | T | F | F |

**19.**

| $p$ | $q$ | $\sim p$ | $\sim p\Rightarrow q$ |
|---|---|---|---|
| T | F | F | T |
| T | F | F | T |
| F | T | T | T |
| F | F | T | F |

**21.**

| $p$ | $\sim p$ | $\sim p\wedge p$ |
|---|---|---|
| T | F | T |
| F | T | T |

**23.**

| $p$ | $q$ | $p\Rightarrow q$ | $p\wedge(p\Rightarrow q)$ |
|---|---|---|---|
| T | T | T | T |
| T | F | F | F |
| F | T | T | F |
| F | F | T | F |

**25.**

| $p$ | $q$ | $r$ | $q\wedge r$ | $p\wedge(q\wedge r)$ | $p\wedge q$ | $(p\wedge q)\wedge r$ | $p\wedge(q\wedge r)\Leftrightarrow(q\wedge r)\wedge r$ |
|---|---|---|---|---|---|---|---|
| T | T | T | T | T | T | T | T |
| T | T | F | F | F | T | F | T |
| T | F | T | F | F | F | F | T |
| T | F | F | F | F | F | F | T |
| F | T | T | T | F | F | F | T |
| F | T | F | F | F | F | F | T |
| F | F | T | F | F | F | F | T |
| F | F | F | F | F | F | F | T |

**27.**

| $p$ | $q$ | $p\vee q$ | $p\wedge(p\vee q)$ | $p\wedge(p\vee q)\Leftrightarrow p$ |
|---|---|---|---|---|
| T | T | T | T | T |
| T | F | T | T | T |
| F | T | T | F | T |
| F | F | F | F | T |

**29.** $p\Rightarrow q$

**31.** $\sim q\vee\sim p$

**33.** $q\Rightarrow p$

## 11.4 Arguments

**1.** Let $p$ and $q$ be the statements, $p$: It is raining, $q$: John is going to school. Assume that $p \Rightarrow \sim q$ and $q$ are true statements. Prove: $\sim p$ is true.
Direct: $p \Rightarrow \sim q$ is true. Also, its contrapositive $q \Rightarrow \sim p$ is true and $q$ is true. Thus, $\sim p$ is true by the law of detachment.
Indirect: Assume $\sim p$ is false. Then $p$ is true; $p \Rightarrow \sim q$ is true. Thus, $\sim q$ is true by the law of detachment. But $q$ is true, and we have a contradiction. The assumption is false and $\sim p$ is true.

**3.** Let $p$, $q$ and $r$ be the statements, $p$: Smith is elected president; $q$: Kuntz is elected secretary; and $r$: Brown is elected treasurer. Assume that $p \Rightarrow q$, $q \Rightarrow \sim r$ and $p$ are true statements. Prove: $\sim r$ is true.
Direct: $p \Rightarrow q$ and $q \Rightarrow \sim r$ are true. So $p \Rightarrow \sim r$ is true by the law of syllogism, and $p$ is true. Thus, $\sim r$ is true by the law of detachment.
Indirect: Assume $\sim r$ is false. Then $r$ is true; $p \Rightarrow q$ is true; $q \Rightarrow \sim r$ is true. So, $p \Rightarrow \sim r$ is true by the law of syllogism. $r \Rightarrow \sim p$, its contrapositive, is true. Thus, $\sim p$ is true by the law of detachment. But $p$ is true, and we have a contradiction. The assumption is false and $\sim r$ is true.

**5.** Not valid.

**7.** Valid.

## 11.5 Logic Circuits

**1.**

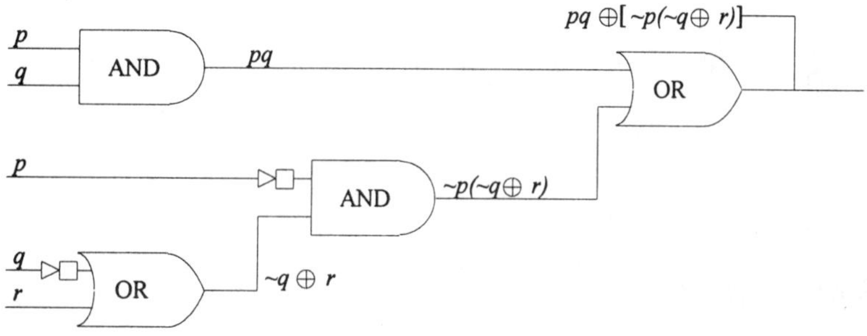


The output $pq \oplus [\sim p(\sim q \oplus r)]$ is 1 when $(p, q, r) = (1, 1, 1)$, $(1, 1, 0)$, $(0, 0, 1)$, $(0, 0, 0)$, or $(0, 1, 1)$.

**3.** The output is $(\sim q \oplus [p(\sim p \oplus q)])q = (\sim q)q \oplus p(\sim p \oplus q)q = p(\sim p \oplus q)q$ $= [p(\sim p) \oplus pq]q = pqq = pq$, which is 1 if and only if $p$ and $q$ are both 1.

**5.**

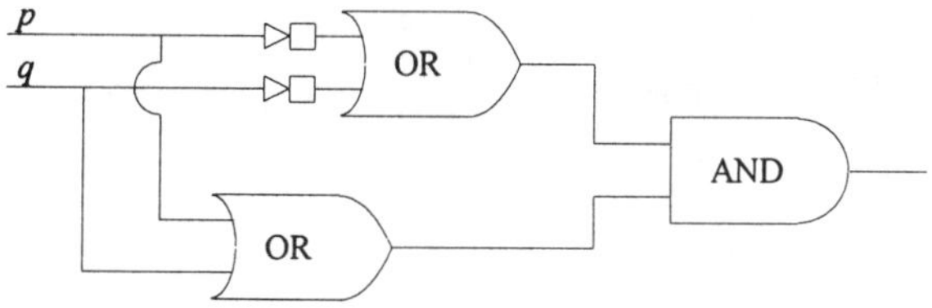

**7.**

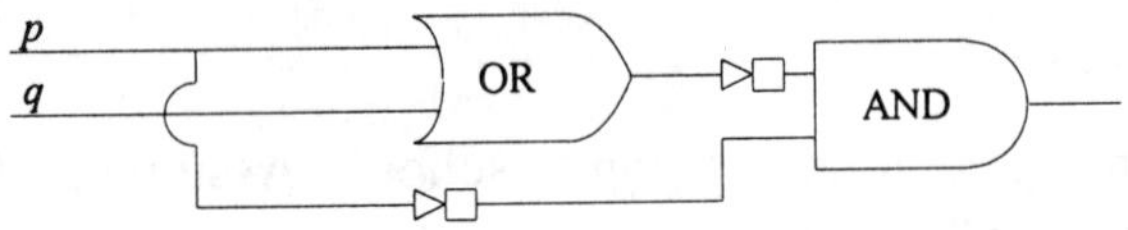


**9.** (For Problem 1)

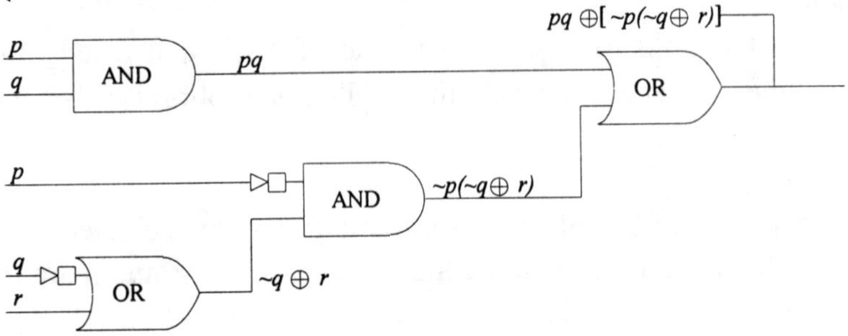


(For Problem 3)

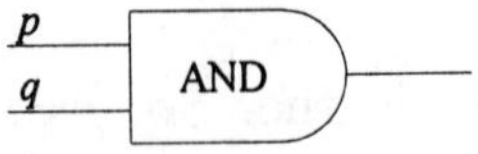


(For Problem 5)

| $p$ | $q$ | $\sim p\oplus \sim q$ | $p\oplus q$ | $(\sim p\oplus \sim q)(p\oplus q)$ |
|---|---|---|---|---|
| 1 | 1 | 0 | 1 | 0 |
| 1 | 0 | 1 | 1 | 1 |
| 0 | 1 | 1 | 1 | 1 |
| 0 | 0 | 1 | 0 | 0 |

$(\sim p\oplus \sim q)(p\oplus q) = [\sim (pq)](p\oplus q)$

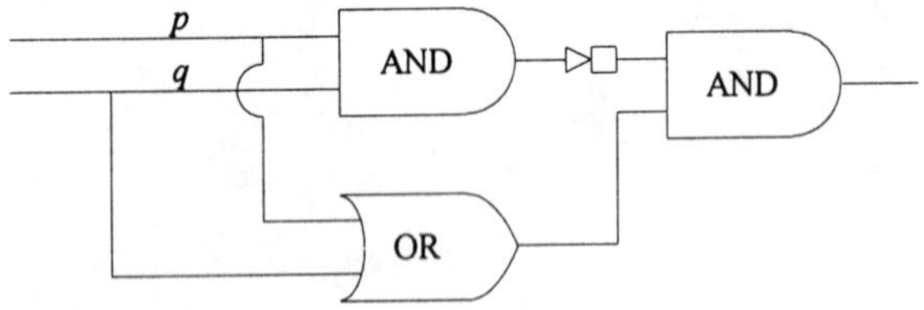


(For problem 7)

$\sim (p\oplus q) \sim p = [(p\oplus q)\oplus p] = \sim [p\oplus q]$

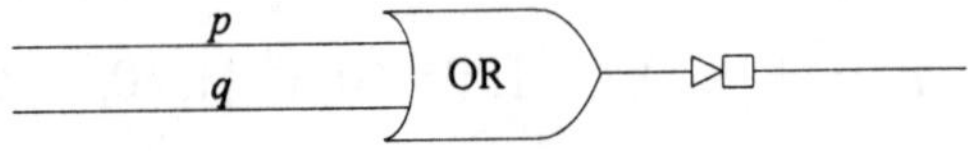

**11.** The truth table for this circuit is either

| $p$ | $q$ | |
|---|---|---|
| 1 | 1 | 1 |
| 1 | 0 | 0 |
| 0 | 1 | 0 |
| 0 | 0 | 1 |

or

| $p$ | $q$ | |
|---|---|---|
| 1 | 1 | 0 |
| 1 | 0 | 1 |
| 0 | 1 | 1 |
| 0 | 0 | 0 |

Thus, two possible circuits are

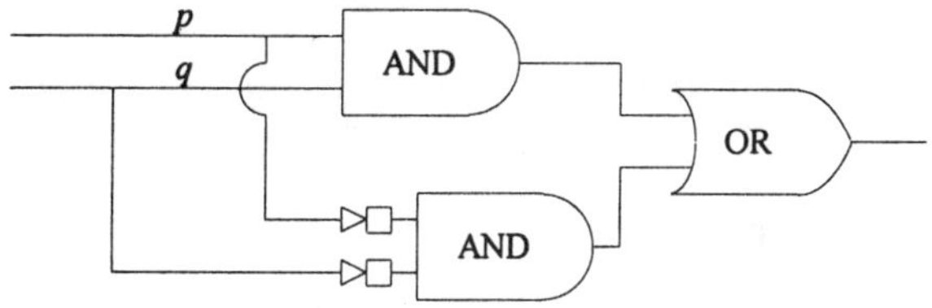


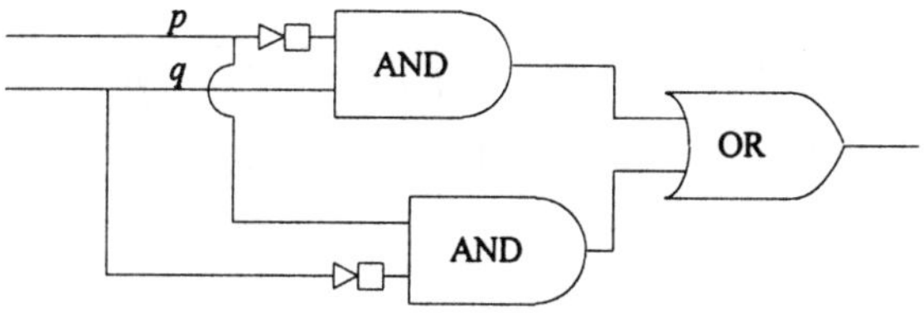


**13.**

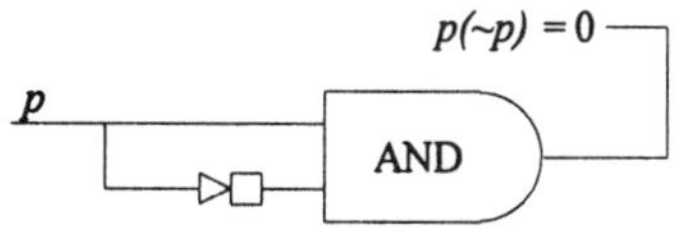


**15.** (a) $p \vee q \equiv \sim(\sim p \sim q)$

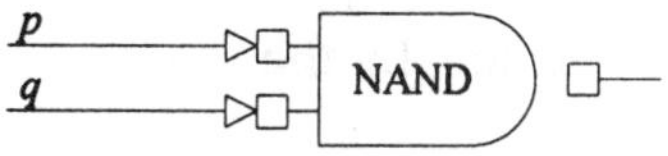


(b)

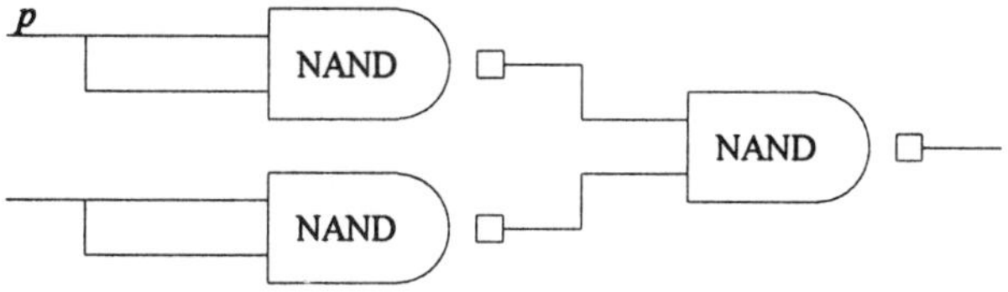


**17.**
$$\begin{aligned} pq \oplus pr \oplus q(\sim r) &= pq(r \oplus \sim r) \oplus pr \oplus q(\sim r) \\ &= pqr \oplus pq(\sim r) \oplus pr \oplus q(\sim r) \\ &= prq \oplus pr \oplus pq(\sim r) \oplus q \oplus \sim r \\ &= pr(q \oplus 1) \oplus (p \oplus 1)q(\sim r) \\ &= pr(1) \oplus (1)q(\sim r) \\ &= pr \oplus q(\sim r) \end{aligned}$$

## Chapter 11 Review

### True or False

**1.** False **3.** True **5.** True

## Fill in the Blank

**1.** $p \vee q$ **3.** logically equivalent **5.** 0; 1

## Review Exercises

**1.** (c) **3.** (a) **5.** Nobody is rich.

**7.** Danny is tall or Mary is not short.

**9.**

| $p$ | $q$ | $\sim p$ | $p \wedge q$ | $(p \wedge q) \vee \sim p$ |
|---|---|---|---|---|
| T | T | F | T | T |
| T | F | F | F | F |
| F | T | T | F | T |
| F | F | T | F | T |

**11.**

| $p$ | $q$ | $\sim p$ | $\sim q$ | $p \vee \sim q$ | $\sim p \vee (p \vee \sim q)$ |
|---|---|---|---|---|---|
| T | T | F | F | T | T |
| T | F | F | T | T | T |
| F | T | T | F | F | T |
| F | F | T | T | T | T |

**13.** $q \Rightarrow p$ **15.** $p \Leftrightarrow q$

**17.** Let $p$ be the statement "I paint the house" and let $q$ be the statement "I go bowling." Assume $\sim p \Rightarrow q$ and $\sim q$ are true. Prove $p$ is true. Since $\sim p \Rightarrow q$ is true, its contrapositive $\sim q \Rightarrow p$ is true. We have $\sim q$ is true and hence, by the law of detachment, $p$ is true.

**19.**

| $p$ | $q$ | $\sim p$ | $\sim p \vee q$ | $p \Rightarrow q$ |
|---|---|---|---|---|
| T | T | F | T | T |
| T | F | F | F | F |
| F | T | T | T | T |
| F | F | T | T | T |

$\sim p \vee q \equiv p \Rightarrow q$

**21.**

| $p$ | $q$ | $(p \oplus q)[\sim(pq)]$ | $p \veebar q$ |
|---|---|---|---|
| 1 | 1 | 0 | 0 |
| 1 | 0 | 1 | 1 |
| 0 | 1 | 1 | 1 |
| 0 | 0 | 0 | 0 |

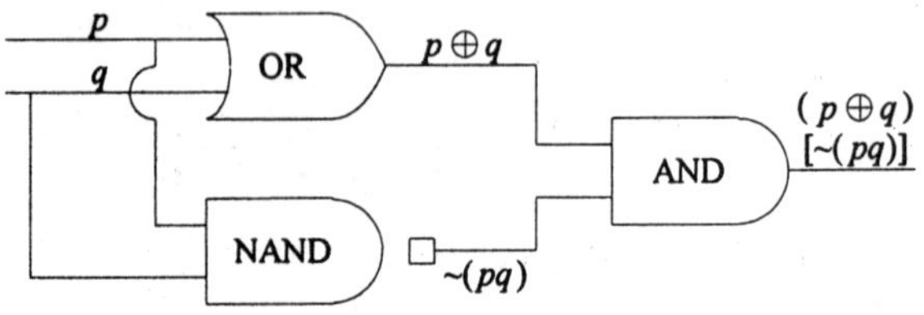

# Chapter 12

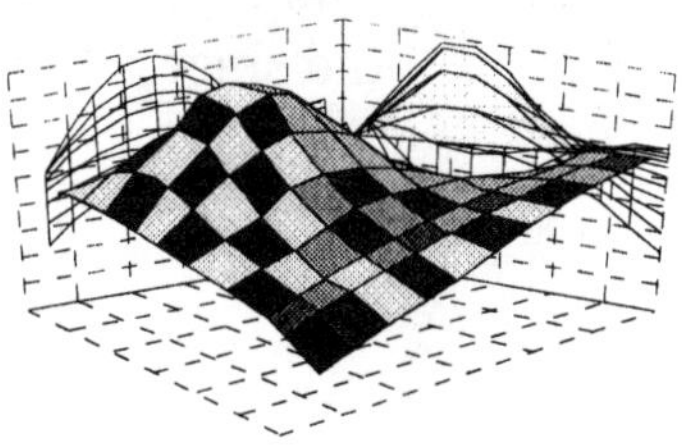

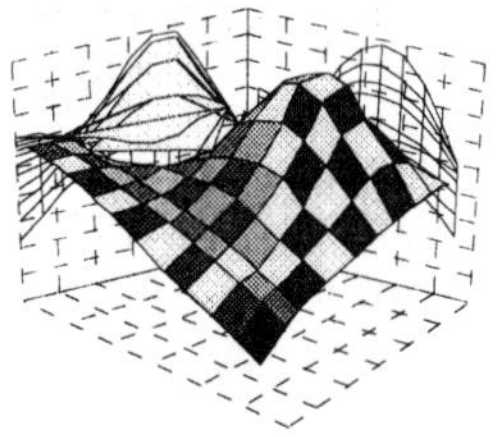

# Relations, Functions and Induction

## 12.1 Relations

**1.** True, false, true, false, true, false, false, false, false.

**3.** [(2, 2), (2, 4), (2, 6), (2, 10), (3, 6), (5, 10)]

**5.** 1, 4, 9, 16, 25, 2, 5

**7.** (a) {(1, 2), (1, 4), (1, 7), (2, 2), (2, 4), (2, 7), (5, 2), (5, 4), (5, 7)}
(b) {(1, 2), (1, 4), (1, 7), (2, 4), (2, 7), (5, 7)}
(c) $R$ is a subset of $A \times B$

**9.** (a) {*aa*, *ab*, *ba*, *bb*}
(b) {(*aa*, *ab*), (*ab*, *aa*), (*ba*, *bb*), (*bb*, *ba*)}

**11.** {(1, 2), (1, 4), (1, 6), (1, 8), (2, 2), (2, 4), (2, 6), (2, 8), (3, 6)}

**13.** {(1, 1), (2, 2), (3, 3), (4, 4), (5, 5), (6, 6), (7, 7), (8, 8), (9, 9), (10, 10), (1, 3), (3, 1), (1, 5), (5, 1), (1, 7), (7, 1), (1, 8), (8, 1), (1, 9), (9, 1), (2, 4), (4, 2), (2, 6), (6, 2), (2, 8), (8, 2), (2, 10), (10, 2), (3, 5), (5, 3), (3, 7), (7, 3), (3, 9), (9, 3), (4, 6), (6, 4), (4, 8), (8, 4), (4, 10), (10, 4), (5, 7), (7, 5), (5, 9), (9, 5), (6, 8), (8, 6), (6, 10), (10, 6), (7, 9), (9, 7), (8, 10), (10, 8)}

**15.** $\{(0, 0), (0, 1), (0, 2), (0, 3), (1, 1), (1, 2), (1, 3), (2, 2), (2, 3), (3, 3)\}$

**17.** $\{(0, 0), (1, 1), (2, 4), (3, 9), (4, 16)\}$

**19.** $\{(1, 1), (2, 1), (5, 2), (7, 3)\}$

**21.** $\{(42, *), (72, H), (47, /), (88, x)\}$

**23.** Reflexive, symmetric and transitive

**25.** Reflexive, symmetric and transitive

**27.** Reflexive, symmetric and transitive

**29.** Reflexive. Not symmetric because $(A, C) \in R$ but $(C, A) \notin R$. Not transitive because $(C, B) \in R$ and $(B, A) \in R$ but $(C, A) \notin R$.

**31.** Since we cannot find *a* and *b* in *A* such that *aRb*, then *R* is symmetric. Similarly, since we cannot find *a*, *b* and *c* in *A* such that *aRb* and *bRc*, then *R* is transitive. And since $(t, t) \notin R$ etc. then *R* is not reflexive.

## 12.2 Functions

**1.** $f(a) = 1$, $f(b) = 4$, $f(c) = 2$. Range $= \{1, 2, 4\}$.

**3.** $f(0) = 5$, $f(5) = 0$, $f(-5) = 10$, $f(10) = -5$, $f(-10) = 15$.

**5.** Does not define a function, because *f* assigns two different values (1 and 3) to *x*.

**7.** Yes, *f* defines a function.

**9.** Yes, *f* defines a function.

**11.** No, *f* is not a function. Domain $\neq A$.

**13.** Yes

**15.** Yes

**17.** One-one, not onto, therefore, not bijective.

**19.** Neither one-to-one nor onto, therefore, not bijective.

**21.** For example $H(10, 10) = 0 = H(01, 01)$. Thus, *H* is not one-to-one.

**23.** No. Odd integers in the range are not associated with any integers in the domain.

**25.** $f^{(-1)}(a) = 3$, $f^{(-1)}(b) = 2$, $f^{(-1)}(c) = 1$

**27.** $g \circ f(1) = g[f(1)] = g(b) = z$ $\quad g \circ f(2) = g[f(2)] = g(a) = x$
$g \circ f(3) = g[f(3)] = g(c) = y$

## 12.3 Sequences

**1.** $s_1 = -1,\ x_2 = \frac{1}{2},\ s_3 = -\frac{1}{3},\ s_4 = \frac{1}{4},\ s_{100} = \frac{1}{100}$

**3.** (a) $b_0 = 0,\ b_1 = \frac{1}{2},\ b_2 = \frac{2}{3},\ b_3 = \frac{3}{4},\ b_4 = \frac{4}{5},\ b_5 = \frac{5}{6}$

(b) $n = 0:\ b_1 - b_0 = \frac{1}{2}$

$n = 1:\ b_2 - b_1 = \frac{1}{6}$

$n = 2:\ b_3 - b_2 = \frac{1}{12}$

**5.** $1, 2, 4, 8, 16, 32, 64, 128$

**7.** $1, 1, 2, 6, 24, 120, 720$

**9.** $M_0 = \begin{bmatrix} 1 & 1 & 0 \\ 1 & 0 & -1 \\ 0 & -1 & 1 \end{bmatrix},\ M_1 = \begin{bmatrix} 1 & 0 & 0 \\ 0 & 1 & 0 \\ 0 & 0 & 2 \end{bmatrix},\ M_2 = \begin{bmatrix} 1 & -1 & 0 \\ -1 & 2 & 1 \\ 0 & 1 & 3 \end{bmatrix},\ M_3 = \begin{bmatrix} 1 & -2 & 0 \\ -2 & 3 & 2 \\ 0 & 2 & 4 \end{bmatrix}$

**11.** $(-1)^n,\ n = 0,\ 1,\ 2,\ \ldots$

**13.** $2n+1,\ n = 0,\ 1,\ 2,\ \ldots$

**15.** $\frac{1}{n+1},\ n = 0,\ 1,\ 2,\ \ldots$

**17.** $s_0 = 0,\ s_1 = 1,\ x_2 = 3,\ s_3 = 7,\ s_4 = 15,\ s_5 = 31,\ s_6 = 63,\ s_7 = 128,\ s_8 = 255$

**19.** $s_0 = 1,\ s_1 = 1,\ x_2 = 2,\ s_3 = 3,\ s_4 = 5,\ s_5 = 8,\ s_6 = 13,\ s_7 = 21,\ s_8 = 34$

## 12.4 Mathematical Induction

**1.** (a) $S(1):\ 1 < 2$. Yes.

(b) $2 < 2^2$. Yes.

(c) $k < 2^k$.

(d) $(k+1) < 2^{k+1}$.

**3.** (a) $1^2 = \frac{1(1+1)(2+1)}{6}$ Yes

(b) $1^2 + 2^2 + 3^2 + 4^2 + 5^2 = \frac{5(5+1)(10+1)}{6}$ Yes

(c) $1^2 + 2^2 + 3^2 + \ldots + k^2 = \frac{k(k+1)(2k+1)}{6}$

(d) $1^2 + 2^2 + 3^2 + \ldots + k^2 + (k+1)^2 = \frac{(k+1)[(k+1)+1][2(k+1)+1]}{6} = \frac{(k+1)(k+2)(2k+3)}{6}$

**5.** (a) $1+2=2^2-1$ Yes

(b) $1+2+2^2+2^3+2^4+2^5=2^{5+1}-1$ Yes

(c) $1+2+2^2+\ldots+2^k=2^{k+1}-1$

(d) $1+2+2^2+\ldots+2^k+2^{k+1}=2^{(k+1)+1}-1$

**7.** Since $1\cdot 2=\frac{1(1+1)(1+2)}{3}$ then $s(1)$ is true and condition I is true. Assume that $s(k)$ is true. Show $s(k+1)$ is true. $s(k+1)$ states:
$1\cdot 2+2\cdot 3+3\cdot 4+\ldots+k\cdot(k+1)+(k+1)[(k+1)+1]=\frac{(k+1)[(k+1)+1][(k+1)+2]}{3}$ (1) using our assumption that $s(k)$ is true the left hand side of

$$(1)=\frac{k(k+1)(k+2)}{3}+(k+1)[(k+1)+1]$$

$$=\frac{k(k+1)(k+2)+3(k+1)(k+2)}{3}$$

$$=\frac{(k+1)+(k+2)(k+3)}{3}=\frac{(k+1)[(k+1)+1][(k+1)+2]}{3}$$

which shows that $s(k+1)$ is true and thus condition II is true.

**9.** Since $\frac{1}{1\cdot 2}=\frac{1}{1+1}$ then $s(1)$ is true and thus condition I is satisfied. Assume that $s(k)$ is true. Show that $s(k+1)$ is also true. $s(k+1)$ states: (1)
$\frac{1}{1\cdot 2}+\frac{1}{2\cdot 3}+\frac{1}{3\cdot 4}+\ldots+\frac{1}{(k+1)[(k+1)+1]}=\frac{k+1}{(k+1)+1}$ Since by assumption $s(k)$ is true then the left hand side of (1) gives

$$\frac{k}{k+1}+\frac{1}{(k+1)[(k+1)+1]}=\frac{k[(k+1)+1]+1}{(k+1)[(k+1)+1]}$$

$$=\frac{k(k+2)+1}{(k+1)[(k+1)+1]}=\frac{k^2+2k+1}{(k+1)[(k+1)+1]}=\frac{(k+1)(k+1)}{(k+1)[(k+1)+1]}$$

$$=\frac{k+1}{[(k+1)+1]}$$

which is the right hand side of (1). Thus $s(k+1)$ is true and so condition II is satisfied

**11.** Since $2=\frac{1(3+1)}{2}$ then $S(1)$ is true and thus condition I is satisfied. Assume that $S(k)$ is true show $S(k+1)$ is also true. $S(k+1)$ states

$$2+5+8+\ldots+[3(k+1)-1]=\frac{(k+1)[3(k+1)+1]}{2}. \quad (1)$$

By our assumption that $S(k)$ is true the left hand side of (1) is then

$$\frac{k(3k+1)}{2}+3(k+1)-1=\frac{k(3k+1)+6(k+1)-2}{2}$$
$$=\frac{3k^2+7k+4}{2}=\frac{(k+1)(3k+4)}{2}$$
$$=\frac{(k+1)(3k+3\quad+1}{2}$$
$$=\frac{(k+1)[3(k+1)+1]}{2}$$

which is the right hand side of (1). Thus $S(k+1)$ is true and so condition II is satisfied.

**13.** Since $2=1\cdot 2$ then $S(1)$ is true and thus Condition I is satisfied. Assume $S(k)$ is true. Show that $S(k+1)$ is true. $S(k+1)$ states:

$$2+4+6+\cdots+2k+2(k+1)=(k+1)(k+2)$$

By our assumption that $S(k)$ is true, the left hand side above becomes:

$$2+4+6+\cdots+2k+2(k+1)=k(k+1)+2(k+1)=(k+1)(k+2)$$

which is the right hand side of the equation above. Thus $S(k+1)$ is true and so Condition II is satisfied.

**15.** Since $-1=-1\cdot\frac{1\cdot 2}{2}$ then $S(1)$ is true and thus Condition I is satisfied. Assume $S(k)$ is true. Show that $S(k+1)$ is true. $S(k+1)$ states:

$$-1+2^2-3^2+4^2-5^2+\cdots+(-1)^k k^2+(-1)^{k+1}(k+1)^2=(-1)^{k+1}\frac{(k+1)(k+2)}{2}$$

By our assumption that $S(k)$ is true, the left hand side above becomes:

$$(-1)^k\frac{k(k+1)}{2}+(-1)^{k+1}(k+1)^2=(-1)^{k+1}\frac{1}{2}\left(-k^2-k+2\left(k^2+2k+1\right)\right)$$
$$=(-1)^{k+1}\frac{1}{2}\left(k^2+3k+2\right)=(-1)^{k+1}\frac{1}{2}(k+1)(k+2)$$

which is the right hand side of the equation above. Thus $S(k+1)$ is true and so Condition II is satisfied.

**17.** Since $\frac{1}{1\cdot 4}=\frac{1}{4}$ then $S(1)$ is true and thus Condition I is satisfied. Assume $S(k)$ is true. Show that $S(k+1)$ is true. $S(k+1)$ states:

$$\frac{1}{1\cdot 4}+\frac{1}{4\cdot 7}+\frac{1}{7\cdot 10}+\cdots+\frac{1}{(3k-2)(3k+1)}+\frac{1}{(3k+1)(3k+4)}=\frac{k+1}{3k+4}$$

By our assumption that $S(k)$ is true, the left hand side above becomes:

$$\frac{k}{3k+1}+\frac{1}{(3k+1)(3k+4)}=\frac{1}{3k+1}\left(\frac{k(3k+4)+1}{3k+4}\right)=\frac{3k^2+4k+1}{(3k+4)(3k+1)}$$
$$=\frac{(3k+1)(k+1)}{(3k+4)(3k+1)}=\frac{k+1}{3k+4}$$

which is the right hand side of the equation above. Thus $S(k+1)$ is true and so Condition II is satisfied.

**19.** Since $\frac{1}{1\cdot 2\cdot 3}=\frac{1\cdot 4}{4\cdot 2\cdot 3}$ then $S(1)$ is true and thus Condition I is satisfied. Assume $S(k)$ is true. Show that $S(k+1)$ is true. $S(k+1)$ states:

$$\frac{1}{1\cdot 2\cdot 3}+\frac{1}{2\cdot 3\cdot 4}+\frac{1}{3\cdot 4\cdot 5}+\cdots+\frac{1}{k\cdot(k+1)\cdot(k+2)}+\frac{1}{(k+1)\cdot(k+2)\cdot(k+3)}=\frac{(k+1)(k+4)}{4(k+2)(k+3)}$$

By our assumption that $S(k)$ is true, the left hand side above becomes:

$$\frac{k(k+3)}{4(k+1)(k+2)}+\frac{1}{(k+1)\cdot(k+2)\cdot(k+3)}=\frac{k(k+3)^2+4}{4(k+1)(k+2)(k+3)}$$
$$=\frac{k^3+6k^2+9k+4}{4(k+1)(k+2)(k+3)}=\frac{(k+1)^2(k+4)}{4(k+1)(k+2)(k+3)}$$
$$=\frac{(k+1)(k+4)}{4(k+2)(k+3)}$$

which is the right hand side of the equation above. Thus $S(k+1)$ is true and so Condition II is satisfied.

**21.** Since $1+3=\frac{1}{2}\left(3^2-1\right)$ then $S(1)$ is true and thus Condition I is satisfied. Assume $S(k)$ is true. Show that $S(k+1)$ is true. $S(k+1)$ states:

$$1+3+3^2+\cdots+3^k+3^{k+1}=\frac{1}{2}\left(3^{k+2}-1\right)$$

By our assumption that $S(k)$ is true, the left hand side above becomes:

$$\frac{1}{2}\left(3^{k+1}-1\right)+3^{k+1}=\frac{1}{2}3^{k+1}-\frac{1}{2}+3^{k+1}=\frac{3}{2}3^{k+1}-\frac{1}{2}=\frac{1}{2}\left(3^{k+2}-1\right)$$

which is the right hand side of the equation above. Thus $S(k+1)$ is true and so Condition II is satisfied.

**23.** Since $1+2\le 3$ then $S(1)$ is true and thus Condition I is satisfied. Assume $S(k)$ is true. Show that $S(k+1)$ is true. $S(k+1)$ states:

$$1+2(k+1)\le 3^{k+1}$$

By our assumption that $S(k)$ is true, the left hand side above becomes:

$$1+2(k+1)=1+2k+2\le 3^k+2<3^k+2\cdot 3^k=3\cdot 3^k=3^{k+1}$$

which is the right hand side of the inequality above. Thus $S(k+1)$ is true and so Condition II is satisfied.

**25.** Since $\frac{1}{2} \le \frac{1}{\sqrt{3+1}}$ then $S(1)$ is true and thus Condition I is satisfied. Assume $S(k)$ is true. Show that $S(k+1)$ is true. $S(k+1)$ states:

$$\frac{1}{2}\cdot\frac{3}{4}\cdot\frac{5}{6}\cdot\cdots\cdot\frac{2k-1}{2k}\cdot\frac{2k+1}{2k+2} \le \frac{1}{\sqrt{3k+4}}$$

By our assumption that $S(k)$ is true, the left hand side above becomes: $\frac{1}{\sqrt{3k+1}}\cdot\frac{2k+1}{2k+2}$. To show that $\frac{1}{\sqrt{3k+1}}\cdot\frac{2k+1}{2k+2} \le \frac{1}{\sqrt{3k+4}}$ we will square both sides and simplify

$\frac{1}{3k+1}\cdot\left(\frac{2k+1}{2k+2}\right)^2 \le \frac{1}{3k+4}$ or $(3k+4)(4k^2+4k+1) \le (3k+1)(4k^2+8k+4)$;

$12k^3+28k^2+16k+4 \le 12k^3+28k^2+20k+4$, or $0 \le 12k$. Thus $S(k+1)$ is true and so Condition II is satisfied.

**27.** Since 3 is a factor of $1^3+2\cdot 1=3$ then $S(1)$ is true and thus Condition I is satisfied. Assume $S(k)$ is true, i.e., $k^3+2k=3\cdot M$ for some positive integer $M$. Show that $S(k+1)$ is true. $S(k+1)$ states that 3 is a factor of $(k+1)^3+2(k+1)$. By our assumption that $S(k)$ is true, the expression above can be written as:

$k^3+3k^2+3k+1+2k+2 = k^3+2k+3k^2+3k+3 = 3\cdot M+3k^2+3k+3 = 3(M+k^2+k+1)$

which is divisible by 3. Thus $S(k+1)$ is true and so Condition II is satisfied.

## 12.5 Recurrence Relations

**1.** 1, 2, 4, 8, 16, 32

**3.** 1, 0, –1, –4, –17, –86

**5.** 2, 4, 12, 48, 240, 1440

**7.** 1, 1, 3, 7, 16, 33

**9.** 1, 2, 2, 4, 8, 32

**11.** 1, 2, 2, 5, 9, 16

**13.** 1, –1, –1, –5, –20, –104

**15.** $r_{10}=89$, $r_{11}=144$, $r_{12}=233$, $r_{13}=377$, $r_{14}=610$

**17.** (a) Length 0: The word with no bits does not contain 00.
Length 1: 1, 0
Length 2: 11, 10, 01
Length 3: 111, 110, 101, 011, 010
Length 4: 1111, 1110, 1101, 1011, 1010, 0111, 0110, 0101

(b) Initial conditions: $s_0=1$, $s_1=2$.
Recurrence relation: $s_n = s_{n-1}+s_{n-2}$ for $n \ge 2$.

**19.** (a) \$1000, \$1100, \$1210, \$1331, \$1464.1

(b) $A_n = A_{n-1} + 0.1A_{n-1}$ (recurrence relations)

$A_0 = 1,000$ (initial condition)

**21.** (a) $p_1 = 2,\ p_2 = 4,\ p_3 = 7,\ p_4 = 11$

(b) Initial condition: $p_1 = 2$.

Recurrence relation: $p_n = p_{n-1} + n$ for $n \geq 2$.

## Chapter 12 Review

### True or False

**1.** False **3.** True **5.** False

### Fill in the Blank

**1.** ordered pairs $(a,b)$ **3.** bijective **5.** Mathematical induction

### Review Exercises

**1.** {(1, 1), (8, 2), (27, 3)} **3.** (a) and (b)

**5.** (a) 1, 0, 3, 2, 5, 4, 7, 6, 9, 8 (b) 0, –1, 2, –3, 4, –5, 6, –7, 8, –9

**7.** 2, 2, 1, 4, 12, 60, 960

# Chapter 13

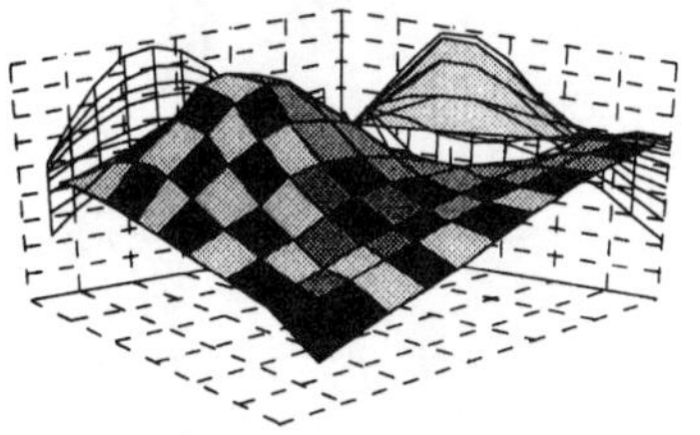

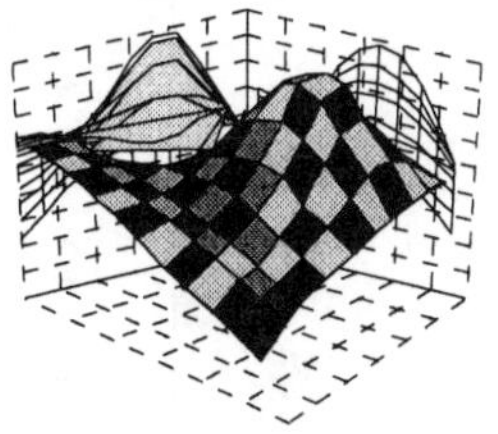

# Graphs and Trees

## 13.1 Graphs

**1.** Vertices: $v_1$, $v_2$, $v_3$, $v_4$
Edges: $e_1$, $e_2$, $e_3$, $e_4$, $e_5$
Loop: $e_1$
Isolated vertex: $v_4$
Parallel edges: $e_3$, $e_4$, $e_5$

**3.** There are two simple graphs which have two vertices.

**5.**

| Vertex | degree |
|---|---|
| $v_1$ | 3 |
| $v_2$ | 0 |
| $v_3$ | 3 |
| $v_4$ | 4 |

| Vertex | degree |
|---|---|
| $v_1$ | 1 |
| $v_2$ | 1 |
| $v_3$ | 2 |
| $v_4$ | 2 |
| $v_5$ | 2 |
| $v_6$ | 4 |
| $v_7$ | 2 |
| $v_8$ | 4 |
| $v_9$ | 2 |
| $v_{10}$ | 2 |

**7.** 9 edges, as following Figure shows.

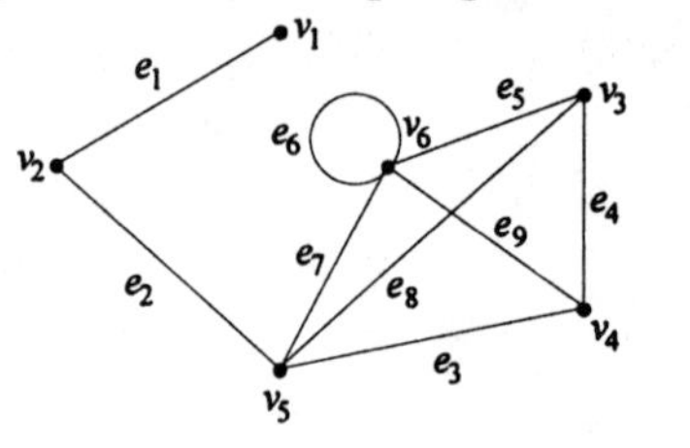


**9.** No, for $\deg(v_4) = 4$ there must be a loop.

**11.** (a)

(b)

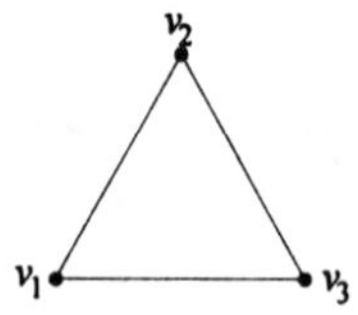


**13.** 15

**15.** Note: there are several correct answers to this problem.

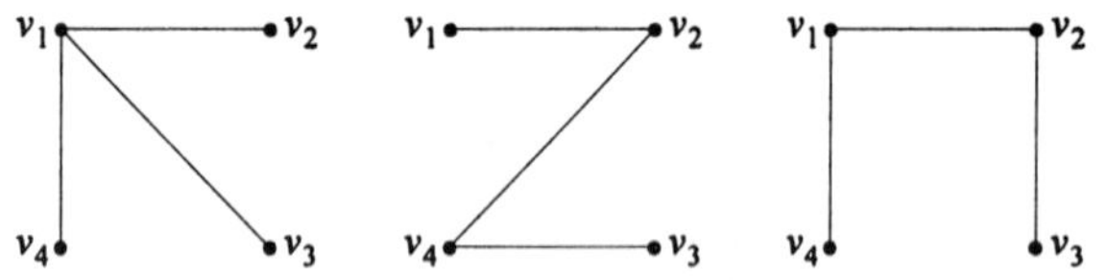


**17.**

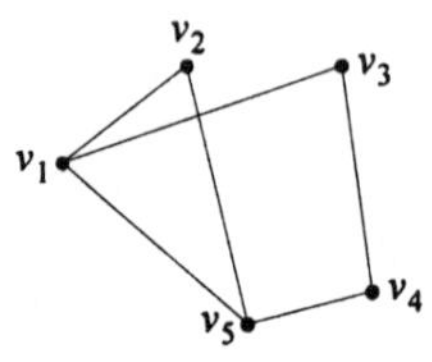


**18.**

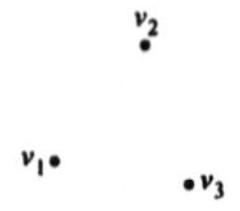


## 13.2 Paths and Connectedness

**1.** (a) Not simple because it contains the circuit $v_4e_4v_5e_6v_2e_7v_4$
(b) Not simple because it contains the circuit $v_1e_5v_5e_6v_2e_1v_1$
(c) Simple path.
(d) A circuit, but not simple circuit because it repeats the vertex $v_2$.
(e) Simple circuit.

**3.** (a), (b) and (c) Simple circuit.
(d) Path, but neither a simple path nor a circuit.

**5.** Note: each part of this problem has several correct answers.
(a) $v_1e_1v_2$; $v_5e_5v_6e_9v_3$; $v_1e_6v_6e_5v_5e_4v_4$; $v_2e_2v_3e_3v_4e_4v_5$
(b) $v_1e_1v_2e_7v_6e_5v_5e_{10}v_3e_9v_6e_6v_1$; $v_1e_1v_2e_8v_5e_5v_6e_9v_3e_2v_2e_7v_6e_6v_1$; $v_1e_1v_2e_7v_6e_9v_3e_{10}v_5e_5v_6e_6v_1$; $v_2e_8v_5e_{10}v_3e_3v_4e_4v_5e_5v_6e_9v_3e_2v_2$
(c) $v_1e_1v_2e_7v_6e_6v_1$; $v_3e_{10}v_5e_4v_4e_3v_3$; $v_2e_8v_5e_{10}v_3e_9v_6e_7v_2$; $v_6e_5v_5e_4v_4e_3v_3e_2v_2e_1v_1e_6v_6$

**7.** $v_1e_1v_2e_2v_3e_9v_5e_4v_4e_3v_3e_8v_6e_6v_1$

**9.** (a)

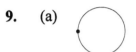

(b)

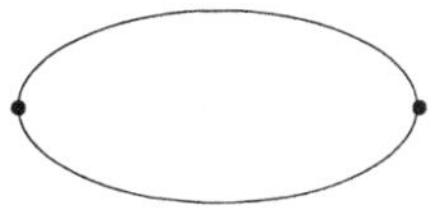

**11.** (a) $e_1, e_2, e_3, e_4$ (b) $e_4$ (c) $e_2, e_4, e_5$

**13.** There are three connected components:

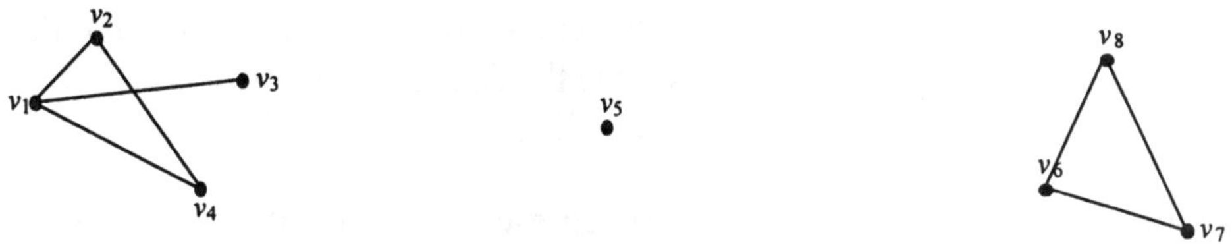


Component I Component II Component III

**15.** (a) No, because $11 > \binom{5}{2} = 10$ (Theorem II)

(b) Since $10 \not> \binom{6}{2}$, Theorem II does not apply. *H* may either be connected or disconnected. For example:

Connected:

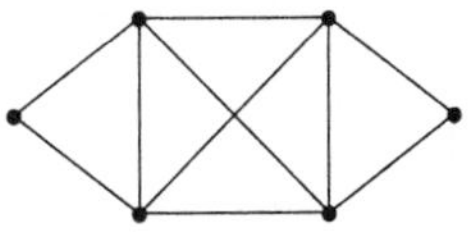

Disconnected:

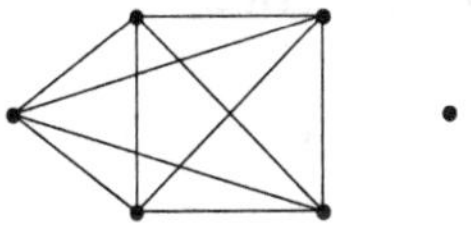

**17.** The shortest path goes from Firsttown through Secondsville to Fifthplace and then to Sixthtown. The total length is 8 + 5 + 4 = 17.

## 13.3 Eulerian and Hamiltonian Circuits

**1.** (a) Every vertex is of even degree. By Theorem III there is an Eulerian circuit. $v_1e_1v_2e_2v_3e_4v_4e_5v_5e_6v_3e_3v_2e_9v_5e_7v_6e_8v_1$ is an Eulerian circuit.
(b) Graph is not connected. Therefore, no Eulerian circuit.
(c) Degree of $v_1 = 3$, not even. Therefore, no Eulerian circuit.
(d) Every vertex is of even degree. Therefore graph contains an Eulerian circuit. The circuit $v_1e_1v_2e_2v_3e_3v_1e_5v_4e_4v_2e_6v_5e_7v_1$ is Eulerian.
(e) Every vertex is of even degree. Therefore graph contains an Eulerian circuit. The circuit $v_1e_1v_2e_{10}v_3e_{11}v_5e_{12}v_1e_{13}v_6e_5v_5e_4v_4e_3v_3e_2v_2e_{14}v_4e_{15}v_6e_9v_2e_8v_7e_7v_6e_6v_1$ is Eulerian.
(f) Degree of $v_1 = 3$, not even. Therefore, graph does not contain an Eulerian circuit.

**3.** Yes, each vertex is of even degree.

**5.**

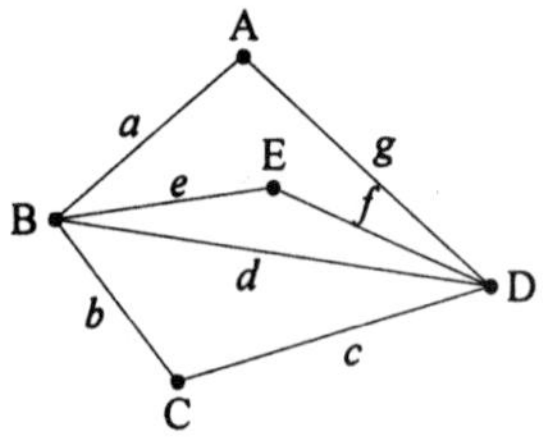


Yes. Let A and C in the figure be the land masses, B, E and D be three islands and *a*, *b*, *c*, *d*, *e*, *f* and *g* be the seven bridges. Note that each vertex has even degree. Therefore, the graph contains an Eulerian circuit the following circuit is the desired round trip: A*a*B*b*C*c*D*d*B*e*E*f*D*g*A.

**7.** No, the corresponding graph has vertex *B* (vertex representing room *B*) of degree 3 which is not even.

**9.** (a) $v_1e_1v_2e_9v_7e_{10}v_8e_3v_3e_4v_4e_5v_5e_6v_6e_7v_1$
(b) $v_1e_1v_7e_2v_2e_4v_8e_5v_3e_6v_4e_7v_5e_8v_6e_9v_1$
(c) $v_1e_1v_2e_2v_3e_3v_4e_{11}v_7e_{10}v_6e_{13}v_5e_5v_1$
(d) $v_4e_3v_3e_9v_8e_{12}v_7e_{11}v_6e_7v_2e_1v_1e_5v_5e_4v_4$

**11.** Use Theorem IV,
(a) $\deg(v_1)+\deg(v_5) = 4 < 6$, number of vertices
(b) $\deg(v_2)+\deg(v_7) = 4 < 8$, number of vertices
(c) $\deg(v_1)+\deg(v_6) = 4 < 7$, number of vertices
(d) $\deg(v_5)+\deg(v_{10}) = 4 < 10$, number of vertices

**13.**

**15.**

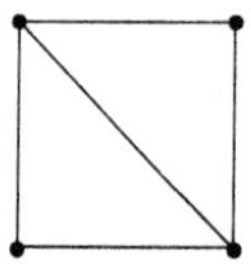

**17.** We have to repeat two of the streets in order to complete circuit: West St. and East St. Therefore the minimal closed circuit is West St. – North Ave – State – East St. – East St. – South Ave – Main St. – West St. The total length of the circuit is 30.

## 13.4 Trees

**1.** 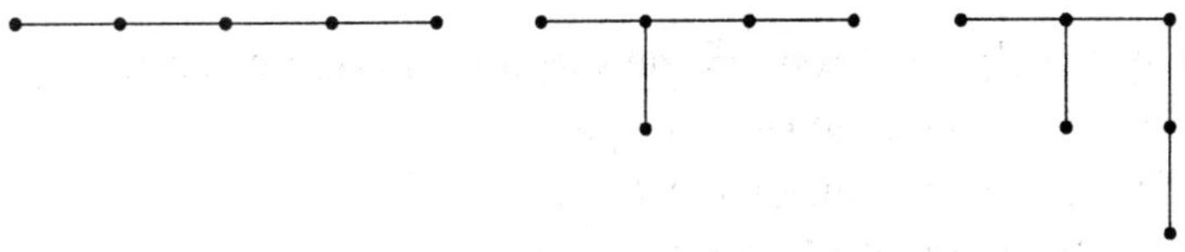

**3.** 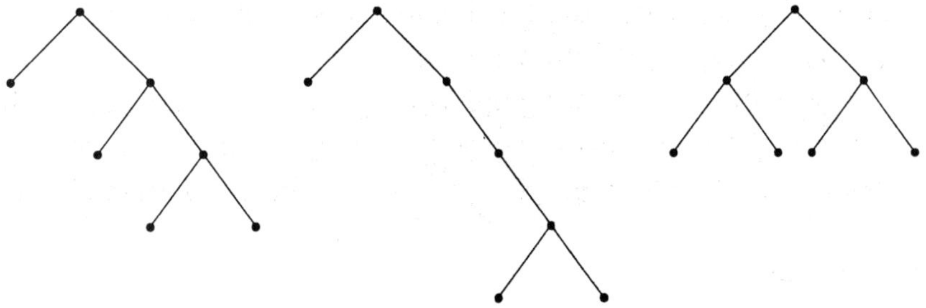

**5.** By Theorem V, such a tree cannot exist

**7.** Cannot exist. A graph with the given specifications must contain a circuit.

**9.** Cannot exist. Such a graph must contain a circuit.

**11.** Cannot exist. Such a graph *must* be a tree.

**13.** Yes. Use Theorem V.

**15.** 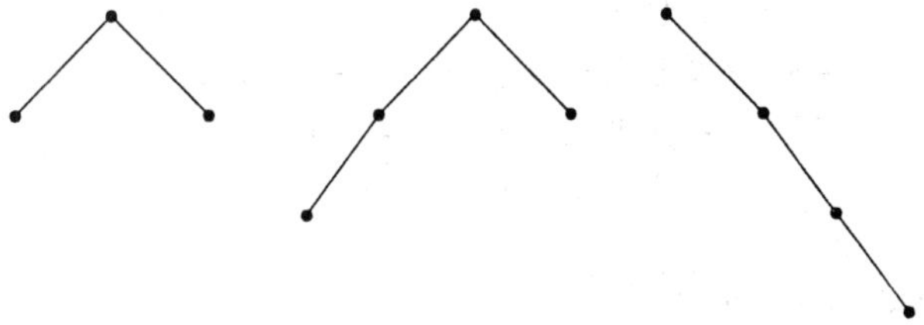

**17.** Leaves: $v_1, v_3, v_4, v_5, v_6, v_7$. Internal vertices: $v_2, v_9, v_8, v_{10}, v_{11}, v_{12}$.

**19.** No. By definition a leaf is connected to the rest of a tree by only one edge.

**21.** 

**23.** (a) $r, p, q, u$ are the descendents of $x$. $u, s, t, w, a, b, c$ are the descendents of $y$.

(b)

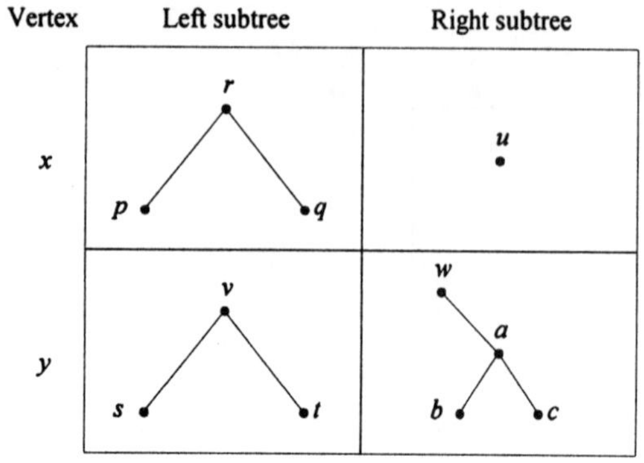


(c)

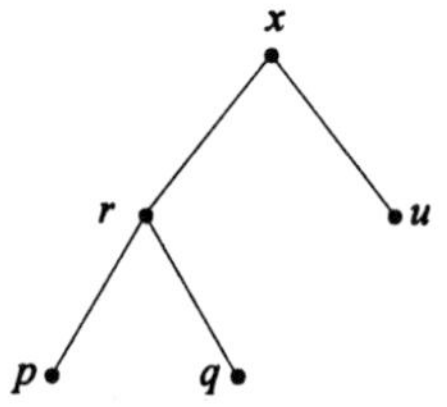


**25.** $a \times b + 1$

**27.** $(x+y) \times z - \frac{a}{b}$

**29.**

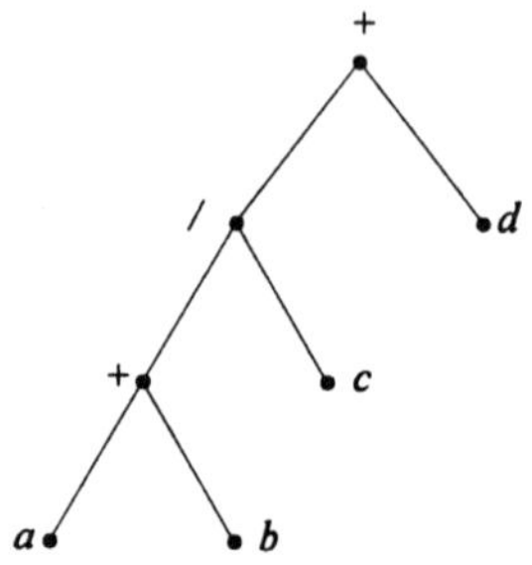


**31.**

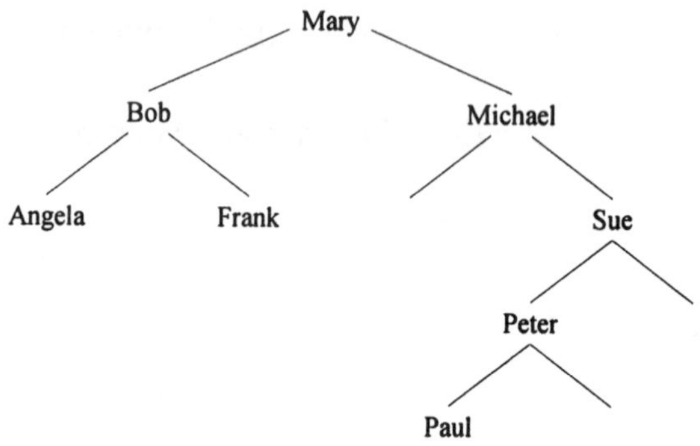


**33.** (a) Yes. (b) Yes

**35.** The minimum spanning tree consists of edges of lengths 2, 5 and the two edges of length 4. The total length of the cable required is then 15.

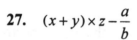

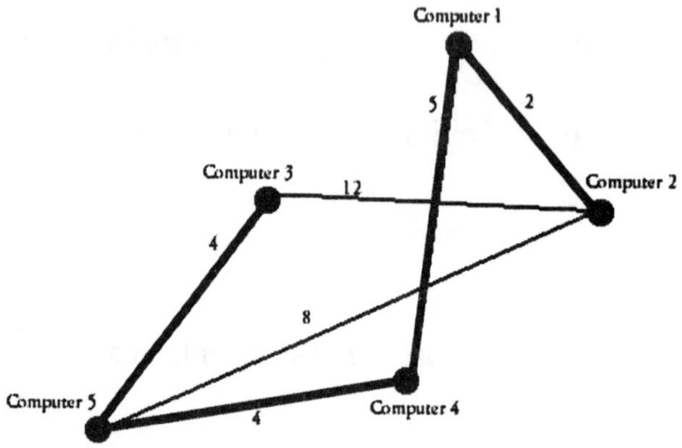

## 13.5 Directed Graphs

**1.** (a)

| Arc | initial point | terminal point |
|---|---|---|
| $e_1$ | $a$ | $a$ |
| $e_2$ | $a$ | $b$ |
| $e_3$ | $a$ | $c$ |
| $e_4$ | $b$ | $c$ |
| $e_5$ | $d$ | $c$ |
| $e_6$ | $c$ | $e$ |
| $e_7$ | $d$ | $e$ |
| $e_8$ | $d$ | $h$ |
| $e_9$ | $c$ | $f$ |
| $e_{10}$ | $f$ | $c$ |
| $e_{11}$ | $f$ | $g$ |
| $e_{12}$ | $f$ | $h$ |
| $e_{13}$ | $h$ | $g$ |

(b)

| Vertex | in degree | Out degree |
|---|---|---|
| $a$ | 1 | 3 |
| $b$ | 1 | 1 |
| $c$ | 4 | 2 |
| $d$ | 0 | 3 |
| $e$ | 2 | 0 |
| $f$ | 1 | 3 |
| $g$ | 2 | 0 |
| $h$ | 2 | 1 |

**3.**

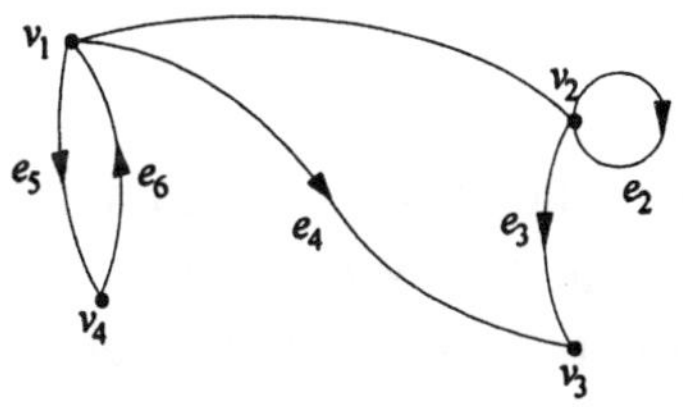


**5.**

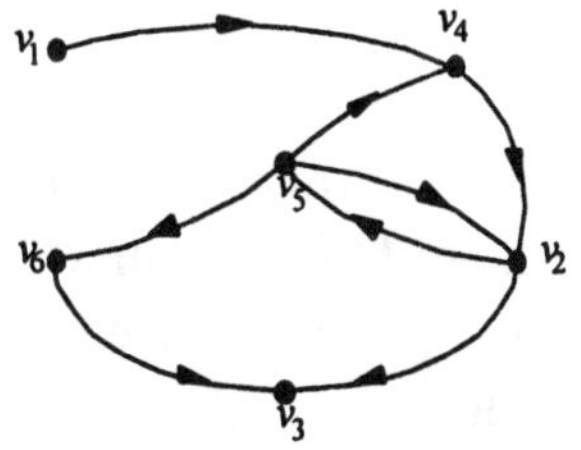


**7.**

**9.** (a) Yes. $v_1e_1v_2e_8v_7e_{10}v_5$ is a directed path from $v_1$ to $v_5$.
(b) No. No arcs go into $v_1$ i.e. $\text{indeg}(v_1) = 0$.

**11.** $v_1e_1v_2e_8v_7e_{10}v_5e_5v_6$, $v_1e_6v_6$, $v_1e_1v_2e_8v_7e_{11}v_4e_{12}v_6$

**13.** No; since outdegree $(v_6) = 0$, then no other vertex is reachable from $v_6$.

15.

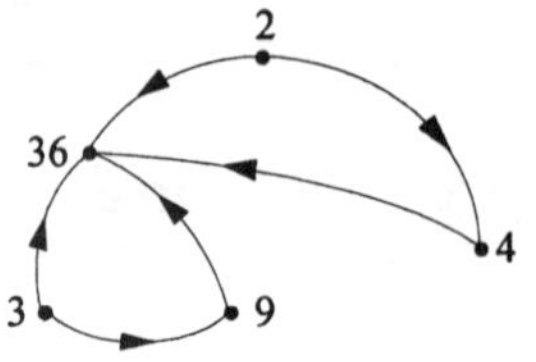


17. Graph is connected and contains no bridges. Thus it is orientable. The graph in the figure is the desired strongly connected digraph.

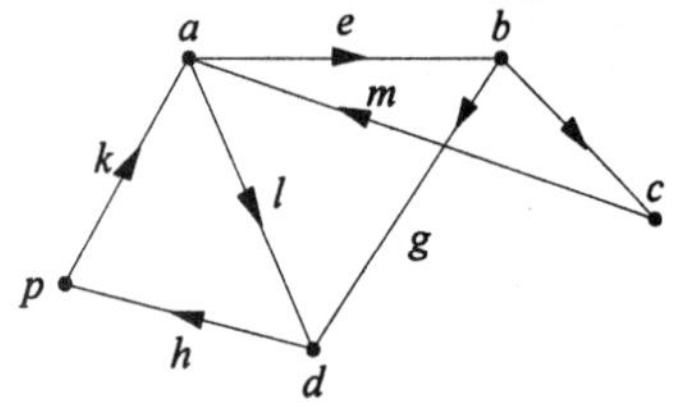


19. The graph is connected and contains no bridges. Therefore, the graph is orientable.

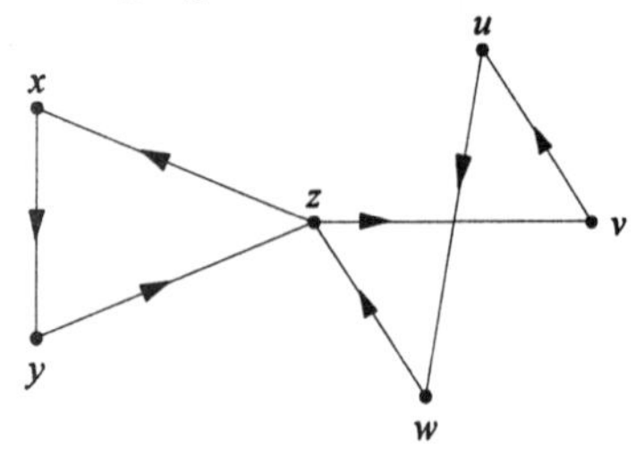


21. In order to reach Los Angeles, we must pass through Denver. The maximum amount of oil we can deliver from Houston directly to Denver, is 5 million gallons. We can also deliver a total of 8 million gallons to Denver, 3 million from Houston to Chicago to Denver, and another 5 from Houston to Atlanta to New York to Chicago to Denver. Therefore, total of $5 + 8 = 13$ million gallons of oil can be delivered to Denver. These 13 million gallons can then be delivered to Los Angeles in two pipelines: one directly from Denver to Los Angeles (10 million gallons) and another 3 million gallons through Seattle. The total capacity is 13 million gallons per hour.

23. In order to reach Los Angeles, we must pass through Denver or Dallas. The number of flights from Dallas to Los Angeles is 6; the number of flights from Denver to Los Angeles is 3. Therefore we can have at most 9 flights from New York to Los Angeles.

## 13.6 Matrix Applications to Graphs and Directed Graphs

**1.** (a)

| | a | b | c | d | e |
|---|---|---|---|---|---|
| a | 0 | 0 | 1 | 0 | 0 |
| b | 0 | 0 | 1 | 0 | 0 |
| c | 1 | 1 | 0 | 1 | 1 |
| d | 0 | 0 | 1 | 0 | 0 |
| e | 0 | 0 | 1 | 0 | 0 |

(b)

| | a | b | c | d | e |
|---|---|---|---|---|---|
| a | 1 | 0 | 0 | 0 | 0 |
| b | 0 | 1 | 1 | 1 | 0 |
| c | 0 | 1 | 0 | 1 | 0 |
| d | 0 | 1 | 1 | 0 | 0 |
| e | 0 | 0 | 0 | 0 | 1 |

(c)

| | a | b | c | d | e |
|---|---|---|---|---|---|
| a | 0 | 0 | 0 | 0 | 0 |
| b | 0 | 0 | 0 | 0 | 3 |
| c | 0 | 0 | 0 | 0 | 0 |
| d | 0 | 0 | 0 | 0 | 2 |
| e | 0 | 3 | 0 | 2 | 0 |

| | a | b | c | d | e |
|---|---|---|---|---|---|
| a | 0 | 1 | 1 | 1 | 1 |
| b | 1 | 0 | 1 | 1 | 1 |
| c | 1 | 1 | 0 | 1 | 1 |
| d | 1 | 1 | 1 | 0 | 1 |
| e | 1 | 1 | 1 | 1 | 0 |

**3.** (a)

| | a | b | c | d | e | f |
|---|---|---|---|---|---|---|
| a | 0 | 0 | 1 | 1 | 0 | 1 |
| b | 0 | 0 | 1 | 0 | 1 | 1 |
| c | 0 | 0 | 0 | 0 | 0 | 0 |
| d | 0 | 0 | 0 | 0 | 0 | 1 |
| e | 0 | 0 | 0 | 0 | 0 | 0 |
| f | 0 | 0 | 0 | 0 | 0 | 0 |

(b)

| | a | b | c | d |
|---|---|---|---|---|
| a | 0 | 1 | 1 | 1 |
| b | 1 | 0 | 1 | 0 |
| c | 1 | 1 | 0 | 0 |
| d | 1 | 0 | 1 | 0 |

(c)

| | a | b | c |
|---|---|---|---|
| a | 0 | 1 | 1 |
| b | 0 | 0 | 1 |
| c | 0 | 0 | 0 |

(d)

| | a | b | c | d |
|---|---|---|---|---|
| a | 0 | 1 | 1 | 1 |
| b | 0 | 0 | 0 | 0 |
| c | 0 | 1 | 0 | 0 |
| d | 0 | 1 | 0 | 0 |

**5.**

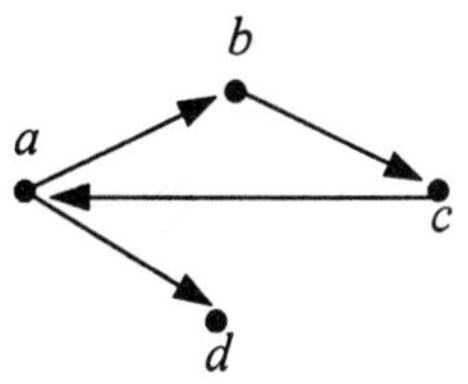


| | a | b | c | d |
|---|---|---|---|---|
| a | 0 | 1 | 0 | 1 |
| b | 0 | 0 | 1 | 0 |
| c | 1 | 0 | 0 | 0 |
| d | 0 | 0 | 0 | 0 |

**7.** The square of the matrix is $\begin{bmatrix} 0 & 0 & 1 \\ 0 & 0 & 0 \\ 0 & 0 & 0 \end{bmatrix}$ so there is only one two-stage dominance, namely a→b→c. The matrix $A + A^2 = \begin{bmatrix} 0 & 1 & 2 \\ 0 & 0 & 1 \\ 0 & 0 & 0 \end{bmatrix}$ is the number of one and two stage dominances.

**9.** Let $A$ be the dominance matrix. So

$$A = \begin{bmatrix} 0 & 0 & 0 & 1 & 1 & 1 & 1 \\ 1 & 0 & 1 & 1 & 0 & 1 & 0 \\ 1 & 0 & 0 & 0 & 1 & 1 & 1 \\ 0 & 0 & 1 & 0 & 0 & 1 & 1 \\ 0 & 1 & 0 & 1 & 0 & 0 & 1 \\ 0 & 0 & 0 & 0 & 1 & 0 & 1 \\ 0 & 1 & 0 & 0 & 0 & 0 & 0 \end{bmatrix}.$$

T o find the winner, we find the two stage dominances by evaluating the square of the dominance matrix, $A^2$. Then we add $A$ and $A^2$,

$$A + A^2 = \begin{bmatrix} 0 & 2 & 1 & 2 & 2 & 2 & 4 \\ 2 & 0 & 2 & 2 & 3 & 4 & 4 \\ 1 & 2 & 0 & 2 & 3 & 2 & 4 \\ 1 & 1 & 1 & 0 & 2 & 2 & 3 \\ 1 & 2 & 2 & 2 & 0 & 2 & 2 \\ 0 & 2 & 0 & 1 & 1 & 0 & 2 \\ 1 & 1 & 1 & 1 & 0 & 1 & 0 \end{bmatrix}.$$

By adding up the entries in each row, we get the number of total dominances:

| Team | a | b | c | d | e | f | g |
|---|---|---|---|---|---|---|---|
| No of dominances | 13 | 17 | 14 | 10 | 11 | 6 | 5 |

Therefore, the winner is team $b$ with the most one and two stage dominances.

**11.** (a)
$$A = \begin{array}{c} \\ a \\ b \\ c \\ d \\ e \end{array}\begin{array}{c} \begin{matrix} a & b & c & d & e \end{matrix} \\ \begin{bmatrix} 0 & 0 & 1 & 0 & 0 \\ 0 & 0 & 1 & 0 & 0 \\ 1 & 1 & 0 & 1 & 1 \\ 0 & 0 & 1 & 0 & 0 \\ 0 & 0 & 1 & 0 & 0 \end{bmatrix} \end{array}$$

(b)
$$A^2 = \begin{array}{c} \\ a \\ b \\ c \\ d \\ e \end{array}\begin{array}{c} \begin{matrix} a & b & c & d & e \end{matrix} \\ \begin{bmatrix} 1 & 1 & 0 & 1 & 1 \\ 1 & 1 & 0 & 1 & 1 \\ 0 & 0 & 4 & 0 & 0 \\ 1 & 1 & 0 & 1 & 1 \\ 1 & 1 & 0 & 1 & 1 \end{bmatrix} \end{array}$$

It represents the number of two-step paths from one vertex to another.

(c) The number of paths of length 2 from $c$ to $c$ is 4.

## Chapter 13 Review

### True or False

**1.** False **3.** True **5.** True

### Fill in the Blank

**1.** vertices; edges **3.** Eulerian **5.** direction

### Review Exercises

**1.** No. Such a simple graph must at most have 6 edges.

**3.** (a) $v_1e_9v_5e_5v_7e_4v_6e_{12}v_3$ (b) $v_1e_9v_5e_5v_7e_4v_6e_{12}v_3e_3v_5e_9v_1$
(c) $v_1e_9v_5e_5v_7e_4v_6e_{11}v_2e_1v_1$

**5.** No. The graph in the figure represents the city where the vertices A and B are the two banks, C, D and E are the three islands and the edges are the bridges. Since deg(B) = 3 which is odd, then the graph does not contain an Eulerian circuit.

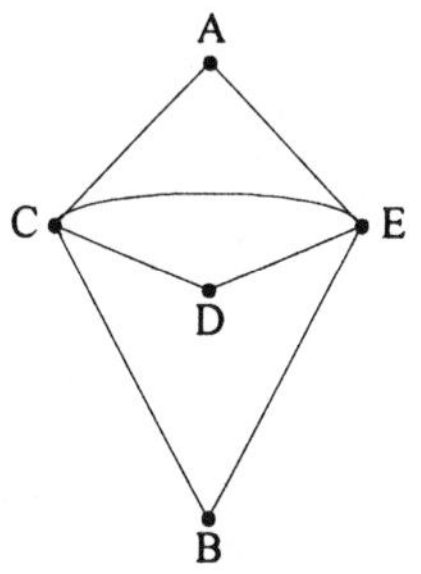


**7.** $v_1e_1v_2e_2v_3e_8v_5e_6v_2e_7v_4e_9v_1e_{10}v_3e_3v_4e_4v_5e_5v_1$ is an Eulerian circuit.

**9.**

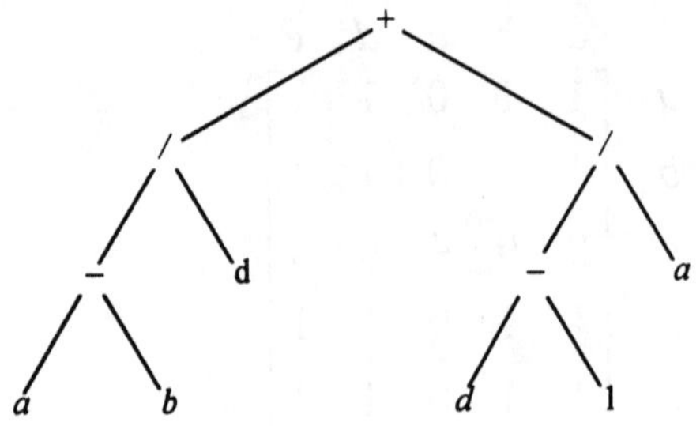


**11.**

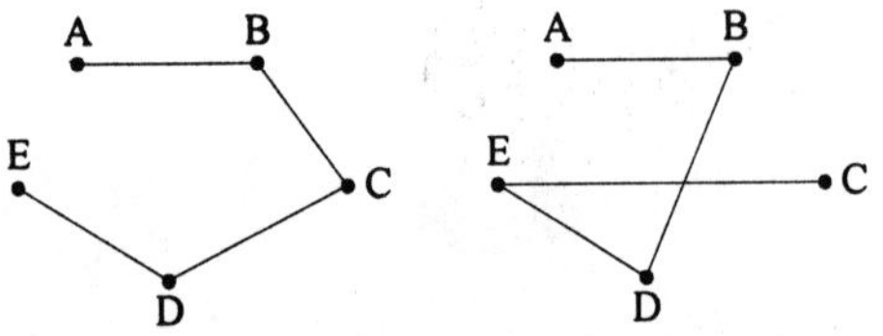


**12.**

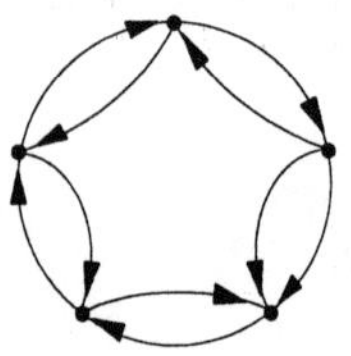

**13.** Yes. Every vertex is reachable from the others.

# Appendix A

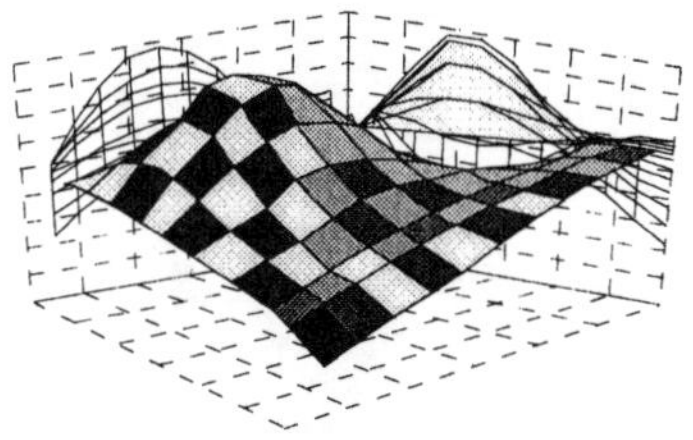

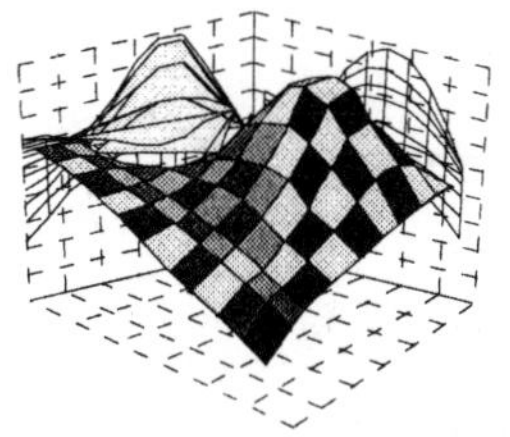

# Review Topics from Algebra and Geometry

## A.1 Basic Algebra

**1.** $x > 0$

**3.** $x < 4$

**5.** $x \le 2$

**7.** $x \ge -1$

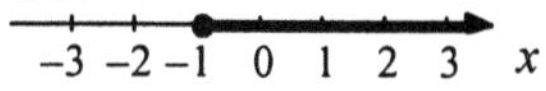


**9.** $x \ge 4$ and $x < 6$

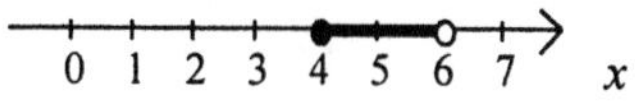


**11.** $x \le 0$ or $x < 6$ (or, more simply, $x < 6$)

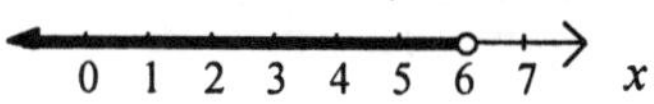


**13.** $x \le -2$ and $x > 1$ (the empty set!)

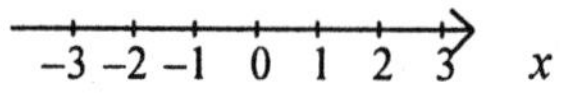


**15.** $x \le -2$ or $x > 1$

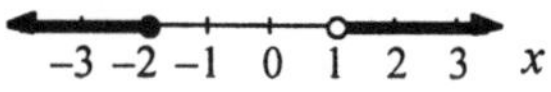


**17.**
$$\begin{aligned} x + 5 &= 7 \\ x &= 7 - 5 \\ x &= 2 \end{aligned}$$

**19.**
$$\begin{aligned} 6 - x &= 0 \\ 6 &= x \\ x &= 6 \end{aligned}$$

**21.**
$$\begin{aligned} 3(2 - x) &= 9 \\ 2 - x &= 3 \\ x &= 2 - 3 \\ x &= -1 \end{aligned}$$

**23.** $4x + 3 = 2x - 5$

$2x = -8$

$x = -4$

**25.** $x + 5 \le 2$

$x \le 2 - 5$

$x \le -3$

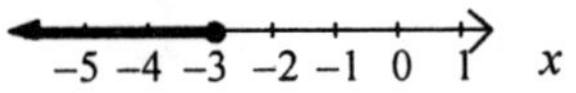

**27.** $3x + 5 \ge 2$

$3x \ge -3$

$x \ge -1$

**29.** $-3x + 5 \le 2$

$-3x \le -3$

$3x \ge 3$

$x \ge 1$

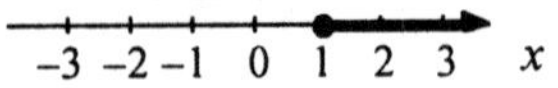

**31.** $6x - 3 \ge 8x + 5$

$6x - 8x \ge 5 + 3$

$-2x \ge 8$

$x \le -4$

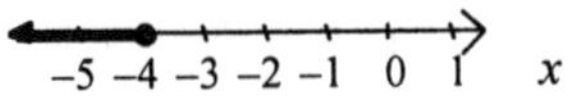

## A.2 Exponents and Logarithms

**1.** $4 \cdot 4 \cdot 4 = 64$

**3.** $\frac{1}{2^3} = \frac{1}{8}$

**5.** 1

**7.** 4

**9.** 3

**11.** 2

**13.** $\left(\sqrt[3]{8}\right)^2 = 2^2 = 4$

**15.** $\frac{1}{16^{3/2}} = \frac{1}{\left(\sqrt{16}\right)^3} = \frac{1}{4^3} = \frac{1}{64}$

**17.** $\frac{1}{(-8)^{2/3}} = \frac{1}{\left(\sqrt[3]{-8}\right)^2} = \frac{1}{(-2)^2} = \frac{1}{4}$

**19.** (a) $3^{2.2} \approx 11.21157846$
(b) $3^{2.23} \approx 11.58725056$
(c) $3^{2.236} \approx 11.66388222$
(d) $3^{\sqrt{5}} \approx 11.66475332$

**21.** (a) $2^{3.14} \approx 8.815240927$
(b) $2^{3.141} \approx 8.821353305$
(c) $2^{3.1415} \approx 8.824411082$
(d) $2^{\pi} \approx 8.824977827$

**23.** (a) $3.1^{2.7} \approx 21.21663834$
(b) $3.14^{2.71} \approx 22.21668955$
(c) $3.141^{2.718} \approx 22.44040295$
(d) $\pi^{e} \approx 22.45915772$

**25.** Solve $3^x = 27$; $3^x = 3^3$; $x = 3$

**27.** Solve $2^x = \frac{1}{2}$; $2^x = 2^{-1}$; $x = -1$

**29.** $N = 2^3 = 8$

**31.** $N = 3^{-1} = \frac{1}{3}$

**33.** $a^3 = 8;\ a^3 = 2^3;\ a = 2$

**35.** $a^2 = 9;\ a^2 = 3^2;\ a = 3$

**37.** $\log_{1.1} 200 = \frac{\log_{10} 200}{\log_{10} 1.1} \approx 55.59025675$

**39.** $\log_{1.005} 1000 = \frac{\log_{10} 1000}{\log_{10} 1.005} \approx 1385.002062$

**41.** $\log_{1.002} 20 = \frac{\log_{10} 20}{\log_{10} 1.002} \approx 1499.363504$

**43.** $\log_{1.0005} 500 = \frac{\log_{10} 500}{\log_{10} 1.0005} \approx 12{,}432.32324$

**45.** $\log_{1.003} 500 = \frac{\log_{10} 500}{\log_{10} 1.003} \approx 2074.641786$

## A.3 Geometric Sequences

**1.** 1, 2, 3, 4, 5

**3.** $\frac{1}{2}, \frac{2}{3}, \frac{3}{4}, \frac{4}{5}, \frac{5}{6}$

**5.** $1, -4, 9, -16, 25$

**7.** $\frac{1}{2}, \frac{2}{5}, \frac{2}{7}, \frac{8}{41}, \frac{8}{61}$

**9.** $-\frac{1}{6}, \frac{1}{12}, -\frac{1}{20}, \frac{1}{30}, -\frac{1}{42}$

**11.** $\frac{1}{e}, \frac{2}{e^2}, \frac{3}{e^3}, \frac{4}{e^4}, \frac{5}{e^5}$ or, approximately 0.367879, 0.270671, 0.149361, 0.0732626, 0.0336897

**13.** 1, 3, 5, 7, 9

**15.** $-2, -1, 1, 4, 8$

**17.** 5, 10, 20, 40, 80

**19.** $3, 3, \frac{3}{2}, \frac{1}{2}, \frac{1}{8}$

**21.** 1, 2, 2, 4, 8

**23.** $A,\ A+d,\ A+2d,\ A+3d,\ A+4d$

**25.** $\sqrt{2},\ \sqrt{2+\sqrt{2}},\ \sqrt{2+\sqrt{2+\sqrt{2}}},\ \sqrt{2+\sqrt{2+\sqrt{2+\sqrt{2}}}},\ \sqrt{2+\sqrt{2+\sqrt{2+\sqrt{2+\sqrt{2}}}}}$

**27.** Common ratio: 2
First four terms: 2, 4, 8, 16
Sum of the first $n$ terms:

$$2\left[\frac{1-2^n}{1-2}\right] = -2\left(1-2^n\right)$$

**29.** Common ratio: $\dfrac{1}{2}$

First four terms: $-\dfrac{3}{2}, -\dfrac{3}{4}, -\dfrac{3}{8}, \dfrac{-3}{16}$

Sum of the first $n$ terms:

$$-\frac{3}{2}\left[\frac{1-\left(\frac{1}{2}\right)^n}{1-\frac{1}{2}}\right] = -3\left[1-\left(\tfrac{1}{2}\right)^n\right]$$

**31.** Common ratio: 2

First four terms: $\dfrac{1}{4}, \dfrac{1}{2}, 1, 2$

Sum of the first $n$ terms:

$$\frac{1}{4}\left[\frac{1-2^n}{1-2}\right] = -\frac{1}{4}\left(1-2^n\right)$$

**33.** Common ratio: $2^{1/3}$

First four terms: $2^{1/3}, 2^{2/3}, 2, 2^{4/3}$

Sum of the first $n$ terms:

$$2^{1/3}\left(\frac{1-2^{n/3}}{1-2^{1/3}}\right)$$

**35.** Common ratio: $\dfrac{3}{2}$

First four terms: $\dfrac{1}{2}, \dfrac{3}{4}, \dfrac{9}{8}, \dfrac{27}{16}$

Sum of the first $n$ terms:

$$\frac{1}{2}\left[\frac{1-\left(\frac{3}{2}\right)^n}{1-\frac{3}{2}}\right] = -\left[1-\left(\tfrac{3}{2}\right)^n\right]$$

# Appendix B

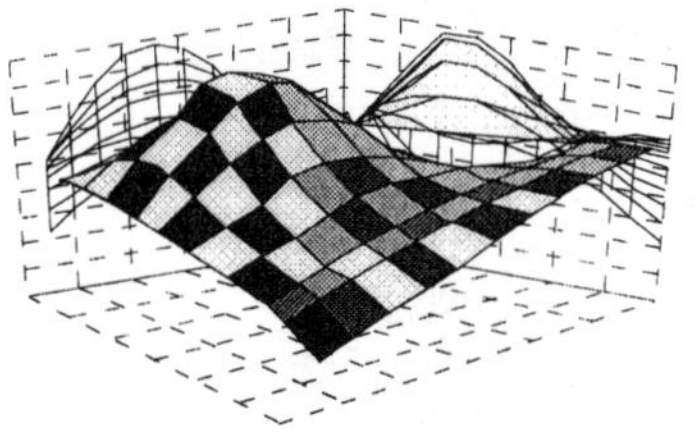

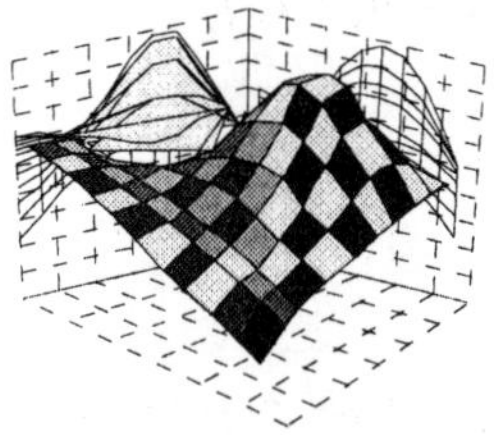

## Using LINDO to Solve Linear Programming Problems

**1.** Maximum $P = 24$, when $x_1 = 0$, $x_2 = 12$, $x_3 = 0$ (and $s_1 = 18$, $s_2 = 0$, $s_3 = 8$). Also, $P = 24$ when $x_1 = 0$, $x_2 = \frac{76}{7}$, $x_3 = \frac{16}{7}$ (and $s_1 = \frac{118}{7}$, $s_2 = 0$, $s_3 = 0$).

**3.** Maximum $P = 15$, when $x_1 = 5$, $x_2 = 0$, $x_3 = 0$ (and $s_1 = 1$, $s_2 = 0$).

**5.** Maximum $P = 40$, when $x_1 = 14$, $x_2 = 0$, $x_3 = 4$ (and $s_1 = 0$, $s_2 = 0$).

**7.** Maximum $P = 6$, when $x_1 = 2$, $x_2 = 4$, $x_3 = 0$ (and $s_1 = 0$, $s_2 = 0$). Also, $P = 6$ when $x_1 = 3$, $x_2 = 0$, $x_3 = 3$ (and $s_1 = 0$, $s_2 = 0$).

**9.** Maximum $P = \frac{76}{5} = 15.2$, when $x_1 = \frac{8}{5} = 1.6$, $x_2 = \frac{24}{5} = 4.8$, $x_3 = \frac{12}{5} = 2.4$ (and $s_1 = 0$, $s_2 = 0$, $s_3 = 0$).

**11.** Maximum $P = 15$, when $x_1 = 0$, $x_2 = 5$, $x_3 = 0$ (and $s_1 = 45$, $s_2 = 0$).

**13.** There is *no* maximum since the feasible region is unbounded in a direction of increasing $P$-values.

**15.** Maximum $P = 30$, when $x_1 = 0$, $x_2 = 0$, $x_3 = 10$ (and $s_1 = 15$, $s_2 = 0$).

**17.** Maximum $P = 42$, when $x_1 = 1$, $x_2 = 10$, $x_3 = 0$, $x_4 = 0$ (and $s_1 = 0$, $s_2 = 0$, $s_3 = 4$).

**19.** Maximum $P = 40$, when $x_1 = 20$, $x_2 = 0$, $x_3 = 0$ (and $s_1 = 0$, $s_2 = 20$, $s_3 = 30$).

**21.** Maximum $P = 50$, when $x_1 = 0$, $x_2 = 15$, $x_3 = 5$, $x_4 = 0$ (and $s_1 = 0$, $s_2 = 0$).

**23.** Minimum $z = \dfrac{305}{4} = 76.25$, when $x_1 = \dfrac{25}{4} = 6.25$, $x_2 = 0$, $x_3 = 0$, $x_4 = 20$, $x_5 = 0$, $x_6 = 50$, $x_7 = 0$ (and $s_1 = 0$, $s_2 = 0$, $s_3 = 0$).